CONDITIONAL INFERENCE AND LOGIC
FOR INTELLIGENT SYSTEMS

CONDITIONAL INFERENCE AND LOGIC FOR INTELLIGENT SYSTEMS

A Theory of Measure-Free Conditioning

I. R. GOODMAN
Naval Ocean Systems Center
San Diego, California
U.S.A.

H. T. NGUYEN and E. A. WALKER
Department of Mathematical Sciences
New Mexico State University
Las Cruces, New Mexico
U.S.A.

1991

NORTH-HOLLAND
AMSTERDAM • NEW YORK • OXFORD • TOKYO

ELSEVIER SCIENCE PUBLISHERS B.V.
Sara Burgerhartstraat 25
P.O. Box 211, 1000 AE Amsterdam, The Netherlands

Distributors for the United States and Canada:
ELSEVIER SCIENCE PUBLISHING COMPANY INC.
655 Avenue of the Americas
New York, N.Y. 10010, U.S.A.

Library of Congress Cataloging-in-Publication Data

Goodman, Irwin R.
Conditional inference and logic for intelligent systems : a theory of measure-free conditioning / I.R. Goodman, H.T. Nguyen, and E.A. Walker.
p. cm.
Includes bibliographical references and index.
ISBN 0-444-88685-0
1. Expert systems (Computer science) 2. Artificial intelligence. 3. Probabilities. 4. Logic, symbolic and mathematical. I. Nguyen, Hung T., 1944- . II. Walker, E. (Elbert), 1930- . III. Title.
QA76.76.E95G67 1991
006.3'3–dc20 91-3852
CIP

ISBN: 0 444 88685 0

Printed in The Netherlands

PREFACE

This book is concerned with a systematic investigation of the concept of "measure-free" conditioning and its associated logic for intelligent systems. Its purpose is to provide a foundation for inference in such systems. The basic problem is the representation and evaluation of implicative statements in natural language, in a way compatible with conditional probability. This longstanding problem involves three distinct disciplines: natural language, logic, and probability. The results are organized in book form here for the first time.

Two audiences are in mind, artificial intelligence researchers who are primarily interested in reasoning under uncertainty in intelligent systems, and mathematicians in the fields of probabilistic modeling, and logic. This diversity of audience requires that some sections be tutorial and elementary in nature.

Specifically, this work bridges the gap between numerically based probabilistic conditioning and the logic underlying implicative statements in natural language. This problem has been addressed in the past, for example, by Boole, DeFinetti, Koopman, Copeland, Schay, and Adams. Those efforts are incomplete, perhaps because of lack of motivation by real world problems. In any case, work in this field has gone unrecognized by the mainstream of researchers, particularly the work of Schay in 1968 on the algebra of conditional events, which remains almost totally uncited in the literature.

The situation is different today. The problem is before us because of the need to provide a firm foundation for probabilistic reasoning in intelligent systems; in particular, how to combine conditional information arising from disparate sources in expert systems and how to compute it probabilistically. This is in line with the Bayesian approach to probabilistic reasoning in intelligent systems (Pearl, 1988). Probability not only has a firm mathematical foundation, but also the conditional probability operator captures a form of non-monotonicity of common sense reasoning.

Our goal is a more complete and satisfactory theory of "measure-free" conditioning. If the concept of "conditional event" can be formalized and a suitable algebra of operators between such events be developed, then the resulting structure will have use in designing inference rules in expert systems. With probability being the method of choice for handling uncertainty despite the plethora of non-probabilistic procedures such as Dempster-Shafer belief functions and Zadeh's fuzzy sets, it is natural to develop a logic of conditional events compatible with conditional probabilities. However, the basic work here can be adapted and extended in various directions, such as to the fuzzy set setting (Chapter 7), as well as to the Dempster-Shafer belief function setting (see, for example, Dubois and Prade (1988)). This development is not to be confused with other "conditional logics", such as that of Nute (1980) and Appiah (1985), which are not compatible with conditional probability, nor with non-commutative extensions of Boolean logic (Guzman and Squier, 1990). Our approach differs also from that of Adams (1975), who takes conditionals as primitives in natural language, while ours are mathematical entities.

This book is primarily concerned with theory. The reader is expected to be familiar with basic probability theory, elementary logic, and elementary facts from ring theory. However, the text is largely self-contained. The hope is that this book will trigger further interest in both the theory and applications of this topic.

In conducting the research leading to this Monograph, we have benefited from discussions with various people. In particular, acknowledgements are expressed to Dr. Philip Calabrese for his thought provoking treatise on measure-free conditional events (Calabrese, 1987), and the lengthy personal communications exchanged on the topic. Thanks are extended to Professors Geza Schay of the University of Massachusetts at Boston, Kevin Hestir and Gerald Rogers of New Mexico State University, and to David Stein of the Naval Ocean Systems Center at San Diego.

The first named author expresses his appreciation for support by Dr. Ralph Wachter of the Office of Naval Research, Dr. Alan Gordon of the Independent Research Office, NOSC, and the backing by the line management of the Naval Ocean Systems Center, in particular John A. Salamann, Jr. and Michael C. Mudurian, both of the Command and Control Department.

The other two named authors extend their thanks to their department head, Professor Carol Walker, for her encouragement.

Finally, we are grateful to Valerie Reed for performing an excellent job in typing this manuscript.

I. R. Goodman
San Diego, California
March 10, 1990

H. T. Nguyen and E. A. Walker
Las Cruces, New Mexico
March 10, 1990

CONTENTS

CHAPTER 0

INTRODUCTION

In this Introduction, we outline the motivation and objectives, as well as the main contributions to the theory of measure-free conditioning.

0.1 Motivation and objectives

This work addresses an anomaly involving probability and logic relative to the interpretation of implicative statements, and the evaluation of those statements compatible with conditional probability. One of our chief motivations is the need to formalize rigorously the connections between conditional probability and the "hidden" logic of implicative statements, such as production rules in expert systems and defaults in common-sense reasoning. The purpose is to provide theoretical results for probabilistic reasoning that will be useful in the design and evaluation of inference rules of such systems.

We now describe the basic problem in some detail. Within the context of logic-based formal methods in artificial intelligence, the space of propositions (facts, evidence, information, and so on) is represented by an algebraic structure R known as a Boolean algebra. The basic connectives among propositions, namely negation, conjunction, and disjunction correspond to operators on R, denoted $'$, $\wedge$, and $\vee$, respectively. When elements of R are uncertain, as is often the case in expert systems, classical two-valued logic has to be replaced by *probability logic*, in which probabilities play the role of truth values. However, our knowledge often contains uncertain conditional information of the form "if b then a", where a and b are elements of R. These conditional propositions are referred to as implicative statements, or conditionals. In expert systems, these are "production rules". In order to make inferences from this type of knowledge, it is necessary to develop an appropriate logic in which these conditionals can be represented and manipulated in order to combine evidence, and in which an entailment relation can be formulated. A quantitative approach to this starts with the quantification of the strength or the "truth" of conditionals. For example, if the conditional "if b then a" is written in the language of Boolean logic, then one can model it by *material implication*, that is

$$b \to a = b' \vee a.$$

If P is a probability measure on R, then $P(b \to a) = P(b' \vee a)$ can be used as such a quantification. However, it is more reasonable to quantify the conditional "if b then a" by the conditional probability $P(a|b)$, which is clearly different from $P(b' \vee a)$. Indeed

$$P(b' \vee a) = P(b' \vee ab)$$

$$= 1 - P(b) + P(ab) \neq P(ab)/P(b).$$

While this is consistent with probability logic for unconditional propositions, that is, for elements of R, one cannot represent the conditional "if b then a" mathematically. Indeed, there is no counterpart of $P(a|b)$ in logic. Logic lacks a conditioning operator corresponding to conditional probability. Since *material implication* $b \to a$ *is not compatible with probability* in the sense that

$$P(b \to a) \neq P(a|b),$$

one might attempt to look for other operations f on R, Boolean or not, such that $P[f(a,b)] = P(a|b)$ for all probabilities P on R and all $a, b \in R$ with $b \neq 0$. Such attempts have been laid to rest by *Lewis' Triviality Result.* (See Chapter 1.) To model "measure-free" conditional events $(a|b)$ compatible with conditional probability, one has to go outside of R. Thus $(a|b)$ cannot be so modeled as an ordinary proposition.

The first question then is to determine a suitable mathematical entity $R|R$ for conditional events $(a|b)$. Once such a model $R|R$ is determined, for each probability P on R, one has P extended to a "semantic evaluation" on $R|R$.

With the space $R|R$ as the counterpart of R in the unconditional case, one then proceeds to define connectives among conditionals, for example conjunctions

$$(\text{if } b \text{ then } a) \wedge (\text{if } d \text{ then } c)$$

whose result is another conditional in $R|R$. Such operations yield an algebra of conditionals, extending the algebra of unconditional events of R. Choosing a correct model $R|R$ for these conditional events, and then choosing suitable logical operations on those conditional events which extend those of R is the backbone of the problem. Once such operations have been found, it is then possible to assign probabilities to compounds of conditionals, since, for example, to evaluate

$$P[(\text{if } b \text{ then } a) \wedge (\text{if } d \text{ then } c)],$$

one merely has to carry out the operation $\wedge$ between the two conditionals, yielding another conditional, and then evaluate P at that conditional. The algebraic structure of $R|R$ together with a probability on R extended to $R|R$ forms the core for the development of conditional probability logic extending that of probability logic.

In summary, the problem we are facing is this. For a Boolean algebra $R(', \wedge, \vee)$,

(1) find a "measure-free" conditioning map f from $R \times R$ to some space $R|R$ so that $P[f(a,b)] = P(a|b)$ defines a function on $R|R$ extending P on R;

(2) define logical operations $', \wedge, \vee$ on $R|R$ extending the corresponding ones on R, and

(3) with conditional probabilities as semantic evaluations, develop a conditional probability logic with syntax $(R|R, ', \wedge, \vee)$.

No satisfactory solution to the problem seems to exist, even in the vast numerically oriented literature treating conditioning in probability and logic. A solution entails the development of "conditional event algebras", and lies outside the scope of conditional probability literature. This aspect has been considered by only a handful of researchers, with no concerted effort being made in that direction. In this monograph, we present a solution to the problem in the form of a conditional events algebra that is new, rigorous, comprehensive, and computationally tractable. The theory of measure-free conditioning presented here can be used both as a basis for treating the problem of combining evidence and as groundwork for further investigations into the connection between probability and logic.

We now return to the topic of inference rules in expert and intelligent systems as one of the main motivating sources for posing the basic problem mentioned above. Automated reasoning in intelligent systems is based on logical entailment (or logical consequences or implication) in some logic. For example, in mechanical theorem proving where first order logic is used, one of the usual ways to draw conclusions is through the use of modus ponens, which simply says that if b implies a and b is true, then a is true. This means that a follows logically from $\{b \to a, b\}$, and this translates into the syntax of first order logic as $(b \to a) \wedge b \leq a$. Note that here $\leq$ is precisely the logical entailment relation of first order logic, and the modeling of "$\to$", conditional information of the form "if b then a", is via material implication mentioned above.

The situation in *reasoning under uncertainty* is more complicated. First, the knowledge base consists of conditional information which is not known with complete certainty. Second, human common sense reasoning is basically "non-monotonic" in nature, whereas first order logic is monotone. This means that one can retract prior conclusions in light of new evidence. From a logical non-numerical approach, the modeling of "if b then a" should be investigated, and a non-monotonic logic for "conditionals" should be found. A well known example is Reiter's (1980) logic of defaults. If we want to treat uncertainty in conditional information in a more quantitative way, various uncertainty measures could be used. The most popular numerical approach is a Bayesian one, in which probabilities assigned to conditionals are conditional probabilities. Suppose we symbolize conditional statements of the form "*if b then a*" or "*most b's are a's*", or "*usually birds fly*" by $(a|b)$. Then the knowledge K is of the form $\{(a_i|b_i) : i = 1,2, ...,n\}$, and the evidence is of the form $E = (e_i)_{i=1,\dots,n}$, $e_i \leq \Omega$. Non-monotonic reasoning is a logical entailment in a non-monotonic logic whose basic objects are of the form $(a|b)$. Note that the elements of E can be identified as $(e_j|\Omega)$. Instead of trying to model $(a|b)$ as a mathematical entity compatible with conditional probability (as a counterpart of non-conditional propositions with respect to unconditional probability), a well known approach (for example, Pearl, 1988) is to rely upon the so-called Adams' logic of conditionals (Adams, 1975), in which conditionals are not modeled mathematically, but are taken as primitives in our natural language, and the probability entailment relation $\Rightarrow$ is defined semantically. The lack of a conditioning operator in logic is mentioned in many places in Pearl's book. Moreover, if a mathematical object $(a|b)$ could be defined, many problems in Adams' book could be clarified. It is interesting to note that in 1968 Shay published a paper providing a proposal for such an object $(a|b)$ and its algebra. Definitely, if objects like $(a|b)$ can be defined, then we can bridge the gap between probability and logic and reasoning can be carried out at the syntax level providing that conditional information can be combined.

Thus the goal is to develop a theory of "conditional events" compatible with conditional probability, analogous to the role played by boolean algebra in the theory of unconditional events and unconditional probability. Perhaps by the very nature of physical systems and statistical problems, the new concept of conditional events might not contribute anything new to them. This might explain why the papers by Copeland published in the Proceedings of the Berkeley Symposium on Mathematical Statistics and Probability (1945,1954), or by Schay (1968) have been largely ignored. This is similar to the case of quantum probability for quantum mechanics but not for ordinary probability models (Gudder, 1988). The need for defining mathematically (measure-free) conditional events appears also in the *Theory of Measurement* (Pfanzagl, 1971, Chapter 12). But unlike Copeland's approach, Pfanzagl proposed to use cosets of Boolean rings to represent conditional events. However, his analysis was restricted only to each fixed (Boolean) quotient ring, so that the algebraic structure of the space of all possible cosets was not investigated. In particular, inference from a collection of conditional events with different antecedents was not formulated. But, as we will see in Chapter 2, the coset form for conditional events is a correct one, and this will be derived axiomatically.

0.2 State-of-the-art

The mathematical problem that we try to analyze in this book has been examined over several decades, but is apparently foreign to probabilists as well as to engineers. Most of the results were published in a scattered, unorganized fashion. However, there are two books on the subject: those of Adams (1975) and Hailperin (1976), which are in logic. See also the book of Pfanzagl (1971, Chapter 12).

Prior to the era of AI, the problem came independently to the attention of the logicians Stalnaker (1968) and Lewis (1976), and as well as to Van Fraasen (1976), Copeland (1941, 1945, 1950, 1954), Koopman (1940, 1941), and DeFinetti (1974). While the discussion of the subject within the logic community remains somewhat active, perhaps because of its philosophical nature, there was no reaction at all in the probability and statistics community. This is exemplified by the largely forgotten Copeland's papers which aimed at providing more basic structures for probability theory and statistics, complementing Kolmogorov's model. The framework that he proposed, that of implicative Boolean algebras, was unsatisfactory, being far too restrictive, and examples and applications were not readily at hand.

At the folklore or unpublished level, all of the attempts to deal with this problem have been shown to be either patently wrong - such as identifying the probability of material implication with conditional probability, or combining antecedents with only union or intersection of antecedents being taken, or using a too restrictive or computationally unfeasible approach (see Chapter 1).

The conditional event "literature" consists of only a couple of dozen papers as opposed to the vast conditional probability literature. Within this meager output, most researchers have reached the point where they have agreed that conditional events should be identified as cosets of principal ideal of events of the original Boolean algebra of events. One exception is Copeland and his colleagues, who used the "implicative" Boolean algebra approach. But this required the original Boolean algebra to be infinite and of a very special sort. Indeed, an "implicative" Boolean algebra R must be isomorphic to R/I for all principal ideals I (Copeland and Harary, 1953a).

Except for Domotor (1969), Pfanzagl (1971) and Calabrese (1987), no justification is proffered by those even proposing cosets of principal ideals as models for conditional events. On the other hand, Hailperin *postulated* that a conditional event should be an element of a Chevalley-Uzkov ring of fractions of a Boolean ring, whose elements he then shows are identifiable with cosets of principal ideals of the original Boolean ring. Thus he could have skipped the ring of fractions step, cosets being a simpler notion. The idea is not so bad: given two elements a and b of a ring, with $b \neq 0$, form a larger ring in which a/b makes sense, that is, in which a is divisible by b. The notion is to model the conditional event $(a|b)$ by the element a/b. For Boolean rings, this cannot be done. Trying to "divide" elements of Boolean rings results in trivialities. For example, in the larger ring,

$$abb(a/b) = aba = ab = ab(a/b) = aa = a.$$

But ab is not necessarily a. Further, using more general "rings of quotients" will also lead nowhere. (See Section 1.2 for more details.)

Among the few who have attempted to define operations among conditional events with different antecedents - the identical antecedent case being similar to the unconditional case - only Schay (1968) has justified his choice of (two proposed systems of) operators, and that indirectly through an abstract characterization theorem. (See Schay (1968), Theorem 5.) However, these operators are chosen initially on an empirical basis, and the characterization theorem appears more as an ad hoc rather than a natural avenue for supporting them.

It will be pointed out in Chapter 3 that both pairs of Schay's conjunction and disjunction operators - and hence Calabrese's operators since they coincide with one system of Schay's, violate the min-conjunction and max-disjunction and related monotonicity properties of probability. Up to now, no one has *derived* operations on conditional events from first principles, and related explicitly the coset form of conditional events to their potential operations. Even further, except for some of Mazurkiewicz's rudimentary results (see Section 1.4), no connections have been established between the coset form and conditional probability assignment of conditional events.

As we will see, there has been a proliferation of definitions for conditional events and of operations between them. This is due perhaps to the fact that each approach is based simply on some intuitive idea or some mathematical analogy rather than a systematic analysis of the problem from basic concepts, or a more axiomatic approach.

In summary, up to now no satisfactory first-principles approach has been taken toward the exposition of a theory of conditional events. Our goal is such a theory.

0.3 Outline of main contributions

With the motivation and objectives described above, our effort will be directed first toward the development of a mathematically rigorous and comprehensive theory of measure-free conditioning. Specifically, a conditioning operator compatible with the probabilistic conditioning operator is introduced into logic. The whole machinery of Boolean logic is extended to "conditional Boolean logic". With this conditional Boolean logic as syntax, the associated conditional probability logic will extend classical probability logic. (See Hailperin (1984) and Nilsson (1986).) Since conditional probability logic is a logic for implicative propositions (such as defaults in common sense reasoning, and productions rules in expert systems), our work makes more rigorous, and goes beyond, that of Adams (1975). Further, it clarifies theoretical issues in algebraic logic in the new direction of non-monotonic logics for AI. The mathematical setting of our conditional extension of first-order logic is an algebraic structure extending the Boolean ring of first order logic, but is not itself a ring. This is however compatible with the goal of achieving non-monotonicity in probability reasoning, more fundamental structures surrounding the theory of probability must be investigated, as has been pointed out by Grosof (1988) and Pearl (1988). Thus, structures more general than Boolean rings must be allowed. This situation is somewhat analogous to that of quantum logic (Gudder, 1988). This need to consider more general algebraic structures can also have some interest for algebraists. For example, combining cosets of different quotient rings of a Boolean ring is possible in a natural way, and the resulting algebraic structure merits attention. The generality in which this phenomenon holds is not clear, although it does extend, for example, to commutative von Neumann regular rings. (See Chapter 8.) A related question of interest here which arises is to characterize commutative partially ordered rings in which cosets of principal ideals are intervals.

The theory of conditioning developed in this book can be used to design inference rules in intelligent machines. Details of these applications to AI should be investigated. At this point, we give some flavor of the theory. We begin by recalling the basics of Adams' logic of conditionals (Adams, 1975), which has been popularized in the AI community by Pearl (1988). Since uncertain implicative propositions in natural language form the core of human and machine knowledge used in reasoning and inference, a logic of these propositions, called conditionals, needs to be developed.

The main thrust of Adams' work is the development of a logic of conditionals, compatible with conditional probability, that is, probabilities of conditionals are taken to be conditional probabilities. See also (Stalnaker, 1968, 1970) and (Lewis, 1976). In classical two-valued logic, the basic structure is a Boolean algebra R of subsets of a universe of discourse Ω. Thus propositions (events) are represented as mathematical entities, namely as elements of R. From this, semantics, or truth values are attached to the "possible worlds". However, Adams, apparently unaware of most of the previous work on the subject, especially that of Shay (1968), took conditionals, as generalizations of ordinary events, as *primitives* in natural language, rather than some entity generalizing elements in a Boolean ring. (It is interesting to speculate on what Adams' book would be like had he known of Schay's work of 1968.) Thus in Adams' conditional extension of classical logic, the collections of conditionals exist only as a formal mathematical structure. However, as human beings, we understand this primitive concept of conditionals, and hence, as in classical logic, proceed to build more complicated conditionals from the simple ones via logical connectives "and", "or", "not", and so on. Now, these "conditional" connectives are extensions of those in ordinary unconditional propositions. As in any extension problem, the solution is not unique. Any proposed extension of the logical operations for conditionals forms only one possible logic amongst all the possible ones. Adams proposed the following ones (1975, pp.46-47). Write "if b then a" as $a|b$. He made the following definitions, perhaps based on intuitive grounds:

$$(a|b)' = (a'|b);$$

$$(a|b) \wedge (c|d) = ((b' \vee a) \wedge (d' \vee c))|(b \vee d);$$

$$(a|b) \vee (c|d) = ((a \wedge b) \vee (c \wedge d))|(b \vee d).$$

These turn out to be precisely Schay's operations (Schay, 1968), which will be discussed in Chapter 3.

The problem with assigning probabilities to compounds of conditionals is discussed using *Lewis' triviality result*, which says that one cannot model conditionals as elements of the Boolean ring R, compatible with conditional probability (Adams, pp. 34-35; Lewis, 1976). Precisely, it says that one cannot associate with "if b then a" an element $\varphi(a,b)$ of R so that for all probability measures P on R,

$$P(\varphi(a,b)) = P(a|b) = P(ab)/P(b).$$

There are some trivial exceptions. A proof of this fact will be reproduced in Chapter 1. This means that the mathematical entity modeling conditionals must properly contain R. This modeling of conditionals is the main thrust of this book.

Another point in Adams' work is his concept of "probabilistic entailment" (Adams, 1975, pp. 56-57). Since reasoning in intelligent systems is based on a logical entailment relation in a given logic, it is not surprising that Pearl (1988) popularized Adams' work because of this concept only. This concept of entailment is particularly suitable for plausible reasoning in a quantitative way, that is, for conditionals $(a|b)$ in which $P(a|b)$ is high, such as "birds fly". Let $K = \{(a_i|b_i) : i = 1, 2, ..., n\}$. Then, by definition, K implies $(c|d)$ if for each $\varepsilon > 0$, there is a $\delta > 0$ such that for any probability measure P on R, if $P(a_i|b_i) > 1 - \delta$ then $P(c|d) > 1 - \varepsilon$. In Chapter 6, we will return to this concept to discuss its practical role in automated reasoning, especially in situations different from plausible reasoning, as well as in the computational aspects of conditional probability logic.

Now consider the problem of assigning a probability to a compound statement of the form S = "if b then a or if d then c", where a, b, c, and d are in R. To do that, we need to model the statement "if b then a" so that its probability is $P(a|b)$ and then we must define the connective $\vee$ (or) appropriately. The hope is that S will again be of the form "if e then f", and $P(S) = P(e|f)$. By Lewis' triviality result, "if b then a" cannot be an element of R, so we are led to look outside R for a model. (See Chapter 2.) In Chapter 3, conditional connectives are derived from algebraic considerations, and in particular, the connective $\vee$ is derived under reasonable assumptions to be

$$(a|b) \vee (c|d) = ((ab \vee cd)|(ab \vee cd \vee bd)),$$

where ab means the intersection or conjunction of a and b, and $a \vee b$ means their union or disjunction. This operator corresponds to Lukasiewicz's three-valued truth table for disjunction.

Another important issue is that of non-monotonic probabilistic reasoning in intelligent systems. Its framework is as follows. Let $T = < K,E >$, where K is a knowledge base and E is a set of evidence. K consists of a collection of implicative propositions symbolized as $(a_i|b_i)$, $i = 1, 2, 3, ..., n$. For example, in the "penguin triangle" example (Pearl, 1988), these are defaults. Note that in the Bayesian approach, where the uncertainty in these defaults is taken into account in a more quantitative way, a default rule of the form "most a's are b's" is modeled semantically as $P(a|b)$ is "high". E is a collection of factual propositions (evidence). Since elements in E can be viewed as implicative statements which are implied by the tautology T (true), the reasoning process will involve a logical entailment relation $\Rrightarrow$ in a conditional logic. Conditionals of interest are of the form $(c|E)$, where E stands for the Boolean conjunction of all elements in E, and c is some event of interest. It is desired to know whether $(c|E)$ follows logically from K. In the case of the penguin triangle example, the ε-semantics of Adams can be used (Pearl, 1988, Ch. 10). It is necessary to be able to handle production rules in expert systems rather than just defaults in plausible reasoning, and also to treat the problem at a syntactic level as in the case of classical first-order logic, where $\Rrightarrow$ is simply the order relation $\leq$ in a Boolean ring. Still this must be done compatible with conditional probability evaluations. The main problem is the representation of K as a whole. Putting all (uncertain) information in K together can be done in two different ways: internal and external. If implicative propositions $(a_i|b_i)$ can be represented as legitimate quantities, as we do in this book, and if logical operations among them are available, then an internal combination of information in K consists simply as taking conjunctions of all the $(a_i|b_i)$. An external combination strategy would consist of forming a "product" of the $(a_i|b_i)$. (See Chapter 3 for details.) To complete the reasoning procedure, a logical entailment relation $\Rrightarrow$ in conditional logic needs to be supplied. It turns out that the order structure of Boolean rings can be extended suitably to provide the desired $\nrightarrow$. Moreover, relative to E, that is, to additional facts or evidence, $\nrightarrow$ is non-monotonic. (See Chapter 8.)

In summary, we first justify the coset form for measure-free conditional events by using an axiomatic approach. A systematic investigation of logical operators among conditionals, including those with different antecedents, is then carried out, resulting in a space of conditional events. Realizing that conditionals have three possible truth values, a systematic study of three-valued logics leads to the conclusion that systems of logical operators among conditionals correspond precisely to various systems of three-valued logics. A conditional probability logic is formulated, extending classical probability logic. In a direction of generalization, we devote a chapter for conditioning in a fuzzy setting.

0.4 Overview of the book

In view of the state-of-the-art presented above, we have looked again at the problem in the last several years (Goodman, 1987, Goodman and Nguyen, 1988). The present book is based essentially on our earlier unpublished work "A Theory of Measure-Free Conditioning" (1987). Some of the results have already appeared in print, and are here augmented by new and improved procedures. In our view, it is not too early to provide a comprehensive presentation of the theory of measure-free conditioning. It is our hope that this book will stimulate further basic research in this area.

The basic program consists of nine parts:

(1) Formulation of the conditional event problem (Chapter 0).

(2) Extensive literature review pointing up the lack of a systematic investigation of the problem (Chapter 1).

(3) Derivation of the necessary form that a conditional event must take, namely that of a coset of a principal ideal in a Boolean algebra of events (Chapter 2).

(4) Derivation of the appropriate operations on conditional events and development of the calculus of these operations and the partial order extending the usual subset relation of ordinary events, together with a justification of the proposed conditional logic via three valued logic (Chapter 3).

(5) Establishment of relevant algebraic properties and a characterization of the algebra of conditional events, and an extension of the Stone Representation Theorem to this conditional setting (Chapter 4).

(6) An analysis of the assignment of conditional probability to conditional events (Chapter 5).

(7) The development of conditional probability logic whose algebraic structure is the conditional event algebra (Chapter 6).

(8) The generalization of results to fuzzy events (Chapter 7).

(9) The investigation of iterated conditioning, and miscellaneous issues (Chapter 8).

CHAPTER 1

A SURVEY OF PREVIOUS WORK ON CONDITIONAL EVENTS

As stated before, investigations of conditional events and their operations have a long history, but are not well known among probabilists, logicians, and computer scientists, who make up a good deal of the AI community. Our literature search revealed that the topic has been considered, independently and at infrequent intervals of time, by logicians and mathematicians, dating back to an idea in Boole's book (1854). Below we present the main approaches that have been taken, as well as duplications of effort that have been made. As we will see, throughout the development of the mathematical theory of conditional events and their calculus, there has been a proliferation of definitions of these objects and of operations among them. This is due to the fact that each approach has been based upon some intuitive idea or some analogy, rather than a systematic analysis from a first principles or axiomatic approach. An axiomatic approach - which we take here - should not only justify rigorously the correct forms for conditional events, but should also shed light on the ones investigated so far. In the same vein, a reasonable conditional logic should be able to be defended axiomatically. See Chapters 2 and 3.

1.1 Implicative Boolean algebras and Lewis' triviality result

The first approach considered for modeling conditional events is that of Copeland (1941, 1945, 1950). See also Copeland and Harary (1953a, 1953b), Balbes (1970), and Jonsson (1954).

Let R be a Boolean ring and let $\rightarrow$ be material implication, that is, $b \rightarrow a$ is the element $b' \vee a$. If P is a probability measure on R, then in general

$$P(b \rightarrow a) \neq P(a|b). \tag{1}$$

The simple-appearing expression in (1) belies an interesting and significant history. First, it can be improved as follows.

$$\begin{aligned} P(b \rightarrow a) &= P(b' \vee a) = P(b' \vee ab) \\ &= P(b') + P(ab) = P(b') + P(a|b)P(b) \\ &= P(b') + P(a|b)[1 - P(b')] \end{aligned}$$

$$= P(a|b) + P(b')[1 - P(a|b)]$$

$$= P(a|b) + P(b')P(a'|b) \geq P(a|b), \tag{2}$$

with equality holding if and only if $P(a'b) = 0$ or $P(b) = 1$, a rather trivial case.

Popper (1963, p. 390, formula 22) was among the first to recognize a form of (2) although earlier in 1956, Copeland (p. 42) implicitly used the inequality as a springboard for his implicative Boolean algebra work. Calabrese independently recognized (2) in 1975 and later in 1987 (p. 201), motivating his development of conditional events outside of the Boolean algebra of unconditional events R. It is tempting to seek another operation $\Diamond$ on R such that $P(a\Diamond b) = P(a|b)$, that is, a binary operation $\Diamond$ on R such that $P(a\Diamond b) = P(a|b)$ is well defined. In other words, one would like to know whether "conditional events" can be modeled as ordinary events, that is, as elements of R. It turns out that, except for trivial cases, the answer is negative (Lewis, 1976; Adams, 1975). Later Calabrese (1987), unaware of Lewis' so called "triviality result", showed, using the normal disjunctive form of Boolean polynomials, that such a $\Diamond$ could not be Boolean, that is, expressible in terms of union, intersection, and complement. Copeland proceeded directly to the search for such a $\Diamond$, and consequently only obtained trivial cases. We now discuss Lewis' Triviality Result, and then outline Copeland's work on implicative Boolean algebras.

Theorem 1 (*Lewis' Triviality Result*). *Let R be a Boolean ring with more than four elements. Then there is no binary operation $\Diamond$ on R such that for all probability measures P on R, and all a, b in R with $P(b) > 0$,*

$$P(a\Diamond b) = P(a|b).$$

Proof. Suppose $\Diamond$ exists. For a probability measure P on R and an element $r \in R$ with $P(r) \neq 0$, denote by P_r be the probability measure on R given by $P_r(x) = P(rx)/P(r)$. Now, if a and b are in R and $P(ab) \neq 0 \neq P(a'b)$, then a and b are P-independent. Indeed, since $P(a)$, $P(a')$, and $P(b)$ are all positive, we have

$$P(a|b) = P(a\Diamond b)$$

$$= P((a\Diamond b)a) + P((a\Diamond b)a')$$

$$= P((a\Diamond b)|a)P(a) + P((a\Diamond b)|a')P(a')$$

$$= P_a(a\Diamond b)P(a) + P_{a'}(a\Diamond b)P(a')$$

$$= P_a(a|b)P(a) + P_{a'}(a|b)P(a')$$

$$= (P_a(ab)/P_a(b))P(a) + (P_{a'}(ab)/P_{a'}(b))P(a')$$

$$= P(a) + 0$$

$$= P(a).$$

Since R has more than four elements, then we can find a and b in R such that $ab \neq 0 \neq a'b$. Indeed, let b be in R, $b \neq 1$, and let $a \in R$ with $a \neq b'$. If $ab = 0$, then $a < b'$ and $ab \neq 0 \neq a'b'$. If $ab \neq 0$, then $ab \neq 0 \neq a'b$ else $b < a$. In any case, we have elements a and b of R with $ab \neq 0 \neq a'b$ and $b \neq 1$. By Stone's Representation Theorem, R is a subalgebra of the algebra $\mathcal{P}(\Omega)$ of all subsets of some set Ω. Let x, y and z be elements of Ω such that $x \in ab$, $y \in a'b$ and $z \notin b$. Let P be a probability measure on $\mathcal{P}(\Omega)$ with $P(\{x\}) = P(\{y\}) = P(\{z\}) = 1/3$. Then P is a probability measure on R such that $P(ab) \neq 0 \neq P(a'b)$. Now $P(a|b) = 1/2$ while $P(a) = 2/3$ or $1/3$, depending on whether or not $z \in a$. Thus $P(a|b) \neq P(a)$, and $\lozenge$ cannot exist. □

The proof above is based on Lewis' original proof (Lewis, 1976). If R has four or fewer elements, then the reader may verify easily the existence of a $\lozenge$ satisfying the condition in the theorem.

As a simple example when such a $\lozenge$ does not exist, let $\Omega = \{x, y, z\}$ and $R = \mathcal{P}(\Omega)$. Define P by $P(x) = P(y) = P(z) = 1/3$. Let $a = \{x\}$ and $b = \{x, y\}$. Then $P(ab) \neq 0 \neq P(a'b)$, and $P(a|b) = 2/3$, while $P(a) = 1/3$. Of course, this is essentially the construction at the end of the proof just given.

When R is finite, there is another approach to Lewis' Triviality Result. If an operation $\lozenge$ on R existed satisfying $P(a\lozenge b) = P(a|b)$, then $P(a|b)$ can have no more than $\#(R)$ values, where $\#(R)$ denotes the number of elements of R. This is simply because $a\lozenge b$ is an element of R. Thus, to prove Lewis' Triviality Result, it suffices to construct on R a probability measure P such that $P(a|b)$ takes more than $\#(R)$ values. We will do a bit more.

Theorem 2. *Let Ω be a finite set, and let R be the Boolean algebra of all subsets of Ω. If Ω has n elements, $n > 0$, and P is any probability measure on R, then there are no more than $3^n - 2^{n+1} + 3$ possible values for $P(a|b)$. Further, then there is a probability measure P on R such that $P(a|b)$ takes on $3^n - 2^{n+1} + 3$ distinct values.*

Proof. Since $P(a|b) = P(ab)/P(b)$, to get the number of possible values of $P(a|b)$ not 0 or 1, we simply have to count the number of pairs (a,b) in R, that is, the number of pairs (a,b) of subsets of Ω, with $0 < a < b$. But this is the number

$$\sum_{m=1}^{2^n} (2^m - 2)\binom{2^n}{m} = (3^n - 1) - 2(2^n - 1) = 3^n - 2^{n+1} + 1,$$

and the first part of the theorem follows.

If μ is any bounded measure on R, then $P = \mu/\mu(\Omega)$ is a probability measure on R, and $P(a|b) = P(ab)/P(b) = \mu(ab)/\mu(b)$. We will prescribe a bounded measure μ on R such that distinct pairs (a,b) of elements with $0 < a < b$ give distinct $\mu(a)/\mu(b)$. A measure is prescribed on R by assigning to each of its singletons, that is to each element of Ω, a positive number. We will have the desired measure, then if there are n positive numbers $\alpha_1, \alpha_2, \ldots, \alpha_n$ such that if I, J, K, and L are subsets of $\{1, 2, \ldots, n\}$ with $\phi \subset I \subset J$ and $\phi \subset K \subset L$, α_I is the sum of those α_i with $i \in I$, and $(I, J) \neq (K, L)$, then $\alpha_I/\alpha_J \neq \alpha_K/\alpha_L$. This construction is relagated to the following lemma. □

Lemma 1. *Let* $n > 0$. *Then there exist positive numbers* $\alpha_1, \alpha_2, \ldots, \alpha_n$ *such that if* I, J, K, *and* L are subsets of $\{1, 2, \ldots, n\}$ *with* $\phi \subset I \subset J$ and $\phi \subset K \subset L$, α_I *is the sum of those* α_i with $i \in I$, *and* $(I, J) \neq (K, L)$, *then* $\alpha_I/\alpha_J \neq \alpha_K/\alpha_L$.

Proof. We get the desired α's inductively. Let α_1 be any positive number greater than 1. (There is some convenience is having $\alpha_1 > 1$.) Having chosen $\alpha_1, \alpha_2, \ldots$, α_{i-1}, for $1 < i \leq n$, let α_i be a positive number with such that

$$\alpha_i > (\alpha_1 + \alpha_2 + \ldots + \alpha_{i-1})^2.$$

For example, if α_1 is taken to be 2, then α_2 may be taken to be greater than $(2)^2 = 4$,. If taken to be 5, say then α_3 then may be taken to be greater than $(\alpha_1 + \alpha_2)^2 = (2 + 5)^2 = 49$, and so on. Note that $\alpha_1 < \alpha_2 < \ldots < \alpha_n$. Now suppose that $\alpha_I/\alpha_J = \alpha_K/\alpha_L$, with the I, J, K, and L having the properties noted above. Let m be the largest index such that m is in one of the sets I, J, K, and L, and is not in all four. There is such an m because $(I, J) \neq (K, L)$. Let $s = \sum_{i > m} \alpha_i$, which is 0, of course, if $m = n$. There are really only three distinct cases to consider:

(a) m is in I, J, and L, and is not in K;

(b) m is J and L, and is not in I or K;

(c) m is in J and not in I or K or L.

Case (a) is the case that m is in exactly three of the sets; case (b) is the case that m is in exactly two of the sets, and case (c) is the case that m is in exactly one of the sets. Write $\alpha_I = u + \alpha_m + s$ or $u + s$, depending on whether or not $m \in I$. Then $\alpha_J = u + v + \alpha_m + s$, since $m \in J$ in all three cases. The number u is just $\sum \alpha_i$, where $i \in I$ and $i < m$. This number may be 0. Now $v = \sum \alpha_i$, where $i \in J, i < m$, and $i \notin I$. Similarly, $\alpha_K = x + s$, since we never have to consider the case where $m \in K$, and finally,

$\alpha_L = x + y + \alpha_m + s$ or $x + y + s$ depending on whether or not $m \in L$. An important point is that $u, v, x,$ and y are sums of α_i's for some i's $< m$, so are small relative to α_m and to s, if $s > 0$.

Suppose that we are in case (a), and that $\alpha_I/\alpha_J = \alpha_K/\alpha_L$. This gives

$$(u + \alpha_m + s)/(u + v + \alpha_m + s) = (x + s)/(x + y + \alpha_m + s)$$

whence

$$(u + \alpha_m + s)(x + y + \alpha_m + s) = (u + v + \alpha_m + s)(x + s).$$

This equality yields

$$s(\alpha_m - v + y) + uy + u\alpha_m + y\alpha_m + (\alpha_m)^2 = xv.$$

This is impossible. All the terms on the left are ≥ 0, and $(\alpha_m)^2 > xv$ since

$$\alpha_m > (\alpha_1 + \alpha_2 + \ldots + \alpha_{m-1})^2 \geq xv.$$

In case (b), we have the equality

$$(u + s)/(u + v + \alpha_m + s) = (x + s)/(x + y + \alpha_m + s),$$

which yields

$$s(y - v) = x(\alpha_m + v) - u(\alpha_m + y).$$

The left side $s(y - v)$ must be 0. Otherwise,

$$s \neq 0 \neq |y - v| > 1,$$

and

$$|x(\alpha_m + v) - u(\alpha_m + y)| < (\alpha_1 + \alpha_2 + \ldots + \alpha_m)^2 < \alpha_{m+1} < s|y - v|.$$

But if $s(y - v) = 0$, then

$$\alpha_m(x - u) = uy - xv,$$

whence $x = u$, from which it follows that $y = v$. But this means that $I = K$ and $J = L$. which is not the case.

In case (c), we have the equality

$$(u + s)/(u + v + \alpha_m + s) = (x + s)/(x + y + s),$$

which yields

$$s(\alpha_m + v - y) = uy - vx - x\alpha_m .$$

If $s > 0$, the left side is positive, and the right side is negative unless $x = 0$. In this case $s(\alpha_m + v - y) > uy$ since $s > (\alpha_1 + \alpha_2 + \ldots + \alpha_m)^2 > uy$. If $s = 0$, then $x\alpha_m = uy - vx$, so $x = 0$. But then $K = \phi$, which is not allowed. □

There are a couple of comments that should be made about getting the desired α_i in the proof of Theorem 2. First, it is clear that they can be chosen to be integers, in which case the resulting probability measure will have all rational values. Secondly, that such α_i exist is no problem. Taking the α_i to be algebraically independent (over the rational numbers) will get the desired distinct $P(a|b)$. Being algebraically independent means that there is no non-trivial polynomial f with rational coefficients such that $f(\alpha_1, \alpha_2, \ldots, \alpha_n) = 0$. If $\alpha_I/\alpha_J = \alpha_K/\alpha_L$ for some $(I, J) \neq (K, L)$, then $\alpha_I\alpha_K - \alpha_J\alpha_L$ is a non trivial polynomial in the α_i's, and is 0. That such algebraically independent sets exist is a well known algebraic fact. A good reference is Hungerford (1974, page 311).

An immediate corollary of Theorem 2 is Lewis' Triviality Result for those R that are the algebra of all subsets of a set with at least 3 elements. Indeed, in that case

$$3^n - 2^{n+1} + 3 > 2^n,$$

and since there is a probability measure P on R such that $P(a|b)$ takes $3^n - 2^{n+1} + 3$ distinct values, there is no binary operation $\lozenge$ on R such that $P(a\lozenge b) = P(a|b)$ for all a and b in R.

Now suppose that R is *any* Boolean algebra with q elements, $q > 4$. Then R is a subalgebra of the algebra $\mathcal{P}(\Omega)$ of all subsets of a finite set Ω with at least three elements. As we have seen, there is a probability measure P on $\mathcal{P}(\Omega)$ taking distinct values $P(a|b)$ not 0 or 1 for every pair (a,b) with $0 < a < b$. The restriction of P to R is a probability measure on R, and taking $b = 1$ yields q distinct values for $P(a|b)$, counting 0 and 1. As in the proof of Lewis' Triviality Result above, there are elements a and b in R with $ab \neq 0 \neq a'b$, and $b \neq 1$. Thus $P(ab|b)$ is yet another value, yielding more than q values of $P(a|b)$ for elements a and b of R. *This implies Lewis' Triviality Result for any finite Boolean algebra with more than four elements.* It is not clear how Theorem 2 can be used to prove Lewis' Triviality Result for infinite R.

The theorem also shows that if R is the algebra of all subsets of a set with n elements, then any model S of the space of conditional events that is compatible with probability must have at least $3^n - 2^{n+1} + 3$ elements, and if R is any finite Boolean algebra such a model must be larger than R.

The key fact in all the above is the existence of a probability measure P on the Boolean algebra of all subsets $\mathcal{P}(\Omega)$ of a finite set Ω. The *existence* of such a P is a trivial consequence of the existence of algebraically independent real numbers, as we have noted. To actually construct a measure, as we did in the lemma, yielding the desired probability involved a bit of arithmetic, but an elementary prescription was given for the numbers needed.

Following is an alternate proof of the existence of such a probability measure on these special finite Boolean algebras. It may have some independent interest. We are going to show that if $R = \mathcal{P}(\Omega)$, with Ω finite, then there is a P_o on R such that P_o is one-to-one on the set

$$\{(a, b) : \emptyset \neq a < b,\ a, b \in R\},$$

that is, whenever $\emptyset \neq a_1 < b_1$, $\emptyset \neq a_2 < b_2$ and $(a_1, b_1) \neq (a_2, b_2)$, then we have $P_o(a_1 | b_1) \neq P_o(a_2 | b_2)$. We will carry out the proof of this by induction on $\#(\Omega)$. Specifically, we are going to show that, for each $n > 1$, there is a probability measure P_n on $R_n = \mathcal{P}(\Omega_n)$, where $\Omega_n = \{\omega_1, ..., \omega_n\}$, such that

$$\#\{P_n(a|b) : a, b \in R_n, P_n(b) > 0\} = 3^n - 2^{n+1} + 3.$$

For $n = 1$, $\Omega_1 = \{\omega_1\}$, and $R_1 = \{\emptyset, \omega_1\}$. Let $P_1 = \delta_{\omega_1}$, that is, $P_1(\emptyset) = 0$, $P_1(\omega_1) = 1$. We have

$$\#\{P_1(a|b) : a = \emptyset, \omega_1; b = \omega_1\} = 2 = 3 - 2^2 + 3\ .$$

For $n = 2$, $\Omega_2 = \{\omega_1, \omega_2\}$, and $R_1 = \{\emptyset, \omega_1, \omega_2, \{\omega_1, \omega_2\}\}$. Denoting as usual the Dirac mass point probability at ω by δ_ω, let $P_2 = (1/3)\delta_{\omega_1} + (2/3)\delta_{\omega_2}$. We have

$$\{(a, b) : \emptyset \neq a < b\} = \{(\omega_1, \{\omega_1, \omega_2\}), (\omega_2, \{\omega_1, \omega_2\})\}$$

and

$$P_2(\emptyset | \omega_1) = P_2(\emptyset | \omega_2) = P_2(\omega_1 | \omega_2) = P_2(\omega_2 | \omega_1)$$

$$= P_2(\emptyset | \{\omega_1, \omega_2\}) = 0\ ;$$

$$P_2(\omega_1 | \omega_1) = P_2(\omega_2 | \omega_2) = P_2(\{\omega_1, \omega_2\} | \{\omega_1, \omega_2\}) = 1\ ;$$

$$P_2(\omega_1 | \{\omega_1, \omega_2\}) = 1/3 \neq 2/3 = P_2(\omega_2 | \{\omega_1, \omega_2\})\ .$$

Thus

$$\#\{P_2(a|b) : a, b \in R, b \neq \emptyset\} = 4 = 3^2 - 2^3 + 3.$$

Note that P_1, P_2 are both one-to-one on R_1, R_2, respectively, and take rational values in $[0, 1]$.

Assume that up to n, there is $P_n : R_n \to [0, 1]$ of the form

$$P_n = \Sigma_{j=1}^{n} (r_j/s_n)\delta_{\omega_j},$$

where the r_j's are distinct positive integers,

$$S_n = \Sigma_{j=1}^{n} r_j,$$

such that

$$\#\{P_n(a|b) : a, b \in R_n, b \neq \emptyset\} = 3^n - 2^{n+1} + 3.$$

Consider $\Omega_{n+1} = \{\omega_1, ..., \omega_n, \omega_{n+1}\}$. Define

$$P_{n+1}(\omega_j) = r_j / s_{n+1}, \ j = 1, ..., n + 1; \ s_{n+1} = \Sigma_{j=1}^{n+1} r_j,$$

and r_{n+1} is to be determined. Let $\emptyset \neq a_1 < b_1, \emptyset \neq a_2 < b_2$ and

$$a_1, a_2, b_1, b_2 \in R_{n+1}, \quad (a_1, b_1) \neq (a_2, b_2).$$

Now

$$\Omega_{n+1} = (b_1 \cup b_2) \cup b_1'b_2'.$$

If $\omega_{n+1} \in b_1'b_2'$, then $\omega_{n+1} \notin a_i, b_i, i = 1, 2;$ and

$$P_{n+1}(a_i|b_i) = P_{n+1}(a_i)/P_{n+1}(b_i) = \Big(\sum_{\{j:\omega_j \in a_i\}} r_j\Big)/\Big(\sum_{\{j:\omega_j \in b_i\}} r_j\Big)$$

$$= P_n(a_i|b_i), \ i = 1, 2.$$

Thus, by the induction hypothesis,

$$P_{n+1}(a_1|b_1) \neq P_{n+1}(a_2|b_2)$$

when

$$\omega_{n+1} \in b_1'b_2',$$

for any choice of integer r_{n+1} different from the r_j's, $j = 1, 2, ..., n$.

Consider next the case where $\omega_{n+1} \in b_1 \cup b_2$. A partition of $b_1 \cup b_2$ is

$$\{a_1a_2, a_1a_2'b_2, a_1b_2', a_1'b_1a_2, a_1'b_1b_2, a_1'b_1b_2', b_1'a_2, b_1'a_2'b_2\} .$$

To express the fact that $\omega_{n+1} \in a_i$ or $\omega_{n+1} \notin a_i$, we write

$$a_i = a_{i,0} \cup \Lambda_{i,1} \cdot \omega_{n+1}$$

for $\Lambda_{i,1} = \Omega_{n+1}$ or $\emptyset$, respectively. Similarly

$$b_i = b_{i,0} \cup \Lambda_{i,2} \cdot \omega_{n+1},$$

with

$$a_{i,0}, b_{i,0} \in R_n.$$

Define

$$\tau_{i,j} = \begin{cases} 1 \text{ if } \Lambda_{i,j} = \Omega_{n+1} \\ 0 \text{ if } \Lambda_{i,j} = \emptyset \end{cases}$$

and

$$\alpha_i = \sum_{\{k: \omega_k \in a_{i,0}\}} r_k ,$$

$$\beta_i = \sum_{\{k: \omega_k \in b_{i,0}\}} r_k .$$

We have

$$P_{n+1}(a_i | b_i) = P_{n+1}(a_i)/P_{n+1}(b_i)$$

$$= \frac{[P_{n+1}(a_{i,0}) + \tau_{i,1}P_{n+1}(\omega_{n+1})]}{[P_{n+1}(b_{i,0}) + \tau_{i,2}P_{n+1}(\omega_{n+1})]}$$

$$= [\alpha_i + \tau_{i,1}r_{n+1}]/[\beta_i + \tau_{i,2}r_{n+1}].$$

Thus

$$P_{n+1}(a_1 | b_1) = P_{n+1}(a_2 | b_2)$$

if and only if r_{n+1} satisfies

$$(**) \qquad Ar_{n+1}^2 + Br_{n+1} + C = 0$$

where

$$A = \tau_{1,1}\tau_{2,2} \quad \tau_{1,2}\tau_{2,1}$$

$$B = \tau_{2,2}\alpha_1 + \tau_{1,1}\beta_2 - \tau_{1,2}\alpha_2 - \tau_{2,1}\beta_1$$

$$C = \alpha_1\beta_2 - \alpha_2\beta_1 .$$

Our proof will be complete if for each pair (a_1, b_1), (a_2, b_2) the coefficients A, B, C cannot be all zero, since then either $A \neq 0$ or $B \neq 0$, and hence the quadratic equation (**) will have at most two solutions. Let J_{n+1} denote the set of all such solutions for all possible pairs (a_1, b_1), (a_2, b_2). Obviously, J_{n+1} is finite. It suffices to choose r_{n+1} to be a positive integer not in J_{n+1} and different from all the r_j's, $j = 1, 2, ..., n$.

The last point to show is:

$$A = B = C = 0 \text{ is impossible.}$$

For this purpose, suppose

$$A = B = C = 0 .$$

Under this assumption, and in view of the hypotheses concerning a_i, b_i, $i = 1, 2$, we note that:

(i) $\beta_i > 0$, $i = 1, 2$.

(ii) If $\alpha_1 = 0$ then $\alpha_2 = 0$ and conversely.

In this case, we have

$$\tau_{11}\beta_2 = \tau_{21}\beta_1$$

since

$$B = 0 .$$

But $\tau_{11}, \tau_{21} > 0$, since otherwise, $a_1 = a_2 = \emptyset$. Thus, $\tau_{11} = \tau_{21} = 1$ and hence $\beta_1 = \beta_2$. But this will imply that $b_1 = b_2$ and $a_1 = a_2$, contradicting the hypotheses. Hence:

$$0 < \alpha_i \leq \beta_i, \quad i = 1, 2.$$

(iii) In fact, $0 < \alpha_i < \beta_i$, $i = 1, 2$. But this is equivalent to

$$\emptyset < a_{i,o} < b_{i,o}, \quad i = 1, 2 .$$

Now, $C = 0$ applied to this case implies

$$\alpha_1/\beta_1 = \alpha_2/\beta_2$$

which is equivalent to $P_n(a_{1o}|b_{1o}) = P_n(a_{2o}|b_{2o})$. By the induction hypothesis,

$$\emptyset < a_{1o} = a_{2o} < b_{1o} = b_{2o} ,$$

that is

$$0 < \alpha_1 = \alpha_2 < \beta_1 = \beta_2 ,$$

which implies, using $B = 0$,

$$a_1 = a_2 \text{ and } b_1 = b_2 ,$$

a contradiction again.

Thus, in summary, $A = B = C = 0$ cannot hold. □

Here is an alternate proof of Lewis' Triviality Result in the general case. For a Boolean ring R, a probability measure P on R, and a mapping $f : R \times R \to R$, define

$$\mathcal{P}_P = \{P_b : P_b(\cdot) = P(\cdot\,|b), b \in R, P(b) > 0\} ,$$

$$\mathcal{U}_P = \{(a, b) : a, b \in R, P(b) > 0, P(a|b) = 0 \text{ or } 1\} ,$$

$$\mathcal{C}_{f,P} = \{(a, b) : a, b \in R, P(b) > 0 , \text{ for all } c \in R \text{ such that } P(bc) > 0 \text{ and } P_c[f(a, b)] = P_c(a|b)\} ,$$

and

$$\mathcal{I}_P = \{(a, b) : a, b \in R, P(b) > 0, (a, b) \notin \mathcal{U}_P, \text{ and } 0 < P(ab) = P(a)P(b) < 1\} .$$

Then

$$\mathcal{C}_{f,P} \subseteq \mathcal{I}_P \cup \mathcal{U}_P.$$

Indeed, let $(a, b) \in \mathcal{C}_{f,P} \setminus \mathcal{U}_P$. Then $0 < P(a|b) < 1$. For $c = 1$, $P(bc) = P(b) > 0$, and we have

$$P[f(a, b)] = P(a|b).$$

Also,

$$P_a[f(a, b)] = P_a(a|b) = 1,$$

and

$$P_{a'}[f(a, b)] = P_{a'}(a|b) = 0,$$

whence

$$P(a|b) = P[f(a, b)] = P[af(a, b)] + P[a' \cdot f(a, b)] = P(a) ,$$

that is, $P(ab) = P(a)P(b)$, which means that $(a, b) \in \mathcal{I}$.

As a consequence, if R is a Boolean ring having at least two elements a, b such that $0 < a < b < 1$, then there is no map $f: R^2 \to R$ compatible with conditional probability. Indeed, if such a map f exists, then choose P to be a probability measure on R such that $0 < P(a) < P(b) < 1$. We have

$$(a, b) \notin \mathcal{U}_P, \ (a, b) \in \mathcal{C}_{f,P} .$$

But

$$0 < P(a)P(b) < P(a) = P(ab) < 1 ,$$

and $(a, b) \notin \mathcal{I}_P$, a contradiction. Thus no such map f exists, and Lewis' Triviality Result follows. □

We turn now to Copeland's work. He looked for a mathematical operator representing the logical connective "if" in R, analogous to division. That is, he apparently had in mind modeling (a|b) with the "fraction" a/b in R. With this he introduced the following notion.

Definition. *An implicative Boolean ring is a Boolean ring R together with an additional binary operation × satisfying the following, for elements a, b, c in R.*

(*i*) $a\times(b\times c) = (a\times b)\times c$,

(*ii*) $a\times(b + c) = a\times b + a\times c$,

(*iii*) $a\times(bc) = (a\times b)(a\times c)$,

(*iv*) *if* $a\times b = a\times c$ *and* $a \neq 0$, *then* $b = c$,

(*v*) $a\times 1 = a$, *and*

(*vi*) *for every a and b with* $b \neq 0$, *there is an element c such that* $ab = b\times c$.

Axioms (*iv*) and (*vi*) enable one to define an operator ◊ by $ab = b\times(b◊a)$, for $b \neq 0$. For $b \neq 0$, the mapping $Rb \to R : rb \to b◊(rb)$ is one-to-one and onto. Indeed, $b◊(rb) = b◊(sb)$ implies that

$$b\times(b◊(rb)) = b\times(b◊(sb)) = rbb = rb = sbb = sb,$$

so that the mapping is one-to-one. Since

$$x(x\times y) = (x\times 1)(x\times y) = x\times(1 \cdot y) = x\times y, \quad x\times y \leq x.$$

Thus $b\times r$ is an element of Rb and $b\times r \to b◊(b\times r) = r$ since $b\times r = (b\times r)b = b\times(b◊(b\times r))$ Thus the mapping is onto. This shows that every implicative Boolean ring $\neq \{0\}$ is infinite, since R and Rb are in one-to-one correspondence and Rb is smaller than R for $b \neq 1$ and for R finite. The mapping above is actually an isomorphism, so R is isomorphic to Rb for every $b \neq 0$. This severely limits the usefulness of implicative Boolean rings. Also, there do not seem to be any appealing examples of them.

By a "probability", Copeland meant any probability measure P on R such that

$$P(a \times b) = P(a)P(b).$$

This makes $\Diamond$ compatible with such probabilities, since

$$P(a|b) = P(ab)/P(b) = P(b \times (b \Diamond a))/P(b) = P(b)P(b \Diamond a)/P(b) = P(b \Diamond a).$$

If all probabilities on R satisfied this condition, then Lewis' Triviality Result would imply that there were no implicative Boolean rings except 0. Thus given a non-trivial implicative Boolean ring, only some probability measures on it are allowable, and Copeland does not elaborate on that point.

Since the intention of Copeland was to stay in the ring R, the problem of "conditional logic", that is, considering operations between conditional events, did not arise. In any case, this approach through implicative Boolean rings seems futile. As Pfanzagl (1971, Chapter 12) has pointed out, if $(a|b) \in R$, then $c \wedge (a|b)$ admits no semantic interpretation.

1.2 Division of events

Although Copeland did mention that his operator "if" is somewhat analogous to division, he did not elaborate further on this connection. It turns out that in Boole's basic work (Boole, 1854), the problem of interpretation of division of propositions was considered in some detail. However, since all elements except 1 in a Boolean ring are zero divisors, there are bound to be some difficulties with this approach. We now outline the idea of Boole and the follow-up work of Hailperin.

Boole's division interpretation

In his basic work (Boole, 1854) which laid down the foundation of symbolic logic, Boole explained vaguely an interpretation for division in a Boolean ring. For elements a and b of R, the element a/b is defined to be an element of R such that $(a/b)b = a$. Now such an element exists only if $a \leq b$. This difficulty can be circumvented by requiring that $a/b = ab/b$. But then, instead of trying to solve the equation $(ab/b)b = ab$ for ab/b, which has many solutions, indeed the whole coset $a + Rb'$, which of course is not an element of R, Boole proceeded differently. Writing down the normal disjunctive form of a binary

Boolean function as

$$f(a,b) = [f(1,1)ab] \vee [f(0,1)a'b] \vee [f(0,0)a'b'] \vee [f(1,0)ab'],$$

he took formally $f(a,b) = a/b$, leading to the expansion

$$a/b = ab \vee (0/0)a'b' = a \vee (0/0)b',$$

since $1/1 = 1$, $0/1 = 0$, and $1/0$ is not defined so that ab' has to be 0, that, is $a \leq b$. It remains to interpret the indefinite "quantity" $0/0$. Here is Boole's description of $0/0$: "The symbol $0/0$ indicates that a perfectly indefinite portion of the class, that is, "some", "none", or "all of its members are to be taken" (Boole, page 92). Translating Boole, $a \vee xb'$ is a candidate for a/b, for $a \leq b$ and for any x in R, and of course $a \vee xb'$ is precisely the coset $a + Rb'$. As in the case of Copeland, no attempt was made concerning logical operations among these "algebraic fractions".

Jevons (1879) objected to Boole's division on the grounds that it lacked clarity. Peirce (1867) retained Boole's operation of division and embellished it. MacFarlane (1879) produced a very readable and improved version of Boole's idea. Unfortunately, this work did not enter the main body of logic. In effect, the vacuum created by the lack of division in Boole-Schroeder logic was filled by the introduction of other operations within logic, such as material implication.

Rigorization of Boole's technique

Hailperin (1976) analyzed thoroughly Boole's original work - especially his long forgotten concepts of logical division and fractions of events. In fact, Hailperin came to the conclusion that Boole's division is viable, provided sufficient rigor is used in developing the idea. This was accomplished by forming a Chevalley-Uzkov "ring of quotients" corresponding to "divide" by an event b (Uzkov, 1949). Indeed, since all the elements of a Boolean ring R are zero divisors (except 1), the standard approach to rings of quotients is not applicable. A way around this situation were given by Uzkov (1949) for commutative rings with unity. If R is such a ring, then the construction of a ring of quotients for R is as follows. Let S be a multiplicatively closed subset of R not containing 0. That is, $S \subseteq R$ and $xy \in S$ whenever $x, y \in S$. An equivalence relation is defined on $R \times S$ by

$$(r, s) \approx (t, u)$$

if and only if there is an element $x \in S$ with

$$x(st - ru) = 0.$$

The set R_S of equivalence classes r/s of $\approx$ is a commutative ring with identity under the operations given by

$$r/s + t/u = (ru + st)/su$$

and

$$(r/s)(t/u) = rt/su.$$

When R is Boolean, then any element $b \in R$ with $b \neq 0$ is a candidate for S above. Thus the ring $R_{\{b\}}$ can be formed, and it is easy to see that

$$R_{\{b\}} \to Rb : a/b \to ab$$

is an isomorphism, and of course

$$Rb \to R|Rb' : ab \to a + Rb'$$

is an isomorphism as well. Thus Hailperin is led to the association of a conditional event $(a|b)$ with the coset $a + Rb'$. Hailperin actually took S to be $R \vee b$, the principal filter associated with b, but this leads to the same ring of quotients, and hence both to $R|Rb'$. Calabrese (1987), without having in mind conditional events as "quotients" in a Boolean ring, proposed an equivalent definition, and hence one equivalent to the coset form. In any event, for Hailperin, "fractional events" became cosets of principal ideals. In any case, it allowed Hailperin to justify Boole's notion of fractional events. He went on to consider these "fractional events" or "conditional events", as the set of all cosets of principal ideals. In ring theory, it is not customary to define operations among cosets of different quotient rings. Because of this fact, Hailperin (p. 112-113) refers simply to the collection of all conditional events as a *partial algebra*, that is, the operations $+$ and $\cdot$ can be only defined on each quotient ring, but not between two cosets from two different quotient rings. This is somehow surprising since it is precisely this point which is important for a logic of conditional propositions. It is here that a good motivation for a new problem in ring theory arises. The problem is this: What are the operations of interest on the union of all quotient rings of R extending those on each fixed one? This question is the topic of Chapter 3. Another point is that in his discussion concerning truth tables, Hailperin (p. 127) did realize that conditional events have three possible truth values. This fact was realized much earlier by DeFinetti (1964) and Schay (1968) who defined conditional events precisely this way, that is from a semantic viewpoint (or equivalently, by extending the concept of ordinary indicator functions of events or sets).

Not only is this approach to conditional events through a three-valued logic equivalent to the coset form of conditional events (see Chapter 2), but it sheds light on how to define logical operations among conditional events, addressing the problem in ring theory mentioned above. Indeed, it is well-known in classical two-valued logic that if

$$f: \{0, 1\}^n \to \{0, 1\},$$

then there is a unique logical operation

$$\varphi_f : R^n \to R$$

such that

$$t[\varphi_f(a_1, ..., a_n)] = f(t(a_1), ..., t(a_n)),$$

where $a_i \in R$, $i = 1, 2, ..., n$ and t stands for "truth value of." (See, for example, Hamilton, 1978). It turns out that this result remains valid in a three-valued logic, as we show in Chapter 3. Thus, not only will each system of operations on conditionals have a logical interpretation, but more importantly, the above extension problem in Boolean ring theory is solvable in view of existing systems of three-valued logics, for example those of Lukasiewicz, Sobocinski, Kleene, and Bochvar (see Rescher, 1969, and our Chapter 3).

To complete a survey of Hailperin's work, it should be also mentioned that in making "Boole's probability rigorous," Hailperin (footnote on p. 191) took the probability of a conditional event, that is, of a coset, to be a conditional probability. This is indeed well-defined, and is precisely a "compatibility condition" with probability leading to an axiomatic theory of conditional events (see Chapter 2). In the same vein, Hailperin (p. 195-197) proceeded to consider the concept of a "conditional events probability realm." This is somehow similar to Renyi's (1970) approach to conditional probability spaces, but in which there is a home for $(a|b)$ in $P(a|b)$. See also our Chapter 5. However, since the space of conditional events was not investigated far enough to reach a reasonable algebraic structure (mainly due to the lack of operations amongst conditional events), no new concepts were introduced beyond that. In Chapter 4, we will show that the space of conditional events is a *Stone algebra*, generalizing Boolean algebras. In other words, the "partial algebra" of Hailperin is in fact an algebra with Lukasiewicz's three-valued logic interpretation.

In light of the attempt to use the theory of "rings of quotients", the following is pertinent, and should lay these attempts to rest. The only quotients that make sense in a Boolean ring are those a/b where $a \leq b$. By a/b, we mean an element in R for which $(a/b)b = a$. Indeed, if $(a/b)b = a$, then multiplying through by b gives $ab = a$, whence $a \leq b$. If $a \leq b$, then taking $a/b = a$ gives an element whose product with b is a.

Further, the ring R cannot be enlarged to another ring so that a/b is defined for a not $\leq b$. Indeed, $b'(a/b)b = b'a = 0$, and this is not the case unless $a \leq b$. So trying to enlarge R so that division of events is possible in that enlargement is futile. Any divisions by elements of R that can be carried out in a larger ring can already be carried out in R. More general statements are true. A "ring of quotients" of a ring R is a ring S and a homomorphism $f: R \to S$ satisfying certain properties. If R is Boolean, then so is the subring $f(R)$ of S. Suppose one wanted to make an element b in R into a unit in S, that is, wanted S to be such that $f(b)$ could be divided into everything in S (or even in $f(R)$). Then in $f(R)$, $[f(1)/f(b)]f(b) = f(1)$, whence, multiplying through by $f(b)$ as above, we get $f(1) = f(b)$. But $f(1) = 1$ in S, f being a homomorphism, so $f(b) = 1$. So the only way to make an element of R into a unit is to make it into the identity element. If a/b is to make sense for every element a (or even just for $a = 1$), that is, if b is to be a unit, then the setting must be such that $b = 1$. What Hailperin did, in effect, was to go to the ring R/Rb', or equivalently, Rb where indeed b is the identity; $b + R/Rb' = 1 + R/Rb'$, and b certainly is the identity of Rb.

Hailperin used a special Chevalley-Uzkov ring of quotients. There are many "rings of quotients" in ring theory, two others being Johnson-Utumi ring of quotients and the "classical ring of fractions." (See for example, Lambek, 1966, for background). This can be seen as follows. We describe these two briefly for commutative rings.

Let R be a commutative ring with unity 1. An ideal I of R is said to be *dense* if, for all $r \in R$, $rI = 0$ implies $r = 0$. A *fraction* is a (module) homomorphism $h: I \to R$ with domain I being a dense ideal. That is, if $x, y \in I$, then

$$h(x + y) = h(x) + h(y),$$

and if $x \in I$, $r \in R$, then

$$h(rx) = rh(x).$$

Let $Hom(I, R)$ be the class of fractions with domain I, and let $F(R)$ be the union of then $Hom(I, R)$ over all dense ideals I. For $f \in Hom(I, R)$ and $g \in Hom(J, R)$, let $f \approx g$ if $f = g$ on $I \cap J$. Then $\approx$ is an equivalence relation on $F(R)$. It is obviously reflexive and symmetric. Transitivity is less obvious. For that, first note that the intersection of dense ideals is dense. Indeed, for I and J dense, $r(I \cap J) = 0$ imples that $rIJ = 0$, so $rI = 0$, whence $r = 0$. Now let $f \in Hom(I, R)$, $g \in Hom(J, R)$ and $h \in Hom(K, R)$ with $f \approx g$ and $g \approx h$. Let $x \in I \cap K$. For $y \in I \cap J \cap K$,

$$f(x)y = f(xy) = g(xy) = h(xy) = h(x)y,$$

so

$$(f(x) - h(x))y = 0,$$

and so

$$f(x) = h(x).$$

Thus $\approx$ is also transitive.

With appropriate operations (see for example, Lambek, 1966), the set $Q(R)$ of equivalence classes is a ring, called the Johnson-Utumi complete ring of fractions of R. Since $Hom(R, R) \to R : f \to f(1)$ is a ring isomorphism, $Q(R)$ contains a copy of R. For a non-zero divisor $r \in R$, Rr is dense, and the $Hom(Rr, R)$ for such r give rise to a subring $CL(R)$ of $Q(R)$ called the classical ring of quotient of R. An element $f \in Hom(Rr, R)$ is identified with the "fraction" x/r, where $f(r) = x$.

When R is a Boolean ring, the only non-zero divisor is 1, thus the only dense principal ideal is R itself, and $CL(R) = Hom(R, R)$ is just R itself. This is the case if R is finite, for example.

If R is the ring of all subsets of a set Ω, then an ideal I is dense if and only if $x = \bigvee_{i \in I} i = 1$, since otherwise $x' \neq 0$ and $x'I = 0$. In this case, $Q(R)$ is also just R itself. To see this, for I dense, $f \in Hom(I, R)$ and $j \in I$,

$$f(j) = f(j) \cdot 1 = f(j)(\bigvee_{i \in I} i) = \bigvee_{i \in I} f(j)i = j(\bigvee_{i \in I} f(i)),$$

so f is just multiplication by $\bigvee_{i \in I} f(i) = x$. This f is equivalent to $h \in Hom(R, R)$ given by $h(r) = rx$, and $Hom(R, R)$ is isomorphic to R as indicated above. More generally, viewing R as a subalgebra of 2^{Ω} for some set Ω, call R *complete* if R is closed under arbitrary unions. Then, as above, I is dense if and only if $\bigvee_{i \in I} = I$, and $Q(R) \approx R$. It turns out that $Q(R) \approx R$ if and only if R is complete (Lambek, 1966). There exist non-complete Boolean rings, so in general $Q(R)$ properly contains R.

1.3 Three-valued logic

Independently of each other, Reichenbach (1948, 1949), Schay (1968), DeFinetti (1972, 1974, Volumes 1 and 2), and Dubois and Prade (1987, 1990) considered the modeling of conditional events from a logical viewpoint. They all viewed a conditional event as an object with three possible "truth" values.

First, Reichenbach considered probability as being determined completely through all standard logical operations over Boolean algebras of propositions and quantified expressions, as well as through the adjunction of a distinct "probability implication" operator P corresponding to $P(\cdot \mid \cdot)$. He also developed a calculus of probabilities (1949,

Chapter 3), and a related probability logic (1949, Chapter 10). Probabilistic conditioning and logical implication were compared in two places: (i) in the discussions of basic axioms for probability (1949, pages 54-57), in which probabilistic conditioning is argued to be a natural, but because of the zero-probability antecedent case, tacitly modified extension of logical implication, and (ii) in the use of P as a "quasi-implication" (1948, table 4b, page 151; table 5, page 168; and pages 166-168). Reichenbach proposed that, in place of classical logical implication, quasi-implication relative to three-valued logic (0 = false, 1 = true, and I = indeterminate) was a more suitable operation. Informally, for α in $\{0, 1, I\}$,

$$0 \overset{P}{\rightarrow} \alpha \text{ and } 1 \overset{P}{\rightarrow} \alpha$$

are defined semantically as $P(\alpha|0) = I$ and $P(\alpha|1) = \alpha$, respectively. Reichenbach also briefly considered an equivalent form of the concept of measure-free conditionals through his "indeterminate" probability implication operator $b \rightarrow a$ via symbolic logic as $(\exists P)(b \rightarrow a)$. (See Reichenbach, 1948, pages, 51, 52, 71, 72).

Schay (1968) asked "could $(a|b)$ be defined in a manner consistent with general usage in probability theory, that is, so that $P(ab)/P(b)$ may be interpreted as the probability of $(a|b)$?" Schay proposed to define $(a|b)$ as a generalized indicator function on Ω (here R is a Boolean ring of subsets of Ω) where

$$(a|b)(\omega) = \begin{cases} 1 \text{ if } \omega \in ab \\ 0 \text{ if } \omega \in a'b \\ u \text{ (undefined) if } \omega \in b'. \end{cases}$$

Note that such functions are clearly in one-to-one correspondence with elements of Rb, or with elements of the quotient ring R/Rb' since such functions specify the subsets b and ab, and conversely.

In discussing conditional probabilities, DeFinetti made a remark about $P(a|b)$, saying that one can even talk about the probability of the "conditional event" $(a|b)$ (DeFinetti, 1974, page 134). He specifies this mathematical object on page 139 of that reference, as a "tri-event", corresponding precisely to Schay's notion. DeFinetti also considered interpreting conditional events through a coset representation (1974, Vol. 1, pages 267-269), but apparently did not connect this with Mazurkiewicz (1956) and others' ideas on the same subject. (See Section 1.4 below.) Furthermore, DeFinetti even considered briefly how one could obtain a "logical sum" of such conditional events (1974, Vol 2, page 310) as sell as how double conditional events could be interpreted (1974, Vol 2, pages 327-328). In a related vein, DeFinetti broached the issue of "counterfactuals" and verifications relative to conditional events, and concluded that compatibility constraints were key to any further analysis of operations on such entities. But, other than brief comments on the potentiality of how a calculus of operations among conditional events could be developed - and was needed - no actual work in this direction was executed.

Bruno and Gilio (1985), inspired by DeFinetti's much earlier work, proposed an abbreviated algebra of measure-free conditional events to produce "conditional hyper-probabilities", which, in turn, were used to obtain some new factorization results for Scozzafava's pseudodensities (1984). However, they did not attach any direct interpretation to conditional events such as being cosets of principal ideals, as presented in our work here. It turns out that their operations are identical to certain of those proposed by Schay (1968) and Calabrese (1987). (See also Section 1.5.) Based on DeFinetti's work, but independent of Bruno and Gilio, Barigelli and Scozzafava (1984) mentioned the lack of apparent attention paid to the domain of conditional probability operators, that is , to conditional events. Their thesis was that careful consideration of such could lead to improved interpretations of frequency data and the elimination of certain confounding problems.

In discussing reasoning with uncertain information, Dubois and Prade (1987, first edition 1985), were led to consider a symbol like $(\cdot\,|\,\cdot)$ for a "non-traditional" logical connective. (It should not be confused with Sheffer's "binary rejection" $(a|b)$, which is defined as $a'b'$). A truth value table for $(a|b)$ is established by observing that the truth values $t(a|b)$ of $(a|b)$ are solutions of the equation $t(ab) = Min\{t(a|b), t(b)\}$ and so are given by $t(a|b) = 1, 0$, or $\{0,1\}$ according as $t(ab) = 1$ or $t(a'b) = 1$, or $t(b) = 0$. Here, $\{0,1\}$ is referred to as an "indeterminate". See also Dubois and Prade (1989, 1990).

Some additional efforts related to conditional events are these: Cox (1961) established an algebra of conditioning for a fixed common antecedent by formally omitting the probability operator everywhere it appears in a conditional or unconditional form. It appears that Cox implicitly recognized the need to establish measure-free conditional events, but did not continue the development. The closest he came to the point of introducing conditional events is in Chapter I.3, corresponding to the measure-free chaining forms. (See Chapter 3 in this book.)

Popper (1961, Appendices IV* and V*) developed a postulate system for probability which included conditional probability perceived as a numerical operator on ordered pairs of primitives subject to certain algebraic relations within the probability arguments. Furthermore, he defined unconditional events within the probabilities through probabilistic equivalence, but unfortunately did not attempt to carry out a similar procedure for the ordered pairs within the probability operator, corresponding to conditional events.

It should be mentioned also the very general work of Foulis and Randall (1971,

1974) concerning the development of measure-free conditioning maps. Future efforts may uncover useful connections between that work and ours.

1.4 Coset form of conditional events

In the attempt to define rigorously the concept of conditional events, compatible with probability theory, except for Copeland, all previous researchers came across, in one form or another, cosets of principal ideals in a Boolean ring. This section is devoted to a survey of more apparent work related to the coset form of conditional events. In historic order, we survey the work of Koopman (1940), Mazurkiewicz (1956), Domotor (1969), Pfanzagl (1971), Hailperin (1986), and Calabrese (1987). Again, it should be noted that all these efforts were carried out independently.

Koopman's program was an investigation into "the axioms and algebra of intuitive probability". The algebra part, that is, the definitions of logical operations among conditional events, was not addressed! The idea of intuitive or qualitative probability is well-known: the primal intuition probability expresses itself in a partial order relation among events. Qualitative (or comparative) probability is motivated by the desire to make numerical probability measures compatible with non-numerical probability comparisons. See, for example, Fine (1973), Fishburn (1983), Villegas (1967), Domotor (1969), and Suppes (1973). Since information is basically conditional, "conditional events" should be the basic building blocks rather than unconditional ones. While Rényi (1970) took this viewpoint, from a numerical approach, extending Kolmogorov's model (see Chapter 5), Koopman first proceeded from a measure-free attack. For a and b in R, an expression denoted by $(a|b)$ is called an "eventuality". In a footnote (1940b, page 270), Koopman mentioned that the notation $(a|b)$ is used in a manner "close" to that of coset $a + Rb'$, without further elaboration. In fact, $(a|b)$ is used as an "alternative notation" for $a + Rb'$. From a qualitative viewpoint, the basic problem is that of comparison in probability of eventualities. That is to say, a system of axioms for a partial order relation among conditionals should first be given. He focuses attention on the set $R|R = \cup_b R/Rb$ of all cosets of all principal ideals, and postulates the existence of "n-scales". From this, upper and lower numerical probabilities of conditional events were shown to exist. There is some analogy here in the work of Dempster (1967) and of Shafer (1976). Conditional probabilities themselves were next defined as common values of upper and lower probabilities (when they happen to be the same). On conditional events with the same antecedent, this conditional probability reduces to an ordinary probability measure. It seems at this point that Koopman wanted simply to show that the numerical probability that he introduced generalized the Kolmogorov model. He made no attempt to introduce operations between conditional events with different antecedents and to consider the behavior of his new probability with respect to such operations. Instead, he gave a system of axioms for orderings among "eventualities," that is conditionals with different antecedents. See Section 3.5 for more discussion on Koopman's system of axioms for intuitive probability.

The major contributions of Mazurkiewicz (1956, Chapter III) were to identify conditional events as cosets of principal ideals and to note the consistency of the assignment of probabilities to these cosets. That is, the assignment $P(a + Rb') = P(a|b)$ is well defined, being $P(ab)/P(b)$, so that the assignment $P(a + Rb) = P(a|b)$ gives a probability measure on the quotient ring R/Rb' of cosets of the principal ideal Rb'.

Domotor (1969) also identified conditionals with cosets of principal ideals, but did not introduce a partial order on it. He embedded R/R into a larger algebraic structure equipped with a vector space structure, identifying R/R with DeFinetti's generalized indicator functions. On this vector space, probability measures are viewed as linear functionals. However, except for vector space operations, no attempt was made to consider extensions of ordinary boolean operators.

In his book on the theory of measurement, Pfanzagl (1971) presented an approach to the simultaneous measurement of utility and subjective probability generalizing Morgenstern-Von Neumann's approach (Von Neumann and Morgenstern, 1947). The syntactic concept of conditional events is essential in his work. Even Pfanzagl was aware of Copeland's implicative algebra (1941), (but not of Koopman's work (1940) in which the coset form was proposed for conditionals!); he did not adopt Copeland's concept of (measure-free) conditional events. Instead, he proposed the coset $a + Rb'$ for $(a|b)$. But he stayed in a fixed (Boolean) quotient ring R/Rb' (that is, for a fixed $b \in R$), so that conditional events with different antecedents were not investigated. He considered conditioning in R/Rb', that is, iterated conditional events of the form $((a|b)|(c|b))$, for b fixed, and this was defined simply as a coset of the Boolean ring R/Rb', that is,

$$((a|b)|(c|b)) = (a|b) + (R/Rb')(c|b)',$$

where

$$(c|b)' = (c'|b),$$

the "negation" of $(c|b)$ in R/Rb'. (All operations involved are coset operations on R/Rb'.) With the mathematical concept of (measure-free) conditional events, the notion of "compound wagers" (conditional bets) can be formulated. For each fixed $b \in R$, the Boolean quotient ring R/Rb' is the collection of all conditional events (or conditionals) with the same antecedent b. The conditional $(a|b) \in R/Rb'$ can be used in the representation of "compound wagers" (conditional bets) (Pfanzagl, 1971, p. 205-207) as follows.

Following Pfanzagl, a "wager" is a situation with a finite number of possible "outcomes," exactly one of which is to occur. Let $A = \{\alpha_1, ..., \alpha_n\}$ be the set of outcomes for a wager W. Which one of the α_i's occurs depends on some uncertain event $a \in R$. A "simple wager" W_a is defined to be a wager with only two possible outcomes, say α, β. Specifically, $w_a = \alpha$ if a occurs and β if a' occurs. In a logical framework, w_a can be viewed as a function on the set of maximal filters Ω of R that is, models of R. (See Chapter 6): For all $\omega \in \Omega$,

$$w_a(\omega) = \begin{cases} \alpha \text{ if } a \in \omega \\ \beta \text{ if } a' \in \omega . \end{cases}$$

A compound wager is defined to be a wager with

$$A = \{\alpha_1, \alpha_2, \alpha_3, \alpha_4\},$$

depending upon $(a|b)$, i.e.

$$w_{(a|b)}(\omega) = \begin{cases} \alpha_1 \text{ if } ab \in \omega \\ \alpha_2 \text{ if } a'b \in \omega \\ \alpha_3 \text{ if } ab' \subset \omega \\ \alpha_4 \text{ if } a'b' \in \omega . \end{cases}$$

For more detail, see Pfanzagl (1971, Chapter 12). See also Neapolitan (1990, p. 57) for conditional bets.

Independently of previous work on the subject, Calabrese (1987) investigated a conditioning operator in logic from an empirical viewpoint. His approach is algebraic, and is based on a relation between logical deducts (or consequences) and filters in a Boolean ring. (See Tarski, 1956.) For each b in R, the set of deducts of b is the filter $R \vee b = \{r \vee b : r \in R\}$. This is precisely the coset $b + Rb'$. Noting

$$R \vee b = \{x \in R : x \geq b\} = \{x \in R : xb = 1 \cdot b\}$$

and replacing 1 by any a in R leads to the class of a-relative deducts

$$(R \vee b)_a = \{x \in R : xb = ab\}.$$

It is easy to verify that $(R \vee b)_a$ is the coset $a + Rb'$. A further critic of Calabrese's work will be found at the end of Section 1.5 below.

1.5 Logical operations among conditional events

Schay (1968), Bruno and Gilio (1985) and Calabrese (1987) contain developments of logical operations among conditional events. (Although Adams (1975) did not define conditional events mathematically, he did propose similar logical operations among them. See Chapter 0.) A comparison with our operations will be given in Chapter 3.

Schay became apparently the first to attempt a full calculus of operations among conditionals, especially among those with different antecendents. Specifically, Schay (1968, page 335) defined five operations as follows: For him the Boolean ring R is explicitly a ring of subsets of some set Ω, and for $a, b \in R$, $(a|b)$ is a generalized indicator function, as defined in Section 1.3, that is $(a|b) : \Omega \to \{0,u,1\}$ with $(a|b) = 0$ on $a'b$, u on b' and 1 on ab. Although not referring explicitly to DeFinetti's idea of conditional event indicator functions, Schay did actually use this idea to help introduce his definitions for logical operations.

Definition (*Schay*). *For a, b, c, and d in R*

$$(a|b)' = (a'|b);$$

$$(a|b) \cap (c|d) = (ac|bd);$$

$$(a|b) \cup (c|d) = (ab \vee cd|b \vee d);$$

$$(a|b) \wedge (c|d) = ((b' \vee a)(d' \vee c)|b \vee d);$$

$$(a|b) \vee (c|d) = (a \vee c|bd);$$

The operations $(', \cap, \vee)$, as well as $(', \wedge, \cup)$ satisfy DeMorgan's Laws, as is easily verified. Further, when $d = b$, $\cup = \vee$ and $\cap = \wedge$, and the operations reduce to the usual set operations on the first component of $(a|b)$, leaving the antecedent fixed. Since $(a|b) = (ab|b)$, one can restrict $a \le b$, with no loss of generality. Then $'$ becomes $(a|b)' = (a'b|b)$.

Schay did investigate the algebraic structure of the space of conditionals which turns out to be equivalent to the set $R|R$ of all cosets of all principal ideals (see Section 2.3). He noted that it is a lattice with respect to $(\le, \cup, \cap)$ and with respect to $(\le, \vee, \wedge)$. He also provided an axiomatic description of his structure, analogous to Stone's Representation Theorem for Boolean algebras. (See especially his Theorem 5.)

Bruno and Gilio (1985), inspired by DeFinetti's work (but independent of Schay) and motivated by some problems in statistics, also developed a calculus of conditional events. Here, as with Schay, $(a|b)$ is identified as a generalized indicator function on Ω. The disjunction and conjunction operations of Bruno and Gilio are identical to Schay's $\cup$ and $\cap$, respectively, and their negation is the same as Schay's. They define an order relation among conditionals by $(a|b) \le (c|d)$ if $a \le c$ and $b \le d$.

Independently of Schay, Adams and Bruno and Gilio, Calabrese used empirical guidelines to propose an algebra for these objects. His operations $'$, $\vee$, and $\wedge$ turn out to be identical to Schay's $'$, $\cup$, and $\wedge$ respectively. Calabrese also investigated an extension of Stone's Representation Theorem, and considered briefly higher order conditioning. (As stated earlier, unaware of Lewis' Triviality Result, Calabrese proved that no mathematical form for conditional events which is compatible with probability, can be given in terms of a Boolean function into R.) A recent discussion of Schay's, Calabrese's, and our work is in Dubois and Prade (1989, 1990). A more systematic comparison is given in Section 3.5.

For ease of reference, we describe below the essentials of Calabrese's work (Calabrese, 1987).

Starting from the viewpoint of a unified algebraic theory of logic and probability, Calabrese argued for an additional operation $(\cdot|\cdot)$ on a set of propositions represented by a Boolean algebra R. This same argument has been advocated much earlier by Copeland (Copeland, 1941). See also Section 1.1. However, unlike Copeland, Calabrese came to realize that a home for conditionals should be outside of R. Although works such as Adams (1975) and Hailperin (1976) were cited in the references of his paper, apparently Calabrese did not notice a certain number of basic facts, namely Lewis' Triviality Result (discussed in Section 1.8 of Adams' book), Adams' proposed logical operations for conditional events (called conditional formulas) (Adams, p. 46-47), and the (equivalent) coset form for conditional events in Hailperin's book. As such, he first reproved a special case of Lewis' Triviality Result, namely that conditional events cannot be represented by binary Boolean operations on R (Calabrese, 1987, Theorem 2.2.1). Calabrese's approach to defining conditional events was based upon the concept of filters in R. Specifically, for $a, b \in R$, the conditional event $(a|b)$ is taken to be the equivalence class of elements of R with respect to the filter $R \vee b$, where by definition $a \approx c$ (under $I = R \vee b$) if and only if $ai = ci$ for some $i \in I$. But it is easy to see that the equivalence class of a under I is precisely the coset $a + I'$ where I' is the ideal defined by $I' = \{i' : i \in I\}$. Indeed, let $a[I]$ denote $\{x : x \in R, x \approx a$ under $I\}$. Observe that $x \in a[I]$ if and only if $x = ri' \vee ai$ for some $r \in R$ and $i \in I$. But

$$x = ri \vee ai = ri' + ai = ri' + a(1 + i') = (a + r)i' + a,$$

so that $a[I] = a + I'$, where I' is an ideal. In particular, for $I = R \vee b$ (the filter generated by b), we have $I' = Rb'$, and hence $(a|b) = a[R \vee b] = a + Rb'$, a coset of the principal ideal Rb'.

Calabrese went on to consider logical operations among conditional events by logical considerations. First, arguing that the statement "if p then (if q then r)" is the same as "if (p and q) then r," he identified $((r|q)|p)$ with $(r|p \wedge q)$. See also

Section 8.1. From that, as an axiom, he defined

$$(r|q)|(p|s)$$

to be

$$(r|q) \wedge (p|s).$$

Also, as an axiom, the disjunction $\vee$ is defined by

$$(p|q) \vee (s|r) = ((p \wedge q) \vee (s \wedge r)|p \vee r)$$

which is precisely Adams' "quasi-disjunction" of conditional formulas (Adams, 1975, p. 47).

Of course, since R/Rb' is a Boolean ring, for each fixed b, the negation for $(a|b)$ should be negation in this Boolean ring, that is to say $(a|b) = (a'|b)$. The conjunction $\wedge$ is derived from $\vee$ via DeMorgan:

$$(q|p) \wedge (s|r) = ((q|p)' \vee (s|r)')'$$

$$= ((q'|p) \vee (s'|r))' = ((q' \wedge p) \vee (s' \wedge r)|p \vee r)'$$

$$= ((p' \vee q) \wedge (r' \vee s)p \vee r)$$

$$= ((p \Rightarrow q) \wedge (r \Rightarrow s|p \vee r),$$

where $\Rightarrow$ denotes material implication. Again, this is Adams' "quasi-conjunction" operation (Adams, 1975, p. 46).

1.6 Notes

Below we include several intuitive or naive approaches to the problem of combining conditional events, and of assigning probabilities to them. Although, they turn out not to be satisfactory, either theoretically or practically, they are presented here for purpose of completeness because of their apparent wide-spread use.

Product space approach.

First consider the case of equal antecedents, and consider $(a|b)$ as primitives in our natural language, as in Adams (1975). If $* : R^2 \to R$ is a Boolean function, then for a probability measure P on R, assign $\hat{P}((a|b)*(c|b)) = P(a*c|b)$. (There is a problem already with $\hat{P}$ being not necessarily well defined.) Now for $(a_i|b_i)$ where the b_i are not necessarily the same, one can try to reduce to the equal antecedent case, by getting a "common denominator". One possibility for a common denominator is $\hat{b} = (b_1, b_2)$ in

R^2. Identify a_1 with $\hat{a}_1 = (a_1, 1)$, a_2 with $\hat{a}_2 = (1, a_2)$, and $(a_i|b_i)$ with $(\hat{a}_i|\hat{b}_i)$. Then we are back in the equal antecedent case, but operating in the product space R^2. Now for a probability measure P on R, one needs a probability measure $\hat{P}$ on R^2 such that

$$\hat{P}(\hat{a}_i|\hat{b}) = P(a_i|b_i), \quad i = 1, 2.$$

But this requires for example,

$$\hat{P}((a_1,1)|(b_1,b_2)) = \hat{P}(a_1b_1,b_2)/\hat{P}(b_1,b_2) = P(a_1b_1)/P(b_1).$$

Taking $\hat{P}$ to be the product measure meets this requirement, but the product measure is unsatisfactory because it implies the independence of the events $(a_1,1)$ and $(1,a_2)$ in R^2. Another possibility is to require that $\hat{P}(\hat{b}) = 1$, and find such $\hat{P}$ with $\hat{P}(a,1) = P(a|b_1)$ and $\hat{P}(1,a) = P(a|b_2)$. Finding such $\hat{P}$'s with given marginals is an extremely difficulty task, but has been solved in the case $\Omega = \mathbb{R}$, the field of real numbers. (Sklar, 1959, 1973). Sklar's Theorem says that if H is an n-dimensional cumulative distribution function with one-dimensional marginal distributions $F_1, F_2, \ldots, F_n$, then there exists an n-dimensional copula C such that for all n-tuples $(x_1, x_2, \ldots, x_n)$,

$$H(x_1, x_2, \ldots, x_n) = C(F_1(x_1), F_2(x_2), \ldots, F_n(x_n)).$$

Conversely, if $F_1, F_2, \ldots, F_n$ are one-dimensional CDF's and C is an n-dimensional copula, then H defined above is an n-dimensional CDF with marginals F_i.

Roughly speaking, for $n = 2$, a copula is a joint distribution for a pair of random variables, each of which is uniformly distributed on $[0, 1]$. More formally, a two dimensional copula is a mapping

$$C : [0,1] \times [0,1] \to [0,1]$$

such that

(i) for all $x \in [0,1]$, $C(x, 0) = C(0, x) = 0$, $C(x, 1) = C(1, x) = x$, and

(ii) for x_i and $y_i \in [0,1]$ with $x_1 \le x_2$ and $y_1 \le y_2$, one has

$$C(x_1, y_1) - C(x_1, y_2) - C(x_2, y_1) + C(x_2, y_2) \ge 0.$$

Note that C is continuous and non-decreasing in each argument, and that

$$Max\{x + y - 1, 0\} \le C(x,y) \le Min\{x,y\} \quad for\ x, y \in [0,1]\}.$$

The bounds above are also copulas, termed minimal and maximal copulas, respectively. For the use of copulas in statistics, see Whitt (1976), Genest and MacKay (1986a, 1986b), and Marshall and Olkin (1988). For the problem of determining joint densities from given conditional densities, see Arnold and Press (1989).

Returning to the problem at hand, it turns out that the above intuitive approach in the case of equal antecedents leads to a very restrictive form of $\hat{P}$. Indeed, consider the case $(\Omega, R) = (\mathbb{R}, \mathcal{B})$, where $\mathcal{B}$ is the Borel σ-field of the reals $\mathbb{R}$. Consider $(a_i|b)$, $i = 1, 2, ..., n$, where $a_i = (-\infty, s_i]$, and denote by $F_{\hat{P}}$ and F_P the CDF's associated with $\hat{P}$ and P, respectively. Then for $\hat{b} = b \times b \times ... \times b$,

$$F_{\hat{P}}(s_1, s_2, \ldots, s_n) = \hat{P}(\times_{i=1}^{n} a_i)$$

$$= \hat{P}[\cap_{i=1}^{n} \hat{a}_i] = \hat{P}[\cap_{i=1}^{n} \hat{a}_i | \hat{b}]$$

$$= \hat{P}[\cap_{i=1}^{n} (a_i | b)] = \mathrm{P}[\cap_{\mathrm{i}=1}^{\mathrm{n}} \mathrm{a_i} | \mathrm{b}]$$

$$= Min_{1 \le i \le n} P((-\infty, s_i] | b) = Min_{1 \le i \le n} F_P(s_i | b),$$

where each marginal CDF of $F_{\hat{P}}$ is $F_P(\cdot | b)$.

Now, when combining n conditionals as above, by Sklar's theorem, one chooses an n-dimensional copula C, independently of the forms of the conditionals. The joint CDF of $\hat{P}$ is then obtained once P and the b_i's are specified. The above form of a maximal copula in the case of equal antecedents contradicts the independent choice of C.

Combination of antecedents approach

One way to combine conditional events $(a|b)$ and $(c|d)$ is via ordinary Boolean operations on both components, for example,

$$(a|b) \vee (c|d) = (a \vee c | b \vee d),$$

and

$$(a|b) \cap (c|d) = (a \cap c | b \cap d).$$

The problem in doing this is that the first is not well defined (for example when $(a|b) = a + Rb'$), and besides violates

$$P[(a|b) \vee (c|d)] \geq P(a|b),$$

while the second violates

$$P[(a|b) \cap (c|d)] \leq P(a|b).$$

In fact, in the latter case, for $0 < a = c = bd$ with $P(bd) < P(b)$ and $P(d)$, one has

$$P((a|b) \cap (c|d)) = P((a|b) \cap (a|d))$$

$$= P((ab|bd) \cap (ad|bd)) = P((bd|bd) \cap (bd|bd))$$

$$= P(bd|bd) = 1,$$

and yet

$$0 < P(a|b) = P(d|b) < 1$$

and

$$0 < P(c|d) = P(b|d) < 1.$$

Thus probabilities do not behave as one would like for either of these operations on conditional events. However, it will be shown later that there is a way to combine conditional events that extends the Boolean operations on ordinary events and so that probabilities do behave properly with respect to that combination. In effect, this approach is a non-cartesian product common denominator one.

Material implication approach

We end this section with some additional remarks about material implication and its relation to conditional events. Material implication is the function $f: R \times R \to R$ defined by $f(a,b) = b' \vee a$, also written $b \to a$. Now material implication satisfies many desirable properties, including the following, which are trivial to verify.

(1) $f(a,b) = f(ab,b)$ (consequent-antecedent invariance);

(2) $f(a,b) = 1$ if and only if $b \leq a$ (tautology);

(3) $f(a,b)b = ab$ (modus ponens);

(4) $f(ac,b) = f(a,b)f(c,b)$, $f(a \vee c,b) = f(a,b) \vee f(c,b)$ (homomorphisms);

(5) $f(a,bc)f(c,b) = f(ac,b)$ (chaining);

(6) $f(a,1) = a$;

(7) $f(f(a,b), c) = f(a,bc)$ (iteration);

(8) $f(b', a') = f(a,b)$ (modus tollens);

(9) $f(a,b)f(c,d) = f(ac, a'b \vee c'd \vee bd)$;

(10) $f(a,b) \vee f(c,d) = f(a \vee c, bd)$;

(11) $f(a,b) + f(c,d) = f((a + c)bd), ab \vee cd \vee bd \vee b'd')$.

Also note that $b \rightarrow a$ is the maximum solution to the equation $xb = ab$. The function f is not one-to-one. For example, for any $s \leq b' \vee a$, $f(s, s \vee a'b) = f(a,b)$. The basic difficulty of material implication is that it is not compatible with probability, that is, $P(a|b) \neq P(b \rightarrow a)$ for all P for which $P(b) \neq 0$. We have seen this before, and of course follows from Lewis' Triviality Result. In fact, as shown in Section 1.1,

$$P(b \rightarrow a) \geq P(a|b).$$

Again, see the discussion in Sections 0.1 and 1.1.

CHAPTER 2

DERIVATION OF CONDITIONAL EVENTS

This chapter is devoted to an axiomatic approach to deriving the mathematical concept of conditional events. From intuitive properties capturing the basic aspects of conditioning and the requirement that conditioning be compatible with probability, we proceed to derive conditioning operators in logic, the values of which are conditional events. A canonical form for conditional events, namely cosets of principal ideals, is obtained. The space of all conditional events so obtained forms the basis of our extension of logic to the conditional case.

2.1 Generalities

In view of Stone's Representation Theorem (see, for example, Halmos, 1963) and in the spirit of symbolic (and algebraic) logic, the basic objects of our analysis are the elements of a Boolean ring $(R, +, \cdot)$. A Boolean ring is a ring with identity such that every element is *idempotent*, that is, for every element a,

$$a^2 = a \cdot a = a.$$

It follows from

$$\begin{aligned}(a + b)^2 &= a + b + ab + ba \\ &= a + b\end{aligned}$$

that $ab = -ba$. Taking $a = b$ gets $a = -a$ so that $a + a = 0$, that is, the ring has characteristic 2, and is also commutative. The *identity* (or *unit*, or *unity*) of R is denoted 1, as usual. Two additional "logical operations" are defined on Boolean rings, $\vee$ (called *or*, or *union*, or *disjunction*) and $'$ (called *not*, or *negation*, or *complement*), by

$$a \vee b = a + b + ab,$$

and

$$a' = 1 + a,$$

respectively. *Conjunction,* or *intersection*, or *and*, sometimes denoted by $\wedge$, is taken to be the multiplication on R. A partial order $\leq$ is defined by $a \leq b$ if $ab = a$. The generic example of a Boolean ring is the set of all subsets of a set Ω, with $+$ and $\cdot$ given by symmetric difference (the "exclusive or") and intersection, that is, by

$$a + b = ab' \cup a'b$$

and

$$a \cdot b = a \cap b,$$

where $'$ is complementation, and $\cup$ and $\cap$ are ordinary union and intersection of sets. The identity 1 is the set Ω and the zero is the empty set ϕ. The partial order then is just ordinary containment of sets. This ring is called the ring of all subsets of Ω, and is particularly pertinent when Ω is a finite set. More generally, a set of subsets R of an arbitrary set Ω which is a ring under the operations given by

$$a + b = ab' \cup a'b$$

and

$$a \cdot b = a \cap b$$

is a Boolean ring, and Stone's Representation Theorem says that every Boolean ring is isomorphic to such a ring of subsets of some set.

An important concept is that of an *ideal* of a Boolean ring R. More generally, an ideal of an arbitrary commutative ring R is a nonempty subset I of R such that $a - b$ is in I for all a and b in I, and $a \cdot b$ is in I for all a in R and b in I. If R is Boolean, then $b = -b$, and the condition that $a - b$ is in I becomes simply that $a + b$ is in I. So an ideal in a Boolean ring R is simply a nonempty subset of R closed under addition and closed under multiplication by elements of R. An ideal is a *principal* ideal if it is of the form

$$Ra = \{ra : r \in R\}.$$

Such ideals will be of particular importance for us.

For an ideal I of R, there is associated a ring R/I, called a quotient ring, whose elements are *cosets*, that is subsets of R of the form

$$a + I = \{a + i : i \in I\},$$

and addition and multiplication are given by

$$(a + I) + (b + I) = (a + b) + I$$

and

$$(a + I) \cdot (b + I) = a \cdot b + I,$$

respectively. It is an easy exercise to show that this makes R/I into a ring, using the properties of an ideal I. Further, R/I is Boolean when R is Boolean. Here, as is the custom, we are using $+$ and $\cdot$ for addition and multiplication in both the rings R and R/I, but the context will make it clear where we are doing our adding and multiplying.

A mapping f from a ring R to a ring S is a *homomorphism* if for a and b in R,

$$f(a + b) = f(a) + f(b)$$

and

$$f(a \cdot b) = f(a) \cdot f(b).$$

For an ideal I of a ring $R, f(a) = a + I$ is a homomorphism from the ring R onto the ring R/I, call the *natural* homomorphism of R onto R/I. Two rings R and S are *isomorphic* if there is a homomorphism from R to S that is one-to-one and onto. If f is a homomorphism from R to S, then

$$Ker(f) = \{a : f(a) = 0\}$$

is called the *kernel* of f, and is an ideal of R. If f is from R *onto* S, then

$$F(\mathrm{a} + Ker(f)) = f(a)$$

is an isomorphism from $R/Ker(f)$ to S. This is the *first isomorphism theorem for rings*. That F is one-to-one from $R/Ker(f)$ onto $f(R)$ is a special case of this. For any mapping f defined on a set $X, x \simeq y$ given by $f(x) = f(y)$ is an equivalence relation. Let $F(x)$ denote the equivalence class to which x belongs, and let the set of equivalence classes be denoted by $X/[f]$. This set of equivalence classes is a partition of X, the map $F : X \to X/[f]$ is the *natural* map from X onto $X/[f]$, and $\eta : X/[f] \to f(X)$ given by $\eta(F(x)) = f(x)$ is one-to-one and onto.

If R is a ring and f is a one-to-one mapping from R onto a set S, then S can be made into a ring isomorphic to R. For example, multiply in S by

$$x \cdot y = f(f^{-1}(x) \cdot f^{-1}(y)).$$

A probability measure P on a Boolean ring R is a function P from R to the closed interval $[0,1]$ such that $P(1) = 1$ and $P(a \vee b) = P(a) + P(b)$ whenever $ab = 0$. This last property is the *finite additivity* of P. There is a stronger property sometimes required, called *σ-additivity,* but we will not need it. An *atom* in a Boolean ring is an element a such that $a \neq 0$ and if $b \leq a$, then $b = a$ or $b = 0$. The ring is *atomic* if every element contains an atom. Finite Boolean rings are always atomic, and the ring of Borel sets of Euclidean space is atomic. If a is an atom in a Boolean ring, then P

defined by $P(b) = 1$ if $a \leq b$ and $P(b) = 0$ otherwise, is a probability measure on R. If a is not an atom, then P is not a probability measure, since for $0 \neq c < a$, $P(ac') = 0 = P(a) = P(ac)$, hence additivity fails. (See, however Section 2.2)

2.2 Conditioning operators

As stated in Chapters 0 and 1, the main goal is to define objects of the form "a given b", denoted $(a|b)$, for a and b elements of a Boolean ring R. Although the operation $(\cdot|\cdot)$ on $R \times R$ is termed measure-free conditioning, the derivation implicitly involved probability measures. We wish to define $(a|b)$ in such a way that for any probability measure P on R, it is possible to assign the conditional probability $P(a|b)$ to $(a|b)$ without ambiguity. In this spirit, the theory of measure-free conditioning developed here is compatible with probability theory. If this compatibility condition is relaxed, then the door is open to other types of conditional objects. Lewis' Triviality Result is established precisely within this probability compatibility condition. The probability compatibility requirement is appealing, for example, in expert systems since the strengh of the production rule "if b then a" is usually quantified by the conditional probability $P(a|b)$. A typical example is the Markov random field model of Lauritzen and Speigelhalter (1988). But measure-free conditional events compatible with probability can be used to investigate other non-probabilitistic conditioning as well. The recent work of Dubois and Prade (1991) is relevant.

In view of previous work on measure-free conditionals, it seems that the coset form is a reasonable one. In the following, we will arrive at this form from an axiomatic approach.

We are going to search for maps $f : R \times R$ onto some space S which captures the basic aspects of conditioning compatible with conditional probability evaluations. A value $f(a,b)$ of f will be called a (measure-free) conditional event. Our strategy is this. The mapping f will be required to satisfy a set of axioms, or requirements. Since S is unknown, we will work on the domain $R \times R$ of f, examining the partition on it induced by the equivalence relation $(a,b) \simeq (c,d)$ if $f(a,b) = f(c,d)$. A version of f is then obtained by assigning to each (a,b) its equivalence class $F(a,b)$. Note that

$$(R \times R)/[f] \to f(R \times R) : F(a,b) \to f(a,b)$$

is a one-to-one correspondence. Thus if all f satisfying the axioms induce the same partition of $R \times R$, then

$$F : R \times R \to (R \times R)/[f]$$

is a canonical form for f. This program will be carried out by examining what each $f(.,b)$ must be like, for each b in R.

Any "conditioning" operator f should at least possess the following three properties.

(1) $f(R,1)$ is a copy of R, that is, $f(.,1)$ is one-to-one. This satisfies the requirement that conditional events should be generalizations of ordinary events.

(2) For (a,b) in $R \times R$, $f(a,b) = f(ab,b)$. This says that conditioning a on b is the same as conditioning ab on b.

(3) For any probability measure P on R, $Q((f(a,b)) = P(a|b)$ defines an extension of P to $W_P = \{f(a,b) : P(b) > 0\}$.

That Q is well defined requires that for $f(a,b)$ and $f(c,d)$ in W_P with $f(a,b) = f(c,d)$, one should have $P(a|b) = P(c|d)$. For Q to be an extension of P also requires that R be contained in W_P. A weaker form of (3) is this.

(3') If $f(a,b) = f(c,b)$ then $P(ab) = P(cb)$ for all P for which $P(b) > 0$.

By Lemma 1 below, (3') is equivalent to

(3") If $f(a,b) = f(c,b)$, then $ab = cb$.

It is reasonable to postulate that conditional events with different antecedents are different, namely

(4) If $f(a,b) = f(c,d)$, then $b = d$.

A weaker form of (4) is

(4') If $f(0,b) = f(c,d)$, then $b = d$, and if $f(b,b) = f(c,d)$, then $b = d$.

In conjunction with (3) or (3'), (4') becomes

(4") If $f(0,b) = f(c,d)$, then $b = d$ and $cd = 0$, and if $f(b,b) = f(c,d)$, then $b = d$ and $cd = d$.

We need some technical lemmas. To construct Dirac-like probability measures on arbitrary Boolean rings, we spell out the following procedure.

Lemma 0. *Let* R *be a Boolean ring. Let* $a_1, a_2, \ldots, a_n$ *be non-zero mutually disjoint elements of* R, *and let* $\alpha_1, \alpha_2, \ldots, \alpha_n$ *be in* $[0,1]$ *with* $\sum \alpha_i = 1$. *Then there is a probability measure* P *on* R *such that if* $\vee_{i \in J} a_i \leq b$, *and* $(\vee_{i \notin J} a_i) = 0$, *then* $P(b) = \vee_{i \in J} \alpha_i$.

Proof. By Stone's Representation Theorem, we may identify R with a subalgebra of the algebra of all subsets of some set Ω. Thus a_i is a subset of Ω. Since each a_i is nonempty, pick $\omega_i \in a_i$, and let $P = \sum_i \alpha_i\delta_i$, where δ_i is the Dirac probability with mass one at ω_i, given by $\delta_i(b) = 1$ if $\omega_i \in b$ and 0 otherwise. It is easy to check that P is a probability measure on R with the desired property.

Lemma 1. *Let R be a Boolean ring. If $P(ab) = P(cb)$ for all probability measures P on R such that $P(b) > 0$, then $ab = cb$.*

Proof. Suppose that $ab \neq cb$. Then either $(ab)(cb') \neq 0$ or $(ab)'(cb) \neq 0$. Suppose $(ab)(cb') \neq 0$. In view of Lemma 0, let $\omega \in (ab)(cb') \neq 0$, $\omega \in \Omega$. Let P be the Dirac probability measure on the set of subsets of Ω with mass 1 at ω. Then $P(ab) = 1$, $P(b) > 0$, and $P(ab) \neq P(cb) = 0$. □

Lemma 2. *Let R be a Boolean ring, and let $a, b. c, d$ be elements of R with $b \neq 0 \neq d$. The following are equivalent.*

(i) $P(a|b) \leq P(c|d)$ for all probability measures P on R for which $P(b) \neq 0 \neq P(d)$.

(ii) Either $ab = 0$, or $d \leq c$, or $ab \leq cd$ and $c'd \leq a'b$.

Proof. Assume *(ii)*. If $ab = 0$ or $d \leq c$, then obviously (i) holds. Suppose that $ab \leq cd$ and $c'd \leq a'b$. Using those two inequalities and the fact that for $t \geq 0$ and $P(y) \geq P(x)$,

$$P(x)/P(y) \leq [P(x) + t]/[P(y) + t],$$

we get

$$P(a|b) = P(ab)/P(b)$$

$$\leq [P(ab) + P(cd) - P(abcd)]/[P(b) + P(cd) - P(abcd)]$$

$$= P(cd)/P(b \vee cd) = P(cd)/P(ab \vee a'b \vee cd)$$

$$= P(cd)/P(a'b \vee cd) \leq P(cd)/P(c'd \vee cd)$$

$$= P(cd)/P(d) = P(c|d).$$

Now assume (*i*), and that neither $ab = 0$ nor $d \leq c$. Then $ab \neq 0$ and $c'd \neq 0$. First, we get $ab \leq cd$. If not, then $(ab)(cd)' \neq 0$. If $(ab)(cd)'d \neq 0$, view R as a subalgebra of the algebra of subsets of a set Ω, and let $\omega \in \Omega$ with $\omega \in (ab)(cd)'d$. The restriction of the Dirac probability measure P on the set of all subsets of Ω given by

$P(\omega) = 1$ has the property that

$$P(b) = P(d) = P(ab) = 1,$$

and $P(cd) = 0$. Thus $P(a|b) = 1$ while $P(c|d) = 0$.

If $(ab)(cd)'d = 0$, let γ and ω be elements of Ω such that $\gamma \in (ab)(cd)'$ and $\omega \in c'd$. Note that $\gamma \neq \omega$. Giving γ and ω each mass $1/2$ yields a probability measure on the set of all subsets of Ω whose restriction P to R has the property that $P(a|b) = P(ab)/P(b) \geq 1/2$, while $P(c|d) = 0$.

The proof that $c'd \leq a'b$ is similar. □

The following corollary is immediate.

Corollary 1. *Let R be a Boolean ring and a, b, c, d be in R with $b \neq 0 \neq d$. The following are equivalent.*

(i) $P(a|b) = P(c|d)$ for all probability measures P on R for which $P(b) \neq 0 \neq P(d)$.

(ii) Either $ab = cd = 0$, or $b \leq a$ and $d \leq c$, or $ab = cd$ and $b = d$.

In view of Lemma 1, we see that if f satisfies (3), then it satisfies (1). Indeed, if $f(a, 1) = f(c,1)$, then $P(a) = P(c)$ for all probability measures P on R, and hence $a = c$. Thus $f(\cdot,1)$ is one-to-one on R, and R is identified with $f(R,1)$. Also it follows from Lemma 1 that if $f(a,b) = f(c,d)$, then $ab = cd$. Thus (2) and (3) are the basic requirements for conditioning operators.

Theorem 1. *If f satisfies (2) and (3), then for each b, $R/[f(.,b)] = R/Rb'$.*

Proof. It suffices to show that Rb' and the kernel of $f(\cdot,b)$ define the same equivalence relation on R. Let a and c be in R. Then $f(a,b) = f(c,b)$ if and only if $f(ab,b) = f(cb,b)$ (by (2)) if and only if $ab = cb$ (by Lemma 1) if and only if $a + Rb' = c + Rb'$. □

Some remarks are in order. First, since $f(R,b)$ and R/Rb' are in one-to-one correspondence, in fact by $a + Rb' \to f(a,b)$, and R/Rb' is a ring, $f(R,b)$ becomes a ring isomorphic to R/Rb'. Second, note that the mapping

$$R \times R \to R/Rb' : (a,b) \to a + Rb'$$

does satisfy (2) and (3). See Section 2.3 for more details.

It remains to describe all f satisfying (2) and (3). Theorem 1 gives a description for such f locally, that is of each $f(\cdot, b)$. Each $f(\cdot, b)$ induces the same partition on R, namely into cosets of R/Rb'. However, two such f do not necessarily induce the same partition on $R \times R$. In fact, let f be defined on $R \times R$ by $f(a,b) = a + Rb'$, and define g by $g = f$ except that $g(a,b) = R$ whenever $a \wedge b = 0$. This does define a function. If $f(a,b) = f(c,d)$ and $a \wedge b = 0$, then $c \wedge d = 0$. Now $f(\cdot,b)$ and $g(\cdot,b)$ both induce the partition of R into cosets of Rb', but do not induce the same partition of $R \times R$, as is obvious. Furthermore, f and g satisfy (2) and (3). The problem is that $g(a,b) = g(c,d)$ does not imply that $b = d$. If this latter condition is satisfied, that is, if we assume property (4) in addition to (2) and (3), then any two such fs are equivalent in the sense that they determine the same partition of $R \times R$. Condition (4) makes the $f(R,b)$'s be mutually disjoint, and (2) and (3) make each be in one-to-one correspondence with R/Rb'. So any f satisfying (2), (3), and (4) is equivalent to the map defined by $(a,b) \to a + Rb'$. Thus we have the following theorem.

Theorem 2. *Let f satisfy* (2), (3), *and* (4). *Then f is equivalent to the map g defined by* $g(a,b) = a + Rb'$. □

If (2) and (3) are assumed, then something a little weaker than (4) will suffice.

Theorem 3. *The the conditions* (2), (3), *and* (4') *imply* (4). *That is* (2), (3), *and* (4') *are equivalent to* (2), (3), *and* (4).

Proof. Assume (2), (3), and (4'). Suppose that $f(a,b) = f(c,d)$. Then there is a probability measure P on R such that $P(b) \neq 0 \neq P(d)$. By Lemma 2, either (i) $b = d$, or (ii) $b \leq a$ and $d \leq c$, or (iii) $ab = cd = 0$. In case (ii),

$$f(a,b) = f(\mathrm{ab},b) = f(\mathrm{b},b) = f(c,d),$$

and (4') implies that $b = d$. In case (iii),

$$f(a,b) = f(0,b) = f(c,d),$$

and (4') implies that $b = d$. Thus $b = d$ in any case, and the theorem follows. □

2.3 Conditional events

The analysis of Section 2.2 has led us to a canonical form for conditional events. This form will be used throughout this book.

Definition. *Let R be a Boolean ring. For a and b in R, the (measure-free) conditional event "a given b", written $(a|b)$ is the coset $a + Rb'$. The space $\cup_{b\in R} R/Rb'$ of all conditional events is denoted by $R|R$. It is sometimes referred to as the conditional extension of logic.*

As we will see, the union $\cup_{b\in R} R/Rb'$ above is a disjoint one. That is, $(R/Rb') \cap (R/Rd') = \phi$ for $b \neq d$. The function $f(a,b) = a + Rb'$ satisfies all the requisite properties discussed in the last section, including the property that $f(a,b) = f(c,d)$ implies that $P(a|b) = P(c|d)$. There are many "conditioning operators" which are not "probability compatible". Examples are

$$f(a,b) = ab,$$

and

$$f(a,b) = (b \rightarrow a) = b' \vee a.$$

More generally, take

$$f(a,b) = ab \vee db'$$

for any d in R. Then for $d = 0$, we get $f(a,b) = ab$, and for $d = 1$, we get $f(a,b) = b' \vee a$. These cannot be compatible with probability by Lewis' Triviality Result in Section 1.1.

We will now look at some of the properties of $(a|b)$. The function $f(a,b) = a + Rb'$ on $R \times R$ will be denoted by $(\cdot|\cdot)$. Thus $(\cdot|\cdot)$ is a function from $R \times R$ onto $\cup_{b\in R} R/Rb' = R|R$.

(1) The function $(\cdot|b)$ is a homomorphism from the ring R onto the quotient ring R/Rb'. This quotient ring consists of all cosets of the form $a + Rb'$, or $(a|b)$, b fixed. These cosets partition R, that is, two cosets $(a|b)$ and $(c|b)$ are equal or disjoint, and every element of R is in some $(a|b)$. In fact, a is in $(a|b)$. Thus to check that two cosets $(a|b)$ and $(c|b)$ are equal, it is enough to get one element in common. Note that $(0|0) = 0 + R = R$ is a coset and leads to the trivial quotient ring R/R, a ring with only one element.

(2) $(\cdot|1)$ is one-to-one on R, and in fact is an isomorphism from R to $(R|1) = R/R0$, which is identified with R itself, cosets of $R0 = \{0\}$ being of the form $a + \{0\} = \{a\}$.

(3) Since

$$a + Rb' = a + ab' + Rb'$$
$$= a(1+b') + Rb'$$
$$= ab + Rb',$$

we get that $(a|b) = (ab|b)$. This is just property (2) in Section 2.2.

(4) In R, a closed interval $[a,b]$ consists of all c such that $a \leq c \leq$ b, and recall that $x \leq y$ if $xy = x$. When we write $[a,b]$, we mean implicitly that $a \leq b$. *Cosets of principal ideals in Boolean rings and closed intervals are the same thing*. In fact, $(a|b) = [ab, b \rightarrow a]$, or $[ab, a \vee b']$, and any closed interval

$$[a,b] = (a|b' \vee a) = (ab|b' \vee a).$$

To see this, for $a + rb'$ in $a + Rb'$,

$$ab(a + rb') = ab,$$

and

$$(a + rb')(b' \vee a) = a + rb'(b' \vee a) = a + rb'.$$

Thus $(a|b) \subset [ab, b' \vee a]$. For $ab \leq c \leq b' \vee a$,

$$c = ab \vee (c(ab)') = ab + ca'b',$$

which is in $ab + Rb' = a + Rb'$. Thus $(a|b) = [ab, b' \vee a]$. For an interval $[a,b] = [ab,b]$,

$$(ab|b' \vee a) = [ab \vee (b' \vee a), (b' \vee a)' \vee (ab(b' \vee a))] =$$

$$[ab, (ba') \vee ((ab) \wedge (b' \vee a))] = [ab, (b \wedge a') \vee (ba)] = [ab,b].$$

This fact that cosets of principal ideals and closed intervals are the same things gives nothing new except the, perhaps important, realization of conditional events as intervals $[a,b]$. Such an interval has a ready interpretation, in fact, a ready meaning - the set of all elements of R between a and b. (Remember , $a \leq b$.) The interval $[a,b]$ is the conditional event "a given $(b' \vee a)$", and the interval $[ab, b' \vee a]$ is the conditional event "a given b", or $(a|b)$. Thinking of a conditional event as an interval has perhaps more intuitive appeal than thinking of it as a coset $a + Rb'$. In any case, it is convenient sometimes to visualize $a + Rb'$ as all sets between ab and $a \vee b'$.

The following property of cosets is fundamental enough for us to be called a theorem.

Theorem 1. *The two cosets* $a + Rb'$ *and* $c + Rd'$ *of* R *are equal if and only if* $ab = cd$ *and* $b = d$.

Proof. If $ab = cd$ and $b = d$, then clearly the two cosets are equal. Now suppose that

$$a + Rb' = c + Rd'.$$

Then

$$ab = cd + rb',$$

so

$$abb = cdb + rb'b = ab = cdb.$$

Thus $ab \leq cd$. By symmetry, $cd \leq ab$, whence $ab = cd$. Now

$$ab + b' = cd + sd' = ab + sd',$$

so $b' = sd'$, whence $b' \leq d'$. By symmetry, $d' \leq b'$, so $b = d$. □

This theorem exhibits all the relevant properties of our conditioning operator $(\cdot|\cdot)$. It asserts that $(a|b) = (c|d)$ if and only if $ab = cd$ and $b = d$. In particular, if $(a|b) = (c|d)$, then

$$P(a|b) = P(ab)/P(d) = P(c|d).$$

If a function f is equivalent to $(\cdot|\cdot)$ in the sense of Theorem 2 of Section 2.2, then f has the property that $f(a,b) = f(c,d)$ if and only if $ab = cd$ and $b =$ d. Further, any function f having this property is equivalent to $(\cdot|\cdot)$.

There are two forms for conditional events that are equivalent to ours that are worth considering. First is the form proposed by Schay (1968) and DeFinetti (1972). Let R be a ring of subsets of some set Ω, and define g on $R \times R$ by $g(a,b)(\omega) = 1$ if ω is in ab, 0 if ω is in $a'b$, and u for ω in b'. That is, $g(a,b)$ is a function from Ω to $\{0,1,u\}$, where the "u" stands for "undefined". Clearly $g(a,b) = g(c,d)$ if and only if $ab = cd$ and $b = d$, so g is equivalent to $(\cdot|\cdot)$.

Another form for conditional events that is equivalent to the coset one adopted here is given by $f(a,b) = (ab,b)$. Clearly, $f(a,b) = f(c,d)$ if and only if $ab = cd$ and $b = d$.

Thus conditional events are pairs (a,b) with $a \leq b$. This form has the appealing interpretation that conditional events are events, (a,b) being viewed as the event a in the subring Rb of R, and this being viewed as different from the event a in the ring R. That is, it is the *pair* (a,b). The space in which it is an event must be kept track of. Another advantage of this realization of conditional events is that pairs are simpler to visualize and to manipulate than cosets.

We remark that $(a|b)$ can be realized as the set of all solutions x of the equation $xb = ab$, which is, of course all those elements between ab and $a \vee b'$. Using other binary operations than $\varphi(a,b) = ab$, and considering the set of all solutions of the equation $\varphi(x,b) = \varphi(a,b)$ gives other formulations of conditional events, and a way to extend the concept to algebraic structures more general than Boolean rings. (See Chapter 8.)

Conditional events $(a|b)$, that is cosets $a + Rb'$, can be expressed in terms of *filters* of R. A *filter* in the Boolean algebra R is a non-empty subset F of R such that if a and b are in F, then $ab \in F$, and if $a \in F$ and $a \leq b$, then $b \in F$. The relation between ideals and filters is that F is a filter in R if and only if $F' = \{1 + x : x \in F\}$ is an ideal in R. Given a filter F, an equivalence relation is defined by $a \approx b$ if there is an element $f \in F$ with $af = bf$. Letting $[a]$ denote the equivalence class containing a, the relation with cosets is expressed in the equation

$$[a] = a + F'$$

For principal ideals, the situation is particularly simple. For for $b \in R$, the set $R \vee b = \{r \vee b : r \in R\}$ is a filter. It just consists of all elements x such that $b \leq x$. Further, $(R \vee b)' = Rb'$, and so in this case,

$$(a|b) = [a] = a + Rb'.$$

Now that we have conditional events $(a|b)$ identified as cosets $a + Rb'$ of R, we must establish logical operations between them, and this will be caarried out in Chapter 3, where the ring operations of R will be extended to its cosets. However, conditional events, as subsets of R, can be combined via union and intersection as well as other ordinary set operations. These, of course, are not extensions of the ring operations of R, but may be of some interest in their own right. The ordinary set operations on conditional events with the same antecedent are of course well understood, since the cosets of an ideal Rb' partition R. For example, the intersection of two such cosets is either empty or the cosets are identical. For cosets with different antecedents, the situation is a bit more complex.

Theorem 2. *The following hold.*

(1) $(a|b) \cap (c|d)$ *is the coset* $((ab \vee cd)|(b \vee d))$ *if* $abd = cbd$*, and is empty otherwise.*

(2) $(a|b) \subset (c|d)$ *if and only if* $cd \le ab$ *and* $a \vee b' \subset c \vee d'$*.*

(3) $(a|b) \cup (c|d)$ *is a coset if and only if one is contained in the other, or*

$$ab \le cd \le a \vee b' \le c \vee d',$$

or

$$cd \le ab \le c \vee d' \le a \vee b'.$$

In the last case, for example,

$$(a|b) \cup (c|d) = (cd|(cd \vee a'b)).$$

Proof. *(1)* If $a + rb' = c + rd'$, then multiplying through by bd gets $abd = cbd$. this latter equality implies easily that

$$ab \vee cd \le (a \vee b') \wedge (c \vee d').$$

The coset $(a|b)$ is the interval $[ab, a \vee b']$ and $(c|d) = [cd, c \vee d']$. It follows that

$$(a|b) \cap (c|d) = [ab, a \vee b'] \cap [cd, c \vee d']$$

$$= [ab \vee cd, (a \vee b') \wedge (c \vee d')]$$

$$= ((ab \vee cd)|(b \vee d)),$$

again using $abd = cbd$.

Viewing cosets as intervals immediately yields (2) and (3). □

CHAPTER 3

LOGICAL OPERATIONS ON CONDITIONAL EVENTS

In this chapter, logical operations between conditional events are defined, extending Boolean operations of the base ring R. As in most extension problems, such an extension is not unique, and the one chosen demands justification. From a semantic viewpoint, the system of logical operations derived here corresponds to Lukasiewicz's three-valued logic. A comparison with other proposed operations is given in Section 3.5. A discussion of the possibiblity of deriving logical operations for conditional events in an axiomatic setting is in Section 3.4. The analysis in this chapter is directed toward Boolean rings, with more general algebraic structures considered in Chapter 8.

3.1 The extension problem

As established in Chapter 2, for $a, b \in R$, by the conditional event "a given b", we mean the coset $a + Rb'$, and use the notation $(a|b)$ for it. Since conditional events are generalizations of events, with $(a|1)$ corresponding to the ordinary event a in R, the logical operations among them should be extensions of the ring operations. That is $(a|1) + (b|1)$ must be $((a + b)|1)$, and so on. There are various ways of doing this. It has been noted at the end of Chapter 2 that ordinary set operations on conditional events are not appropriate. The space $R|R$ of conditional events is the disjoint union $\cup R/Rb'$, with the union over all $b \in R$. We have in each R/Rb' the usual quotient ring operations which come from the operations of R. What is needed are operations combining cosets from different quotient rings, that is, combining elements from R/Rb' and R/Rd' with $b \neq d$, and of course with the result of such a combination being a coset of a principal ideal. This is not a standard ring theory operation, and has been largely avoided. For example, Hailperin (1976) just called $R|R$ a partial universal algebra (see, for example Grätzer, 1968), and considered logical operations only between elements of the same quotient ring. That is clearly unsatisfactory. We will define operations between any two cosets of principal ideals, and investigate the resulting algebraic structure of $R|R$ in Chapter 4.

For any ring R, its operations + and $\cdot$ induce corresponding operations on subsets of R. Namely, for subsets A and B of R,

$$A + B = \{a + b : a \in A, b \in B\},$$

and

$$AB = \{ab : a \in A, b \in B\}.$$

We have two other commonly used operations for Boolean rings, $a \vee b = a + b + ab$, and $a' = 1 + a$. These extend to set operations as well, namely

$$A \vee B = \{a + b + ab : a \in A, b \in B\}, \text{ and}$$

$$A' = \{a' : a \in A\}.$$

A convenient fact, and one easily checked, is that for subsets A and B of R, DeMorgan's laws hold:

$$(AB)' = A' \vee B',$$

and

$$(A \vee B)' = A'B'.$$

We have already been using set addition in writing down cosets: $a + Rb'$ means $\{a\} + Rb'$, which is $\{a + rb' : r \in R\}$. Now coset addition in each quotient ring R/Rb' is just this set addition. Cosets of R/Rb' are added by the formula

$$(a + Rb') + (c + Rb') = (a + c) + Rb',$$

but this coincides with the set addition above since

$$(a + rb') + (c + sb') = (a + c) + (r + s)b'$$

is in $(a + c) + Rb'$, and the other inclusion is equally as easy to check. Further, this addition is well defined - set addition is certainly well defined, and if

$$a + Rb' = x + Rb'$$

and

$$c + Rb' = y + Rb',$$

then

$$(a + c) + Rb' = (x + y) + Rb'.$$

These remarks for coset addition are valid for any ring and any ideal I, not just Boolean rings and principal ideals Rb'.

It is not generally true that coset multiplication is set multiplication. That is, it is not true for all rings that set multiplication of $(a + I)$ and $(b + I)$ is $ab + I$, the product of the two cosets $(a + I)$ and $(b + I)$. However, it is true for Boolean rings, and in fact for commutative von Neumann regular rings. For Boolean rings, an arbitrary element of $(a + I)(b + I)$ is

$$(a + i)(b + j) = ab + aj + bi + ij$$

with i and j in I. This is clearly in $ab + I$. On the other hand for

$$ab + k \in ab + I,$$

taking $i = ka'$ and $j = k + kba'$ puts $ab + k$ in the form $ab + aj + bi + ij$. So coset multiplication in Boolean rings is just set multiplication. This unusual fact suggests that perhaps set addition and multiplication are appropriate operations on any pair of elements of $R|R$. Similar remarks hold for the set operations $'$ and $\vee$ on $R|R$.

3.2 Conditional logical operations

First we will show that $R|R$ is closed under the set operations $'$, $+$, multiplication or $\wedge$, and $\vee$. This will give us an "algebra" of conditional events, and its properties will be subsequently investigated. It is convenient to note first that for ideals I and J of a Boolean ring R, the product $IJ = \{ij : i \in I, j \in J\}$ is indeed an ideal. Clearly, $I \cap J$ is an ideal, and $I \cap J \subset IJ$. For x in $I \cap J$, $x = x \cdot x$ is in IJ, so $IJ = I \cap J$. Since sums of ideals are ideals, the following theorem then implies that sums, products, and disjunctions $\vee$ of two cosets are cosets.

Theorem 1. *Let R be a Boolean ring and let I and J be ideals of R. Then*

(1) $(a + I)' = (a' + I)$,
(2) $(a + I) + (b + J) = (a + b) + I + J$,
(3) $(a + I)(b + J) = ab + bI + aJ + IJ$,
(4) $(a + I) \vee (b + J) = a \vee b + b'I + a'J + IJ$.

Proof. For (1),

$$(a + I)' = \{(a + i)' : i \in I\} =$$

$$\{(1 + a) + i : i \in I\} = a' + I.$$

For (2),

$$(a + I) + (b + J) = \{a + i + b + j : i \in I, j \in J\} =$$

$$\{(a + b) + i + j : i \in I, j \in J\} = (a + b) + I + J.$$

Part (3) is more difficult. An element in $(a + I)(b + J)$ is of the form

$$(a + i)(b + j) = ab + bi + aj + ij,$$

which is clearly in $ab + bI + aJ + IJ$. The other inclusion is the difficult part. First, we will establish it for principal ideals. So let $I = Rx$ and $J = Ry$. An element in $(a + Rx)(b + Ry)$ is of the form

$$(a + rx)(b + sy) = ab + asy + brx + rsxy,$$

and an element of $ab + bRx + aRy + RxRy$ is of the form

$$ab + btx + auy + vxwy.$$

Let $z = vwxy + txb + uya + ab$. Then letting

$$r = (z - a)xy + tx(1 - y),$$

and

$$s = (1 - b)xy + uy(1 - x)$$

puts $ab + asy + brx + rsxy$ in the form $ab + btx + auy + vxwy$. This is a bit tedious but straightforward to check. Thus we have

$$(a + I)(b + J) = ab + bI + aJ + IJ$$

for principal ideals. For arbitrary ideals I and J, we need

$$ab + bI + aJ + IJ \subset (a + I)(b + J)$$

That is, we need an element of the form $ab + bu + av + wx$ to be of the form

$$ab + bi + aj + ij,$$

where i, u, and w are in I, and j, v and x are in J. Now

$$Ru + Rw = R(u + w + uw),$$

as is easily checked, and similarly $Rv + Rx = R(v + x + vx)$. Let

$$e = u + w + uw$$

and

$$f = v + x + vx.$$

Then $ab + bu + av + wx$ is in $ab + bRe + aRf + ReRf$, which, by the principal ideal case, is $(a + Re)(b + Rf)$, and which, in turn, is contained in $(a + I)(b + J)$. This proves part (3).

For (4) we use DeMorgan's laws and (3). We have

$$(a + I) \vee (b + J) = [(a' + I)(b' + J)]'$$

$$= 1 + a'b' + b'I + a'J + IJ = a \vee b + b'I + a'J + IJ. \qquad \square$$

Several comments are in order. We have, for example, the formula

$$(a + I)(b + J) = ab + bI + aJ + IJ.$$

There is no question of the product $(a + I)(b + J)$ being well defined. It is just the product of the two sets $a + I$ and $b + J$. If representatives are changed, that is, if a is replaced by x and b by y such that $a + I = x + I$ and $b + J = y + J$, then

$$(a + I)(b + J) = (x + I)(y + J) = xy + yI + xJ + IJ.$$

Similar remarks hold for the other operations. Since addition, multiplication and disjunction on R are commutative and associative, their extensions to subsets of R are commutative and associative. Multiplication and disjunction are also idempotent, that is, $xx = x = x \vee x$. In particular, these three binary operations on the set of all cosets of R are both commutative and associative. Thus we can perform these operations on any (finite) number of cosets with the result independent of order or association.

The operations $'$, $+$, $\vee$, and $\wedge$ or multiplication were defined by extending the corresponding operations of R to subsets of R. Back in R, the operations satisfy the following relations:

(1) $x' = 1 + x$,

(2) $x + y = x'y \vee xy'$,

(3) $(xy)' = x' \vee y'$ and $(x \vee y)' = x'y'$,

(4) $x(y \vee z) = xy \vee xz$,

(5) $x \vee (yz) = (x \vee y)(x \vee z)$,

(6) $x(y + z) = xy + xz$,

(7) $x \vee y = x + y + xy$.

For cosets of R, only (1) through (5) hold, and those are easily checked using the theorem above. We have already noted that DeMorgan's laws (the properties in (3)) hold for any subsets of R. An example of the failure of (6) for cosets of principal ideals is given below, and (7) does not hold for the cosets $a + R$ and $1 + Ra'$, where $a \neq 0$. In that case, we have

$$(a + R) \vee (1 + Ra') = a \vee 1 + Ra',$$

$$(a + R) + (1 + Ra') + (a + R)(1 + Ra') = a \vee 1 + R,$$

and

$$a \vee 1 + Ra' \neq a \vee 1 + R,$$

since $a \neq 0$.

Finally, Theorem 1 holds for commutative von Neumann rings with little change in the proof. See Chapter 8.

We turn now to specializing these results to the case where the ideals are principal. In that case, we have the three binary operations, and the unary operation $'$ on $R|R$. We now change to the notation $(a|b)$ for $a + Rb'$. The following theorem states the basic facts about the operations on the space $R|R$ of principal ideals.

Theorem 2. *The following hold.*

(1) $(a|b)' = (a'|b)$,
(2) $(a|b) + (c|d) = (a + c|bd)$,
(3) $(a|b)(c|d) = (ac|a'b \vee c'd \vee bd)$,
(4) $(a|b) \vee (c|d) = (a \vee c|ab \vee cd \vee bd)$.

Proof. The proof of (1) is easy. For (2),

$$(a|b) + (c|d) = (a + Rb') + (c + Rd') = a + c + Rb' + Rd'.$$

Now we have observed in the proof of the previous theorem that $Rx + Ry = R(x \vee y)$. Thus

$$Rb' + Rd' = R(b' \vee d') = R(bd)'.$$

For (3),

$$(a|b)(c|d) = ac + cRb' + aRd' + Rb'Rd',$$

using (3) of the previous theorem. We need the ideal $Rb'c + Rad' + Rb'd'$ to be the principal ideal $R(a'b \vee c'd \vee bd)'$. It is the principal ideal $R(b'c \vee ad' \vee b'd')$. It is routine to show that

$$(a'b \vee c'd \vee bd)' = b'c \vee ad' \vee b'd'.$$

For (4),

$$(a|b) \vee (c|d) = a \vee c + c'Rb' + a'Rd' + Rb'd',$$

using (4) of the previous theorem. We need the ideal

$$Rb'c' + Ra'd' + Rb'd'$$

to be the principal ideal $R(ab \vee cd \vee bd)'$, and it is the ideal $R(b'c' \vee a'd' \vee b'd')$. Again, it is routine to check that

$$(ab \vee cd \vee bd)' = b'c' \vee a'd' \vee b'd'. \qquad \square$$

Note that $(0|1)$ is the zero of $R|R$, that is, is the additive identity, and that $(1|1)$ is the multiplicative identity. That is, $(0|1)$ is the only element such that

$$(0|1) + (a|b) = (a|b)$$

for all $(a|b)$, and $(1|1)$ is the only element such that for all $(a|b)$,

$$(1|1)(a|b) = (a|b).$$

Indeed,

$$(0|1) + (a|b) = (0 + a)|1 \cdot b = a|b,$$

and

$$(1|1)(a|b) = a|(1'1 \vee a'b \vee b) = a|b.$$

If $(x|y)$ were another zero, then

$$(0|1) + (x|y) = (0|1) = (x|y).$$

Similarly, $(1|1)$ is the only multiplicative identity for $R|R$.

Elements in $R|R$ do not have negatives, in general. If

$$(a|b) + (c|d) = (a + c)|bd = (0|1),$$

then $bd = 1$, so $b = d = 1$. So the $(a|b)$ with negatives are exactly the $(a|1)$, whose negative is itself. Further, multiplication of sets does not distribute over addition of sets, even for cosets of principal ideals. For example,

$$(1|b)((1|d) + (1|f)) \neq (1|b)(1|d) + (1|b)(1|f),$$

the first being $(0|df)$, and the second being $(0|bdf)$. Just pick b, d, and f so that

$df \neq bdf$. However, multiplication does distribute over $\vee$, and $\vee$ over multiplication, as we have observed above for any cosets.

In any case, the "algebra" $R|R$ is far from being a ring under the operations of multiplication and addition. It does, however contain isomorphic copies of all the R/Rb', since

$$(a|b) + (c|b) = (a + c)|b,$$

and

$$(a|b)(c|b) = (ac|b).$$

The operations in $R|R$ have many interesting properties and interrelations. We record some of the more fundamental ones here. Their proofs are straightforward. In the following, we will use just x for the element $(x|1)$ in $R|R$.

Theorem 3. (*Bayes*) *Let $a_1 + a_2 + \ldots + a_n = 1$ be a partition of 1. In particular, the a_i are mutually disjoint. Then for b in R,*

(*1*) $b = (b|a_1)a_1 + (b|a_2)a_2 + \ldots + (b|a_n)a_n$,

(*2*) $(a_j|b) = (((b|a_j)a_j)|b)$,

(*3*) $(a_j|b)b = (b|a_j)a_j = a_jb$,

(*4*) $(a_j|b) = ((b|a_j)a_j)|((b|a_1)a_1 + (b|a_2)a_2 + \ldots + (b|a_n)a_n)$. □

In particular, from (4) we get

$$b = (b|a)a + (b|a')a'$$

and

$$(a|b) = ((b|a)a)|((b|a)|a + (b|a')a').$$

Recall that logical (material) implication $b \rightarrow a$ in R is defined to be $b' \vee a$. We denote $(b \rightarrow a)(a \rightarrow b)$ by $a \leftrightarrow b$. These operations extend to $R|R$ in the same manner as the others. We define

$$(c|d) \rightarrow (a|b) = \{y \rightarrow x : y \in (c|d), x \in (a|b)\},$$

and

$$(c|d) \leftrightarrow (a|b)$$

in the obvious way.

In the following theorem, parts (1) through (5) give *connections of* $\vee$ *and* $\wedge$ *with logical implication*, parts (6) and (7) are *absorbing properties*, while part (9) is a *decomposition property*. Again, the verifications are straightforward.

Theorem 4. *The following hold.*

(1) $b \to a = (a|b) \vee b' = (b'|a') \vee a$,

(2) $(a|b) = ((b \to a)|b) = (b \to a)(b|b)$,

(3) $(a|b) = (b'|a')(0|0) \vee (b' \to a)$,

(4) $(c|d) \to (a|b) = (c|d)' \vee (a|b) = ((cd \to ab)|(c'd \vee ab \vee bd))$,

(5) $(c|d) \leftrightarrow (a|b) = ((c|d) \to (a|b))((a|b) \to (c|d))$
$= ((ab \leftrightarrow cd)|bd) = ((a|b) + (c|d))'$,

(6) $(a|b) = (a|b)((a|b) \vee (c|d))$,

(7) $(a|b) = (a|b) \vee (a|b)(c|d)$,

(8) $(a|b) = (a|1) + (0|b)$. □

3.3 An order relation and related concepts

The Boolean ring R has a partial order $\leq$ given by $a \leq b$ if $ab = a$. Being a partial order means that $\leq$ is reflexive, anti-symmetric, and transitive. That is, $a \leq a$, $a \leq b$ and $b \leq a$ imply that $a = b$, and finally, $a \leq b$ and $b \leq c$ imply that $a \leq c$. The partial order does respect multiplication and $\vee$, in the sense that if $a \leq b$, then $ac \leq bc$ and $a \vee c \leq b \vee c$. Further, $a \leq b$ implies that $b' \leq a'$. These properties are trivial to check.

We now define a partial order on $R|R$ in the analogous way, and note some of its properties. In particular, it will extend the partial order on R, identifying R with the elements of the form $(r|1)$. Note that, in his discussion on qualitative probability, Savage (1972, p. 44) mentioned the lack of qualitative counterpart of $P(a|b) \leq P(c|d)$. It is necessary, even from a qualitative viewpoint, to compare "interconditionals," that is, conditionals with different antecedents. See also Koopman (1940), and our Chapter 5.

Definition. *For* $(a|b), (c|d) \in R|R$,

$$(a|b) \leq (c|d)$$

if

$$(a|b) = (a|b)(c|d).$$

The relation $\leq$ is indeed a partial order on $R|R$. Since

$$(a|b)(a|b) = (a^2|a'b \vee a'b \vee b^2) = (a|b),$$

we have $(a|b) \leq (a|b)$, so $\leq$ is reflexive. If $(a|b) = (a|b)(c|d)$ and $(c|d) = (c|d)(a|b)$, then certainly $(a|b) = (c|d)$, so that $\leq$ is symmetric. Finally, to get transitivity for $\leq$, if $(a|b) = (a|b)(c|d)$ and $(c|d) = (c|d)(e|f)$, then

$$(a|b)(e|f) = ((a|b)(c|d))(e|f) =$$

$$(a|b)((c|d)(e|f)) = (a|b)(c|d) = (a|b).$$

The partial order above depends only on the multiplicaiton in $R|R$ being idempotent, commutative, and associative. Finally, it should be noted that if

$$(a|b) \le (c|d),$$

then

$$(a|b) \vee (e|f) \le (c|d) \vee (e|f)$$

and

$$(a|b)(e|f) \le (c|d)(e|f),$$

while it is not true that $(a|b) \le (c|d)$ implies that

$$(a|b) + (e|f) \le (c|d) + (e|f).$$

We now give some useful alternate conditions equivalent to being $\le$.

Theorem 1. *The following are equivalent, and hence are all equivalent to* $(a|b) \le (c|d)$.

(*1*) $(a|b) = (a|b)(c|d)$,
(2) $(a|b)' \ge (c|d)'$,
(*3*) $ab \le cd$ *and* $c'd \le a'b$,
(*4*) $(c|d) = (c|d) \vee (a|b)$.

Proof. First we prove that (1) implies (3). If $(a|b) = (a|b)(c|d)$, then

$$(a|b) = (ac|a'b \vee c'd \vee bd),$$

and we have

$$ab = ac(a'b \vee c'd \vee bd) = abcd,$$

so that $ab \le cd$. Also $b = a'b \vee c'd \vee bd$, whence $c'd \le b$, and so $ac'd \le ac'b$. But $ab \le cd$ gets $ac' = 0$, so $ac'd = 0$. Thus $c'd \le a'$, and already we have $c'd \le b$. Thus $c'd \le a'b$, and so (1) implies (3). Assume (3). Then $ab \le cd$ and $c'd \le a'b$. To get

$$(a|b) = (a|b)(c|d) = (ac|(a'b \vee c'd \vee bd),$$

we need first that $ab = ac(a'b \vee c'd \vee bd)$. The last is $acbd$, which is indeed ab since $ab \le cd$. Finally, we need $b = (a'b \vee c'd \vee bd)$. Now

$$(a'b \vee c\, d \vee bd) = (a' \vee d)b \vee c'd.$$

But $c' \vee d \leq b$, and $abd' \leq cdd' = 0$, so $b \leq a' \vee d$. It follows that (3) implies (1).

Part (2) is equivalent to (1), using (3), and (4) is equivalent to (1) using DeMorgan's laws and (2). □

Note that (3) implies that $\leq$ is monotone in the first argument, that is, $(a|b) \leq (c|b)$ if $a \leq c$. More generally, if $(a|b) \leq (c|d)$ then $(a|b) \leq ((c \vee x)|d)$, as follows readily from (3). This is not true for the second argument. For example, $(a|b)$ and $(a|bc)$ are not comparable, in general. It is not true that $ab \leq abc$, so $(a|b)$ is not $\leq (a|bc)$, and it is not true that $a'b \leq a'bc$, so that $(a|bc)$ is not $\leq (a|b)$.

Theorem 2. *The following hold.*

(1) $0 \leq ab \leq (a|b) \leq (b \to a) \leq 1$.

(2) $ab \leq (a|b)(b|a) \leq (a \leftrightarrow b)$.

(3) *If* $a_1 \leq a_2 \leq \ldots \leq a_n$, *then*

$$(a_1|a_2)(a_2|a_3)\ldots(a_{n-2}|a_{n-1})(a_{n-1}|a_n) = (a_1|a_n).$$

(4) $(a|bc)(b|c) - (ab|c)$. □

Items (1) and (2) above give some connections between material implication and $\leq$, with (2) being an immediate consequence of (1). Items (3) and (4) are called "chaining" conditions, and (4) is a consequence of (3).

It is possible to give a formal characterization for our operations $\cdot$ and $\vee$ on $R|R$. A systematic investigation of the rationale of our operations will be given in Sections 3.4 and 3.5.

Theorem 3. *The mapping* $\varphi : R|R \to R$ *defined by* $\varphi(a|b) = b \to a = b' \vee a$ *is a* $(\vee,\wedge)$*-homomorphism from* $R|R$ *onto* R. *That is,*

$$\varphi[(a|b) \vee (c|d)] = \varphi(a|b) \vee \varphi(c|d)$$

and

$$\varphi[(a|b)(c|d)] = \varphi(a|b)\varphi(c|d).$$

Proof. First, φ is well defined, and clearly φ maps $R|R$ onto R. Now

$$\varphi[(a|b) \vee (c|d)] = \varphi(a \vee c|a \vee c \vee bd) = (a \vee c \vee bd)' \vee a \vee c,$$

taking $a \leq b$ and $c \leq d$ without loss of generality.

$$\varphi(a|b) \vee \varphi(c|d) = (b' \vee a) \vee (d' \vee c).$$

Thus we need

$$(a \vee c \vee bd)' \vee a \vee c = (a'c'(b' \vee d')) \vee a \vee c = b'c' \vee a'd' \vee a \vee c$$

to be $b' \vee a \vee d' \vee c$, which it clearly is. Similarly,

$$\varphi[(a|b) \wedge (c|d)] = \varphi(a|b) \wedge \varphi(c|d). \qquad \square$$

Theorem 4. *Let* $\circ$ *and* $\cup$ *be any operations on* $R|R$ *extending* $\cdot$ *and* $\vee$ *on* R. *Suppose that the mapping given by* $(a|b) \to b'a$ *is a* $(\circ, \cup)$*-homomorphism. Suppose further that*

$$(a|b)\circ(c|d) = (abcd|\alpha(a,b,c,d))$$

and

$$(a|b) \cup (c|d) = (ab \vee cd|\beta(a,b,c,d)),$$

where $abcd \leq \alpha(a,b,c,d)$ *and* $(ab \vee cd) \leq \beta(a,b,c,d)$. *Then* $\circ = \cdot$ *and* $\cup = \vee$.

Proof.

$$\varphi[(a|b)\circ(c|d)] = \varphi(abcd|\alpha(a,b,c,d)) =$$

$$(b' \vee a)(d' \vee c) = \alpha(a,b,c,d)' \vee abcd =$$

$$\alpha(a,b,c,d)' + abcd\,.$$

Let $r = a'b \vee c'd \vee bd$. Then

$$(b' \vee a)(d' \vee c) = r' + rac = r' + abcd,$$

whence $\alpha(a,b,c,d)' = r$. Thus $\circ = \cdot$. Similarly,

$$\varphi[(a|b) \cup (c|d)] = \beta(a,b,c,d)' + ab \vee cd =$$

$$b' \vee a \vee d' \vee c = b' \vee ab \vee d' \vee cd = b' \vee d' \vee ab \vee cd =$$

$$(b' \vee d')(ab)'(cd)' + ab \vee cd,$$

the last two summands being disjoint. It follows that

$$\beta(a,b,c,d) = [(b' \vee d')(ab)'(cd)']' = bd \vee abcd,$$

and that $\cup = \vee$. $\square$

Other operations for combining evidence

For inference purposes, it is sometimes appropriate to combine several pieces of conditional information, that is, conditional events, using appropriate operations. If two conditional events $(a|b)$ and $(c|d)$ arise from the same Boolean ring, then we have various ways to do that now: multiply them, or use $\vee$, or use $+$ in $R|R$, or use other operations in $R|R$. Then, for example, given a probability measure P on R, one could calculate the probability of the resulting conditional event. But what if the events a and b came from the Boolean ring R, and c and d came from the Boolean ring S? How do we get a single conditional event capturing the essence of the two conditional events $(a|b)$ and $(c|d)$? One way is to do it is as for ordinary events. If R and S are Boolean rings, then the Cartesian product $R \times S = \{(r,s) : r \in R, s \in S\}$ is a Boolean ring under the componentwise operations. That is, just operate componentwise. Now if r is an event in R and s is an event in S, then (r, s) is an event in, that is, is an element of, the Boolean ring $R \times S$. Since $R \times S$ is a Boolean ring, we can form $(R \times S)|(R \times S)$. The objects of interest are the two conditional events $(a|b)$ and $(c|d)$, or the pair $[(a|b), (c|d)]$, with a and b in R, and c and d in S, say. This pair is an element of the set

$$(R|R) \times (S|S) = \{(x,y) : x \in R|R, y \in S|S\}$$

of all pairs of $R|R$ and $S|S$. But this set is in natural one-to-one correspondence with $(R \times S)|(R \times S)$ via the mapping

$$[(a|b), (c|d)] \to ((a,c)|(b,d)).$$

The upshot is that the pair $(a|b)$ and $(c|d)$ of conditional events is associated with a conditional event, namely one in the space $(R \times S)|(R \times S)$. For any probability measure P on $R \times S$, one may assign the probability of $[(a|b), (c|d)]$ to be $P[(a,c)|(b,d)]$, which makes sense.

Another way to combine evidence of the form above is this. Regard the Boolean rings R and S as rings of subsets of Ω_1 and Ω_2, respectively. Let

$$\mathrm{C} = \{a \times b : a \in R, b \in S\},$$

that is, the set of all Cartesian products of elements of R by elements of S. Thus each element of C is a subset of $\Omega_1 \times \Omega_2$, or $\mathrm{C} \subset \mathcal{P}(\Omega_1 \times \Omega_2)$, the Boolean ring of all subsets of $\Omega_1 \times \Omega_2$. Now C is not a subring of the ring $\mathcal{P}(\Omega_1 \times \Omega_2)$, as can be seen by observing the the basic relations between union, intersection, and complement are given by the formulas

$$(a \times b)' = (a' \times \Omega_2) \cup (\Omega_1 \times b'),$$

$$(a \times b) \cap (c \times d) = (ac \times bd),$$

and

$$(a \times b) \cup (c \times d) = (a \times b)' \cap (c \times d)'.$$

However, there is a unique smallest subring $R \bullet S$ of $\mathcal{P}(\Omega_1 \times \Omega_2)$ containing C, namely the intersection of all those subrings containing C. The operations of the ring are then just the usual set theoretic operations of $\mathcal{P}(\Omega_1 \times \Omega_2)$. For $a, b \in R$ and $c, d \in S$, we define

$$(a|b) \times (c|d) = \{e \times f : e \in (a|b), f \in (c|d)\}.$$

But observe that

$$(e \times f) \cap (b \times d) = eb \times fd = ab \times cd.$$

Hence

$$e \times f \in (ab \times cd|b \times d) \in (R \bullet S)|(R \bullet S).$$

That is,

$$(a|b) \times (c|d) = (ab \times cd|b \times d).$$

Note that if P is a probability measure on $R \bullet S$ then

$$P[(a|b) \times (c|d)] = P[(ab \times cd)|(b \times d)].$$

As an illustration of the possibility to use this type of operation $\times$ among conditionals in the problem of combining evidence, consider the well-known "penguin triangle" problem in AI, as discussed for example in (Pearl, 1988).

Let

$$f = \text{flying animals}$$

$$b = \text{birds}$$

$$\text{p} = \text{penguins},$$

so that $(f|b)$ = "birds fly", $(f'|p)$ = "penguins do not fly". For an analysis of this type of information, see (Zadeh, 1985). It is appropriate here to combine the peices of evidence $(f|b)$ and $(f'|p)$ via the operation $\times$ among conditionals. This is in line with familiar situations in statistics. Now

$$(f|b) \times (\text{f}'|p) = (fb \times \text{f}'p)|b \times p).$$

Thus $P(fb \times f'p | b \times p)$ should be close to 1 for any reasonable probability P on $R \bullet S$.

3.4 Connections with three-valued logic

So far we have studied the logical operations on the space $R|R$ from a syntax viewpoint. In this section, we will investigate the semantic relation with three-valued logics. To that end, we first discuss that relationship between Boolean algebras and classical two valued logic. *We then show that an analogous relationship exists between $R|R$ and three-valued logic.* Since Boolean polynomials play a fundamental role, we begin with a discussion of them and their properties that are pertinent to our situation. This discussion is informal, but should be sufficient for our purposes.

An *elementary Boolean polynomial in the n variables* $X_1, X_2, ..., X_n$ is $Y_1Y_2...Y_n$, where $Y_i = X_i$ or X_i'. The symbol X_i' should be thought of as the complement of X_i, and the elementary polynomial $Y_1Y_2...Y_n$ should be thought of as the product, or conjunction of the Y_i. We are using juxtaposition to indicate this conjunction, rather than inserting the conjunction symbol $\wedge$. *There are* 2^n *of these elementary Boolean polynomials*. A *Boolean polynomial in the n variables* $X_1, X_2, ..., X_n$ is an expression of the form $E_1 \vee E_2 \ldots \vee E_m$, where the E_i are distinct elementary Boolean polynomials. Thus a Boolean polynomial is the (formal) disjunction of elementary ones. The empty disjunction is allowed and is denoted 0. The order of the E_i in the disjunction is immaterial. (As an aside, the set of Boolean polynomials in the n variables $X_1, X_2, ..., X_n$ may be thought of as the Boolean algebra of all subsets of the set of elementary Boolean polynomials in those variables.)

Here are some examples for the case n = 3. There are 2^3 elementary Boolean polynomials, namely

$$X_1X_2X_3,\ X_1'X_2X_3,\ X_1X_2'X_3,\ X_1X_2X_3',\ X_1'X_2'X_3,\ X_1'X_2X_3',\ X_1X_2'X_3',\ X_1'X_2'X_3'.$$

The expression

$$f = X_1X_2X_3 \vee X_1X_2'X_3 \vee X_1X_2X_3' \vee X_1'X_2X_3'$$

is a Boolean polynomial in three variables having four elementary terms.

Generally, a Boolean polynomial in the variables $X_1, X_2, ..., X_n$ is any expression formed from the X_i using $\wedge$, $\vee$ and $'$. For example, in the case $n = 3$,

$$f = X_1X_2 \vee X_1)(X_2'X_3 \vee X_1X_2) \vee (X_3' \vee X_1)'X_2X_3'$$

is such an expression. However, manipulating this expression as if the X_i were elements of a Boolean algebra, and $\wedge$, $\vee$, and $'$ were the usual operations on it, one may bring f into the form of a disjunction, or union, of elementary Boolean polynomials, and this form is unique. This is the well known fact that every Boolean polynomial can be written in its disjunctive normal form. We regard any two Boolean polynomials the same is they have the same disjunctive normal form. This is the same thing as requiring that two are the same if they induce the same Boolean function $R^n \to R$. The disjunctive normal form of a Boolean polynomial is not usually the simplest form of that polynomial. For example, if $n = 3$, the Boolean polynomial

$$X_1X_2X_3 \vee X_1X_2'X_3' \vee X_1X_2'X_3 \vee X_1X_2X_3',$$

which is in disjunctive normal form, may be more simply represented as X_1. Our unions of elementary Boolean polynomials are of course in disjunctive normal form.

The following proposition is clear.

Lemma 1. *There are* 2^{2^n} *Boolean polynomials in n* variables.

The connection between Boolean polynomials and mappings is this. *If f is a Boolean polynomial in n variables and R is a Boolean ring, then f induces a map* $f : R^n \to R$ *by evaluation.* Note that we use the same symbol f to denote a Boolean polynomial as well as the function it induces on any Boolean algebra R^n. This is convenient, and should cause no confusion. For example, if n = 3 and

$$f = X_1X_2X_3 \vee X_1X_2'X_3 \vee X_1X_2X_3' \vee X_1'X_2X_3',$$

then the mapping $f: R^3 \to R$ is given by the formula

$$f(a_1, a_2, a_3) = a_1a_2a_3 \vee a_1a_2'a_3 \vee a_1a_2a_3' \vee a_1'a_2a_3'.$$

Definition 1. *A function* $R^n \to R$ *is called Boolean if it is induced by a Boolean polynomial in n variables.*

There are a couple of pertinent elementary facts about about these evaluation maps and elementary polynomials. First notice that there is a natural one-to-one correspondence between elementary polynomials in n variables and n-tuples of 0's and 1's. For $n = 3$, $X_1X_2'X_3$ corresponds to $(1, 0, 1)$, for example. An elementary polynomial takes the value 1 on that n-tuple of 0's and 1's to which it corresponds, and takes the value 0 on all other n-tuples of 0's and 1's. *Thus, given an n-tuple a of 0's and 1's from a Boolean algebra R, there is exactly one elementary Boolean polynomial e in n variables for which* $e(a) = 1$, *and that e has value 0 on all other such n-tuples.* This fact is the basis of the following lemma.

Lemma 2. *For any Boolean algebra* R *and any* n, *distinct Boolean polynomial in* n *variables induces distinct maps* $R^n \to R$.

Proof. Suppose f and g are Boolean polynomials in n variables and $f \neq g$. Then there exists an elementary polynomial e in n variables such that e is a term of f and not a term of g (say). Now $e(a) = 1$ for exactly one n-tuple a of 0's and 1's, and $f(a) = 1$ while $g(a) = 0$. Thus the polynomials f and g induce distinct mappings from R^n to R. □

Lemma 3. *Let* $\{0, 1\}$ *be the two-element Boolean algebra. Then every map* $f : \{0, 1\}^n \to \{0, 1\}$ *is induced by a Boolean polynomial in n variables.*

Proof. There are 2^{2^n} Boolean polynomials in n variables and 2^{2^n} maps. Use Lemmas 1 and 2. □

The previous lemma says that given any map

$$g : \{0, 1\}^n \to \{0, 1\},$$

there is a Boolean polynomial f in n variables inducing that map g. That Boolean polynomial is easy to construct, given g. For each n-tuple a from $\{0, 1\}$ for which $g(a) = 1$, there is exactly one elementary Boolean polynomial e for which $e(a) = 1$. The Boolean polynomial inducing g is the union of those e. Thus, if

$$g : \{0, 1\}^n \to \{0, 1\}$$

is presented by a table

$$\begin{array}{cc} (a_1, a_2, ..., a_n) & g(a_1, a_2, ..., a_n) \\ \cdot & \cdot \\ \cdot & \cdot \\ \cdot & \cdot \end{array}$$

then the Boolean polynomial inducing it is the union of the elementary Boolean polynomials $Y_1Y_2...Y_n$ where $Y_i = X_i$ if $a_i = 1$ and $Y_i = X_i$ if $a_i = 0$ and $g(a_1, a_2, ..., a_n) = 1$. For example, for $n = 3$, if the table is

a_1	a_2	a_3	$g(a_1, a_2, a_3)$
0	0	0	0
0	0	1	1
0	1	0	0
1	0	0	1
0	1	1	0
1	0	1	0
1	1	0	1
1	1	1	0

then the Boolean polynomial is $X_1'X_2'X_3 \vee X_1X_2'X_3' \vee X_1X_2X_3'$.

Corollary. *If* R *is any Boolean algebra and if* $f\colon R^n \to R$ *is a Boolean function, then* f *is completely determined by its action on* $\{0, 1\}^n$.

Let R and S be sets, and let $t : R \to S$ be any function. Then t induces a function $t^n : R^n \to S^n$ by the formula

$$t^n(r_1, r_2, \ldots, r_n) = (t(r_1), t(r_2), \ldots, t(r_n)).$$

Now suppose that R and S are Boolean algebras and t is a homomorphism. That is, t is a function such that

$$t(r \vee s) = t(r) \vee t(s),$$

$$t(r \wedge s) = t(r) \wedge t(s),$$

and

$$t(r') = t(r)'$$

for r, s in R. In particular, $t(0) = 0$ and $t(1) = 1$, as may be checked. If f is any Boolean polynomial in n variables, then since t is a homomorphism, we have immediately that for $(r_1, r_2, \ldots, r_n) \in R^n$,

$$tf(r_1, r_2, \ldots, r_n) = f(t(r_1), t(r_2), \ldots, t(r_n)).$$

This may be rephrased as follows.

Proposition 1. *Let* R *and* S *be Boolean rings and* $t : R \to S$ *a homomorphism. Let* f *and* g *be a Boolean polynomials in* n *variables. Then the diagram*

$$\begin{array}{ccc} R^n & \xrightarrow{f} & R \\ {\scriptstyle t^n}\downarrow & & \downarrow{\scriptstyle t} \\ S^n & \xrightarrow{g} & S \end{array}$$

commutes if and only if $f = g$.

Proof. Suppose that $f = g$. Since t is a homomorphism, we have

$$t(f(a_1, ..., a_n)) = f(t(a_1), ..., t(a_n)) = f(t^n(a_1, ..., a_n)),$$

whence the diagram commutes. Now suppose that f and g are Boolean polynomials such that the diagram commutes. Since $t(0) = 0$ and $t(1) = 1$, f and g must induce the same map on $\{0, 1\}^n$, which is contained in both R^n and S^n. Thus by Lemma 2, $f = g$. □

Now we specialize the results above to the case where S is the two element Boolean algebra $\{0, 1\}$. In that case, *the homomorphism t is called a truth evaluation on R.* In the diagram

$$\begin{array}{ccc} R^n & \xrightarrow{f} & R \\ {\scriptstyle t^n}\downarrow & & \downarrow{\scriptstyle t} \\ \{0,1\}^n & \xrightarrow[\Psi_f]{} & \{0,1\} \end{array}$$

if f is any Boolean polynomial, then the map Ψ_f induced by the Boolean polynomial f, is called the *truth function, or truth table* of the Boolean function f. It of course depends also on the homomorphism t, that is on a truth evaluation on R. More generally, *any* function $\Psi : \{0, 1\}^n \to \{0, 1\}$ is called a *truth function, or truth table.* So for the case $S = \{0,1\}$, the result may be stated as follows:

Theorem 1. *Let R be a Boolean algebra, let t be a truth evaluation on R, and let $f : R^n \to R$, be a Boolean function. Then there is exactly one truth function*

such that

$$\Psi_f \circ t^n = t \circ f,$$

namely $\Psi_f = f$. *Conversely, given a truth function* Ψ, *that is any mapping*

$$\Psi : \{0,1\}^n \to \{0,1\},$$

there is exactly one Boolean function $f : R^n \to R$ such that

$$\Psi \circ t^n = t \circ f,$$

namely that given by the Boolean polynomial f inducing Ψ. In particular, there is a one-to-one correspondence bwtween truth functions $\{0,1\}^n \to \{0,1\}$ and Boolean functions $R^n \to R$. □

Finally, it should be noted that above, given f, the construction of Ψ_f is immediate. If f is given as a Boolean polynomial, then there is no computation to be made for the construction: Ψ_f is that same Boolean polynomial. In any case, f is determined by its action on $\{0, 1\}^n$ inside R^n, and given any function from $\{0, 1\}^n$ to $\{0, 1\}$, we have specified earlier how to write down the Boolean polynomial inducing that function. So the construction of Ψ_f from f is routine. Now given $\Psi : \{0, 1\}^n \to \{0, 1\}$, write down the Boolean polynomial inducing Ψ, and that gives the unique f such that $\Psi \circ t^n = f \circ t$. So not only do the requisite f's and Ψ_f's exist, we have an explicit procedure for constructing them.

We are now going to generalize the results above to the conditional case. In particular, $R|R$ will play the role of R. First, we must decide on, and develop the relevant properties of, the analogs of Boolean polynomials for the conditional case. That is, which maps $(R|R)^n \to R|R$ should play the role that Boolean maps $R^n \to R$ play? Elements of $R|R$ are of the form $(a|b)$, with $a, b \in R$. This representation is unique if a is taken to be contained in b, that is if $ab = a$. Any mapping $(R|R)^n \to R|R$ takes an element of the form $(a_1|b_1, a_2|b_2, \ldots, a_n|b_n)$ to one of the form $(a|b)$. Again, a is not unique, but ab is, and thus ab should be a function of the $2n$ variables

$$(a_1b_1, a_2b_2, \ldots, a_nb_n, b_1, b_2, \ldots, b_n).$$

We require that this function be induced by a Boolean polynomial f *of $2n$ variables.* Similar requirements are mandated for the existence of a Boolean polynomial g of $2n$ variables yielding b. But the situation is not as simple as in the classical case. Different Boolean polynomials can induce the same mappings on $2n$-tuples of the form $(r_1, r_2, \ldots, r_n, r_{n+1}, \ldots, r_{2n})$, where $r_i \leq r_{i+n}$. A moment's reflection shows that two such polynomials induce the same mapping on such $2n$-tuples if and only if their elementary terms are the same except for those of the form

$$Y_1Y_2 \ldots X_i \ldots Y_n Y_{n+1} \ldots X_{i+n}' \ldots Y_{2n}.$$

These are precisely those elementary terms that are 0 on $2n$-tuples of the form

$(r_1, r_2, \ldots, r_n, r_{n+1}, \ldots, r_{2n})$ where $r_i \leq r_{i+n}$. We call a Boolean polynomial in $2n$ variables *reduced* if it contains no elementary terms of the form displayed above. It should be clear that in our considerations here, only reduced Boolean polynomials need be considered. Thus we are requiring that a function $(R|R)^n \to R|R$ be given by two reduced Boolean polynomials f and g of $2n$ variables. The polynomial f will consist of some of the elementary terms of g, so that $f \leq g$ in that sense. Such a pair of Boolean polynomials will be denoted $f|g$, and is called a *conditional Boolean polynomial* of $2n$ variables. For any Boolean algebra R, a conditional Boolean polynomial of $2n$ variables induces a function $(R|R)^n \to R|R$ by the formula

$$(f|g)(a_1|b_1, a_2|b_2, \ldots, a_n|b_n) =$$

$$f(a_1b_1, a_2b_2, \ldots, a_nb_n, b_1, b_2, \ldots, b_n)|g(a_1b_1, a_2b_2, \ldots, a_nb_n, b_1, b_2, \ldots, b_n).$$

Lemma 4. *There are* 3^{3^n} *conditional Boolean polynomials of* $2n$ *variables.*

Proof. A conditional Boolean polynomial is of the form $f|g$, with f and g reduced and the elementary terms of f among those of g. The number of reduced *elementary* Boolean polynomials of $2n$ variables is 3^n. To see this, note that for such a polynomial, there are 2^n choices for its first n entries. For those entries that are X_i', there is only one choice for the $(i + n)$-th entry, namely X_i. For those entries that are X_i', there are two choices for the $(i + n)$-th entry, namely X_i or X_i'. So there are 2^i elementary terms in which i of the first n entries are X_i's. It follows that there are indeed

$$\sum_{i=0}^{n} 2^i \binom{i}{n} = 3^n$$

elementary reduced Boolean polynomial $2n$ variables. For each such g with i elementary terms, one has the choice of 2^i fs. Thus there are

$$\sum_{i=0}^{3^n} 2^i \begin{bmatrix} i \\ 3^n \end{bmatrix} = 3^{3^n}$$

possible $f|g$'s, and the proof is complete. □

Any Boolean algebra R contains the two element Boolean algebra $\{0, 1\}$. *We denote this two element Boolean algebra by* V. Thus inside $R|R$ is

$$V|V = \{(0|1), (1|1), (0|0)\},$$

and so inside $(R\,|\,R)^n$ is $(V\,|\,V)^n$. *Now* $V\,|\,V$ *will play the role here that* V *did in the classical two valued case. The elements* $(0\,|\,1)$, $(1\,|\,1)$, $(0\,|\,0)$ *will be identified with the truth values* 0 *(false),* t *(true), and* u *(undecided), respectively.*

Lemma 5. *Let* R *be any Boolean algebra. Distinct conditional Boolean polynomials induce distinct functions* $(R\,|\,R)^n \to R\,|\,R$.

Proof. This follows from the observation that distinct reduced Boolean polynomials induce distinct mappings on the set of sequences $(r_1, r_2, \ldots, r_n, r_{n+1}, \ldots, r_{2n})$ of 0's and 1's with $r_i \leq r_{i+n}$.

Lemma 6. *Every function* $(V\,|\,V)^n \to V\,|\,V$ *is induced by exactly one conditional Boolean polynomial in* $2n$ *variables.*

Proof. There are 3^{3^n} such functions. Use Lemmas 4 and 5.

Definition 2. *A function* $(R\,|\,R)^n \to R\,|\,R$ *is a conditional Boolean function if it is induced by a conditional Boolean polynomial.*

For a conditional Boolean polynomial $f\,|\,g$ of $2n$ variables, the function $(R\,|\,R)^n \to R\,|\,R$ it induces will also be denoted $f\,|\,g$. *Such a Boolean function* $f\,|\,g : (R\,|\,R)^n \to R\,|\,R$ *is determined by its action on* $(V\,|\,V)^n$. This follows from Lemma 6.

Let R and S be Boolean algebras, and let $t : R \to S$ be a homomorphism. Then t induces a function $R\,|\,R \to S\,|\,S$, which we also denote by t, by the formula

$$t(a\,|\,b) = (t(ab)\,|\,t(b)).$$

Now t is well defined since $t : R \to S$ is a homomorphism so that $t(ab) \leq t(b)$. The function $t : R\,|\,R \to S\,|\,S$ induces in the usual way the function $t^n : (R\,|\,R)^n \to (S\,|\,S)$. The following proposition generalizes Proposition 1 to the conditional case.

Proposition 2. *Let* R *and* S *be Boolean algebras, let* $t : R\,|\,R \to S\,|\,S$ *be induced by a homomorphism from* $t : R \to S$, *and let* $f\,|\,g$ *and* $h\,|\,k$ *be conditional Boolean polynomials in* $2n$ *variables. Then the diagram*

$$\begin{array}{ccc} (R\,|\,R)^{n} & \xrightarrow{\;f\,|\,g\;} & R\,|\,R \\ {\scriptstyle t^{n}}\big\downarrow & & \big\downarrow{\scriptstyle t} \\ (S\,|\,S)^{n} & \xrightarrow{\;h\,|\,k\;} & S\,|\,S \end{array}$$

commutes if and only if $f|g = h|k$.

Proof. Suppose that $f|g = h|k$. Then

$$t[(f|g)((a_1|b_1), \ldots, (a_n|b_n))]$$

$$= t[f(a_1b_1, \ldots, a_nb_n, b_1, \ldots, b_n)|g(a_1b_1, \ldots, a_nb_n, b_1, \ldots, b_n)]$$

$$= (t(f(a_1b_1, \ldots, a_nb_n, b_1, \ldots, b_n))|t(g(a_1b_1, \ldots, a_nb_n, b_1, \ldots, b_n)))$$

$$= (f(t(a_1b_1), \ldots, t(a_nb_n), t(b_1), \ldots, t(b_n))|g(t(a_1b_1), \ldots, t(a_nb_n), t(b_1), \ldots, t(b_n)))$$

$$= (f|g)[(t(a_1b_1)|t(b_1)), \ldots, (t(a_na_n)|t(b_n))]$$

$$= (f|g)[t(a_1b_1|b_1), \ldots, t(a_nb_n|b_n)]$$

$$= (f|g)[t^n((a_1b_1|b_1), \ldots, (a_nb_n|b_n))],$$

and the diagram commutes. Conversely, if the diagram commutes, then since t is the identity on $V|V$, viewed as contained in both $R|R$ and $S|S$, the conditional Boolean polynomials must agree on $V|V$, whence they are equal by Lemma 6. □

For the case $n = 2$, a conditional Boolean polynomial $f|g$ gives a binary operation on $R|R$ and one on $S|S$, and the commutativity of the diagram just says that $t : R|R \to S|S$ is a homomorphism with respect to those operations. *Thus t is a homomorhism for any binary operation induced on $R|R$ and $S|S$ by any conditional Boolean polynomial $f|g$.*

Let t be a truth evaluation on the Boolean algebra R. That is, t is a homomorphism from R to the two element Boolean algebra $\{0, 1\} = V$.

Definition 3. *A truth evaluation on $R|R$ is a function $t : R|R \to V|V$ induced by a truth evaluation t on R by the formula*

$$t(a|b) = (t(ab)|t(b)).$$

Note that we are using t both for the truth evaluation on R and the truth evaluation it induces on $R|R$. Viewing $R|R$ as containing R, the truth evaluation on $R|R$ induced by a truth evaluation t on R is an extension of t to all of $R|R$. Viewing $V|V$ as a subset of $R|R$, a truth evaluation t on $R|R$ is the identity function on $V|V$.

Since $V|V = \{(0|1), (1|1), (0|0)\}$ *has three elements, each conditional event* $(a|b)$ *has one of three possible truth values,* $(0|1)$ *(false, or* 0*),* $(1|1)$ *(true, or* 1*), and* $(0|0)$ *(undecided, or* u*). The truth value* $t(a|b)$ *of* $(a|b)$ *is thus called true if* $t(ab) = 1$*, false if* $t(a'b) = 1$*, and undecided if* $t(b') = 1$*.*

We pause here to discuss these three possible truth values, their justification, motivation, and history. The connection of conditional events and three-valued logic, at an informal level, appeared in DeFinetti (1964). Following his discussion on conditional prevision and probability, in which the concept of conditional events was mentioned (DeFinetti, 1974, vol I, p.134), he brought out the connection as follows. In the conditional event $(a|b)$, there are three cases to consider, ab, ab', and b', corresponding to "thesis", "anti-thesis", and "anti-hypothesis", respectively. The event a enters the picture only through its intersection with b. Thus $(a|b)$ can be written in its "reduced" form $(ab|b)$. For DeFinetti, $(a|b)$ is a formal object with no strict mathematical meaning. He stated that "one might consider $(a|b)$ as a tri-event with values $(1|1) = 1$, $(0|1) = 0$, and $(0|0) = \phi$, where 1 = true, 0 = false, and ϕ = void, according as it leads to a "win", or a "loss", or a "calling off" of a possible conditional bet."

A similar idea appeared in Schay (1968). Generalizing indicator functions of ordinary events, Schay defined conditional events $(a|b)$ as functions, defined on a sample space Ω, and taking three possible values $\{0, 1, u\}$, with u denoting "undefined". This approach is similar to the one taken in fuzzy set theory (Zadeh, 1965). The truth space $\{0, 1, u\}$ is standard in three-valued logic. (See Rescher, 1969.) However, in the calculations to be presented in this chapter, DeFinetti's notation will be used, and we will justify the meaning given to the symbols $(1|1)$, $(0|1)$, and $(0|0)$. (See also, Boole, 1854, pp. 89-97, and Hailperin, 1876, pp. 123-137.)

In classical two-valued logic, the truth values of a Boolean expression such as $b \rightarrow a$, or equivalently $b' \vee a$, are determined from those of the variables a and b. The truth space $\{0, 1\}$ is a Boolean ring which can be viewed as being contained in every Boolean ring R, so that the determination of the possible truth values of a Boolean expression is equivalent to that determination for the case $R = \{0, 1\}$. That is, the determination of the possible truth values can be made by substituting only the values 0 and 1's into the expression. Consider now a conditional event $(a|b)$. It is not a Boolean expression, but one can formally apply this evaluation process to get the possible "truth values" of $(a|b)$. This is what DeFinetti did. Doing this for $(a|b)$ yields the three possiblilities $(1|1)$, $(0|1)$, and $(0,0) = (1|0)$. Using our modeling of conditional events as cosets of principal ideals,

$$(1|1) = 1 + \{0, 1\}0 = 1 + \{0\} = \{1\},$$

$$(0|1) = 0 + \{0, 1\}0 = 0 + \{0\} = \{0\}, \text{ and}$$

$$(0|0) = 0 + \{0, 1\}1 = 0 + \{0, 1\} = \{0, 1\}.$$

The first two we identify with "true" and "false", respectively, but there is a third possible "truth value" $(0|0) = \{0,1\}$, which can be interpreted as "undecided" since we cannot reasonably choose one of the values "true" or "false" for $(a|b)$ when both a and b are 0.

Now back to our more mathematical truth evaluations $t : R|R \to V|V$. In the diagram

$$\begin{array}{ccc} (R|R)^n & \xrightarrow{\quad f|g \quad} & R|R \\ \downarrow t^n & & \downarrow t \\ (V|V)^n & \xrightarrow[\quad \Psi_{f|g} \quad]{} & V|V \end{array},$$

if $f|g$ is a conditional Boolean polynomial and t is a truth evaluation on $R|R$, the map $\Omega_{f|g}$ induced by that polynomial on $(V|V)^n$ is called the *truth function or truth table of* $f|g$. It of course depends on the truth evaluation t. More generally, *any* function $\Psi : (V|V)^n \to V|V$ is called a *truth function or truth table.* Here is our main theorem for the conditional case. It follows immediately from the previous Proposition 2 and Lemma 6.

Theorem 2. *Let R be a Boolean algebra, let t be a truth evaluation on $R|R$, and let $f|g : (R|R)^n \to R|R$, be a conditional Boolean function. Then there is exactly one truth function*

$$\Psi_{f|g} : (V|V)^n \to V|V$$

such that

$$\Psi_{f|g} \circ t^n = t \circ (f|g),$$

namely $\Psi_{f|g} = f|g$. *Conversely, given a truth function* Ψ, *that is, any mapping*

$$\Psi : (V|V)^n \to V|V,$$

there is exactly one conditional Boolean function $f|g : (R|R)^n \to R|R$ *such that*

$$\Psi \circ t^n = t \circ (f|g),$$

namely that given by the conditional Boolean polynomial $f|g$ *inducing* Ψ. *In particular, there is a one-to-one correspondence between truth functions* $(V|V)^n \to V|V$ *and conditional Boolean functions* $(R|R)^n \to R|R$. □

In the theorem, given $f|g$, how can one actually construct $\Psi_{f|g}$? Given Ψ how can one actually construct $f|g$? If $f|g$ is given, it is almost always given in the form of a conditional Boolean polynomial, in which case simply take $\Psi_{f|g} = f|g$. In any case, the action of the function $f|g$ on $V|V$ is given, and that action determines $f|g$. So from a function $(V|V)^n \to V|V$, we need to construct the conditional Boolean polynomial inducing it. Thus we need to construct two Boolean polynomials inducing two given Boolean functions $V^{2n} \to V$. We have seen earlier how to do this explicitly. Now, conversely, this is the same problem as constructing from $\Psi : (V|V)^n \to V|V$ the requisite $f|g$. So carrying out these constructions is just a problem in constructing Boolean polynomials inducing given functions $V^{2n} \to V$. We will have occasion to carry out some of these constructions in Section 3.5 for the cases $n = 1$ and $n = 2$.

In case $n = 2$, each conditional Boolean polynomial gives a binary operation on $R|R$, and in particular on $V|V$, and we have a one-to-one correspondence between binary operations (given by conditional Boolean polynomials) on $R|R$ and (binary) truth functions on $V|V$. The case $n = 1$, of course, gives a unary operation on $R|R$, or just a mapping from $R|R$ into itself, and there is a one-to-one correspondence between unary operations on $R|R$ (given by conditional Boolean polynomials) and unary truth functions on $V|V$. The space $V|V = \{(0|1), (1|1), (0|0)\}$ is called the *truth space*. We sometimes label its elements 0, 1, and u for $(0|1)$, $(1|1)$, and $(0|0)$, respectively, thinking of 0 as *false*, 1 as *true*, and u as *undecided*. Various authors have defined logical connectives, or operators $\vee$, $\wedge$, and $'$ on $R|R$, and there are several well known sets of truth tables for the truth space $V|V$. Given logical operators $\vee$, $\wedge$, and $'$ on $R|R$, there are corresponding truth tables for them. These truth tables may or may not be reasonable ones from a logical point of view. It is typical that a three-valued logic is specified by giving five truth tables, one for each of the connectives $\vee$, $\wedge$, $'$, $\to$, and $\leftrightarrow$. In any case, truth tables for them give rise to algebraic operations on $R|R$, and with these operations, $R|R$ may or not be an interesting or tractable algebraic system. This one-to-one correspondence between truth tables (for $V|V$) and operations on $R|R$ is of interest, with this latter structure providing a syntactic home for a given three-valued logic. We will look at several such correspondences in Section 3.5.

In discussing conditional Boolean polynomials, we have stuck to those $f|g$ in *reduced* form. That is, f and g are Boolean polynomials in normal disjunctive form with no terms of the form

$$Y_1, Y_2, , , , X_i, \ldots, Y_n, Y_{n+1}, \ldots, X_{i+n}', \ldots Y_{2n},$$

and every elementary term of f is one of g. Usually, a Boolean polynomial can be written in much more compact form than its normal disjunctive form. For this reason, and for computational purposes, we indicate how to associate a conditional Boolean polynomial with $f|g$ for *any Boolean polynomials* f and g. To do this, just put f and g in their disjunctive normal forms, discard from each their elementary terms of the form displayed above, and from f those elementary terms not in g. This last step is the same as "intersecting" f with g. In fact, one could intersect f and g first, and then put fg and g in their normal disjunctive forms, discarding those terms of the form displayed above. This gives a pair $f|g$ in reduced form, and starting from any pair, it should be clear that it is associated with exactly one $f|g$ in reduced form. Further, any pair $f|g$ induces a function

$$f|g : (R|R)^n \to R|R$$

just as in the case of reduced forms, and *two* $f|g$*'s induce the same function if and only if they have the same reduced form.* We will call two $f|g$'s *equivalent* if they have the same reduced form, or what is the same thing, if they induce the same mapping just indicated.

The procedure outlined above is useful in verifying that two pairs $f|g$ are equivalent. We illustrate with an example. Let $n = 2$, and consider the two conditional polynomials

$$f|g = (X_1' \vee X_2 \vee X_3'X_4')|(X_1'X_3 \vee X_2X_4 \vee X_3X_4 \vee X_3'X_4')$$

and

$$h|k = (X_1X_2 \vee X_1'X_3 \vee X_2X_3' \vee X_3'X_4')|(X_1'X_3 \vee X_1X_4 \vee X_2X_3' \vee X_3'X_4').$$

Now it is an easy calculation to get

$$fg = X_1'X_3 \vee X_2X_4 \vee X_3'X_4'$$

and

$$hk = X_1X_2X_4 \vee X_1'X_3 \vee X_2X_3' \vee X_3'X_4'.$$

Still, it is not clear at all that

$$fg|g = (X_1'X_3 \vee X_2X_4 \vee X_3'X_4')|(X_1'X_3 \vee X_2X_4 \vee X_3X_4 \vee X_3'X_4')$$

and

$$hk|k = (X_1X_2X_4 \vee X_1'X_3 \vee X_2X_3' \vee X_3'X_4')|(X_1'X_3 \vee X_1X_4 \vee X_2X_3' \vee X_3'X_4')$$

are equivalent. The disjunctive normal form of $fg = X_1'X_3 \vee X_2X_4 \vee X_3'X_4'$ is

$$X_1'X_2X_3X_4 \vee X_1'X_2'X_3X_4 \vee X_1'X_2X_3X_4' \vee X_1'X_2'X_3X_4'$$

$$\vee X_1X_2X_3X_4 \vee X_1'X_2X_3X_4 \vee X_1X_2X_3'X_4 \vee X_1'X_2X_3'X_4$$

$$\vee X_1X_2X_3'X_4' \vee X_1'X_2X_3'X_4' \vee X_1X_2'X_3'X_4' \vee X_1'X_2'X_3'X_4',$$

which, after discarding duplicate terms and those of the forms $X_1WX_3'Y$ and WX_2YX_4', becomes

$$X_1'X_2X_3X_4 \vee X_1'X_2'X_3X_4 \vee X_1'X_2'X_3X_4'$$

$$\vee X_1X_2X_3X_4 \vee X_1'X_2X_3'X_4 \vee X_1'X_2'X_3'X_4'.$$

Similarly, the normal disjunctive form of $hk = X_1X_2X_4 \vee X_1'X_3 \vee X_2X_3' \vee X_3'X_4'$ is

$$X_1X_2X_3X_4 \vee X_1X_2X_3'X_4$$

$$\vee X_1'X_2X_3X_4 \vee X_1'X_2'X_3X_4 \vee X_1'X_2X_3X_4' \vee X_1'X_2'X_3X_4'$$

$$\vee X_1X_2X_3'X_4 \vee X_1'X_2X_3'X_4 \vee X_1X_2X_3'X_4' \vee X_1'X_2X_3'X_4'$$

$$\vee X_1X_2X_3'X_4' \vee X_1'X_2X_3'X_4' \vee X_1X_2'X_3'X_4' \vee X_1'X_2'X_3'X_4'.$$

Again, discarding duplicate terms and those of the forms $X_1WX_3'Y$ and WX_2YX_4' yields

$$X_1X_2X_3X_4 \vee X_1'X_2X_3X_4 \vee X_1'X_2'X_3X_4$$

$$\vee X_1'X_2'X_3X_4' \vee X_1'X_2X_3'X_4 \vee X_1'X_2'X_3'X_4',$$

which is the same form as that of fg. Similarly, g and k have the same such forms, so that $f|g$ and $h|k$ are equivalent. They represent the same conditional Boolean polynomials, and induce the same conditional Boolean functions.

In summary, the situation is this. There is a one-to-one correspondence between (conditional) logical operations on $R|R$ and truth functions on the truth space $\{0, 1, u\}$. Note however that, unlike the case of R, there is a variety of three-valued logics. See, for example, Rescher (1969) for background. Also, note the difference with the Boolean case: since both R and $\{0, 1\}$ are Boolean rings, truth evaluations are specified as homomorphisms; the situation in three-valued logics is somewhat different. Indeed, as far as three-valued logics are concerned, all logicians insist on the choice of some system of "truth tables" for basic connectives between implicative propositions without syntax considerations. This is not surprising since the concrete space $R|R$ of implicative propositions, as a mathematical entity, was never considered at the level of Boolean rings for unconditional propositions. Now, since $R|R$ is shown to be the space of all cosets of principal ideals of R, it is possible to investigate its algebraic structures induced by semantic considerations.

In the case of $R|R$, which has no a priori algebraic structure, we have only at our disposal truth evaluations $t : R|R \to \{0, 1, u\}$ defined previously. The objective is to establish an analogous commutative diagram for the conditional case. This type of diagram will provide algebraic structures for $R|R$ from given semantics and vice versa. If $\{0, 1\}$ is the truth space in classical two-valued logic, then formally $(\{0, 1\}|\{0, 1\})$ is the truth space for elements of the conditional space $R|R$. From the above identification, we see that three-valued logic is natural for conditional events. This is in line with earlier considerations of DeFinetti (1964) and Schay (1968). It is interesting to note that the symbols $(0|1)$, $(1|1)$, $(0|0)$ appeared also in Boole's Laws of Thoughts (Boole, 1854), apparently in his attempt to provide a disjunctive normal form for ratios of propositions. See also Hailperin (1976).

Another constructive proof of Theorem 2 will now be given. First, in view of Stone's Representation Theorem, we regard the Boolean ring R as a field of subsets of some set Ω. As such, truth evaluations can be expressed in terms of indicator functions. Recall that the generalized indicator function of $(a|b)$, for $a, b \in R$, is defined as:

$$\varphi(a|b) : \Omega \to \{0, u, 1\}$$

$$\varphi(a|b)(\omega) = \begin{cases} 1 & \text{if } \omega \in ab \\ 0 & \text{if } \omega \in a'b \\ u & \text{if } \omega \in b' \ . \end{cases}$$

Assuming $a \leq b$, $(a|b)$ partitions Ω as $a, a'b, b'$, so that if we let

$$w_j(a|b) = \begin{cases} a(=ab) & \text{if } j = 1 \\ a'b & \text{if } j = 0 \\ b' & \text{if } j = u, \end{cases}$$

where $j \in \{0, u, 1\}$, then

$$\varphi(a|b)(\omega) = j \text{ if and only if } \omega \in w_j(a|b).$$

For $n \geq 1$, $\varphi_n : (R|R)^n \to (\{0, u, 2\}^n)^\Omega$ is defined componentwise, that is

$$\varphi_n(\underline{a}|\underline{b})(\omega) = \varphi_n((a_1|b_1), ..., (a_n|b_n)(\omega) = (\varphi(a_1|b_1)(\omega), ..., \varphi(a_n|b_n)(\omega)).$$

For f a conditional Boolean map $(R|R)^n \to R|R$, that is, $f = (\alpha|\beta)$ where α, $\beta : R^{2n} \to R$, are Boolean polynomials in $2n$ variables, we write

$$f(\underline{a}|\underline{b}) = (\alpha(a_1, ..., a_n, b_1, ..., b_n)|\beta(a_1, ..., a_n, b_1, ..., b_n)$$

$$= (\alpha(\underline{a}, \underline{b})|\beta(\underline{a}, \underline{b})).$$

We have

$$\varphi(f(\underline{a}|\underline{b}))(\omega) = \begin{cases} 1 & \text{if } \omega \in \alpha(\underline{a}, \underline{b})\beta(\underline{a}, \underline{b}) \\ 0 & \text{if } \omega \in (\alpha(\underline{a}, \underline{b}))'\beta(\underline{a}, \underline{b}) \\ u & \text{if } \omega \in (\beta(\underline{a}, \underline{b}))' . \end{cases}$$

Theorem 2 can be rephrased as follows. Each conditional Boolean map $f : (R|R)^n \to R|R$ has a unique "truth table" $\psi_f : \{0, u, 1\}^n \to \{0, u, 1\}$, that is

$$\forall(\underline{a}|\underline{b}) \in (R|R)^n, \ \forall\omega \in \Omega,$$

$$\varphi(f(\underline{a}|\underline{b}))(\omega) = \psi_f(\varphi_n(\underline{a}|\underline{b})(\omega)). \qquad (*)$$

Conversely, for each $\psi : \{0, u, 1\}^n \to \{0, u, 1\}$, there exists a unique conditional Boolean map $f : (R|R)^n \to R|R$ such that $\psi_f = \psi$, that is, (*) holds.

The construction goes as follows. The disjunctive normal form of $\alpha(\underline{a}, \underline{b})$ is

$$\vee \ \alpha(\delta_1, ..., \delta_n, \gamma_1, ..., \gamma_n)a_1^{\delta_1}...a_n^{\delta_n}b_1^{\gamma_1}...b_n^{\gamma_n}$$

where $\vee$ is over all $\delta_i, \gamma_i, i = 1, 2, ..., n$ with $\delta_i, \gamma_i \in \{0, 1\}$. Since $a_i \leq b_i$, for all

$i = 1, 2, \ldots, n$, there are only three pairs $(0, 1), (0, 0), (1, 1)$ for each (δ_i, γ_i), thus, letting

$$j_i = \begin{cases} 1 \text{ if } (\delta_i, \gamma_i) = (1, 1) \\ 0 \text{ if } (\delta_i, \gamma_i) = (0, 1) \\ u \text{ if } (\delta_i, \gamma_i) = (0, 0) \, , \end{cases}$$

and, for

$$\underline{j} = (j_1, j_2, \ldots, j_n),$$

$$w_{\underline{j}}(\underline{a}|\underline{b}) = w_{j_1}(a_1|b_1) \ldots w_{j_n}(a_n|b_n) \, ,$$

we have

$$\alpha(\underline{a}, \underline{b}) = \bigvee_{\underline{j} \in \{0, u, 1\}^n} \alpha(\delta_1, \ldots, \delta_n, \gamma_1, \ldots, \gamma_n) w_{\underline{j}}(\underline{a}|\underline{b}) = \bigvee_{\underline{j} \in J(\alpha)} w_{\underline{j}}(\underline{a}|\underline{b})$$

where

$$J(\alpha) \subseteq \{0, u, 1\}^n \, ,$$

$$J(\alpha) = \{\underline{j} : \alpha(\delta_1, \ldots, \delta_n, \gamma_1, \ldots, \gamma_n) = 1\} \, .$$

Note that $\underline{j} = (j_1, j_2, \ldots, j_n)$ with j_i corresponds to (δ_i, γ_i). Define $\psi_f : \{0, u, 1\}^n \to \{0, u, 1\}$ by

$$\psi_f(\underline{j}) = \begin{cases} 1 \text{ if } \underline{j} \in J(\alpha) \\ 0 \text{ if } \underline{j} \in J^c(\alpha) \cap J(\beta) \\ u \text{ if } \underline{j} \in J^c(\beta) \end{cases}$$

where $J^c(\alpha)$ denotes the set-complement of $J(\alpha)$ in $\{0, u, 1\}^n$ and similar notation applies to $J(\beta)$.

Note that $f(\underline{a}|\underline{b})$ might have another representation form, say $(\Lambda(\underline{a}, \underline{b}) | \beta(\underline{a}, \underline{b}))$, but then $\alpha(\underline{a}, \underline{b}) \cap \beta(\underline{a}, \underline{b}) = \Lambda(\underline{a}, \underline{b}) \cap \beta(\underline{a}, \underline{b})$, implying that

$$J(\alpha) \cap J(\beta) = J(\Lambda) \cap J(\beta),$$

so that ψ_f is well-defined.

For ψ_f defined above, (*) holds. Indeed, for $(\underline{a}|\underline{b}) \in (R|R)^n$ arbitrary but fixed,

$\varphi(f(\underline{a}|\underline{b}))(\omega) = 1$ if and only if $\varphi(\alpha(\underline{a}, \underline{b})|\beta(\underline{a}, \underline{b}))(\omega) = 1$ if and only if $\omega \in \alpha(\underline{a}, \underline{b})$ (assuming $\alpha \leq \beta$) if and only if

$$\omega \in \bigvee_{\underline{j} \in J(\alpha)} w_{\underline{j}}(\underline{a}|\underline{b})$$

if and only if

$$\omega \in w_{\underline{j}}(\underline{a}|\underline{b})$$

for some

$$\underline{j} \in J(\alpha)$$

if and only if

$$\omega \in w_{j_i}(a_i|b_i),\ \forall i = 1, 2, \ldots, n$$

(where $\underline{j} = (j_1, \ldots, j_n)$) if and only if

$$\varphi(a_i|b_i)(\omega) = j_i,\ \ \forall i = 1, 2, \ldots, n$$

if and only if

$$1 = \psi_f(\underline{j}) = \psi_f(\varphi(a_1|b_1)(\omega), \ldots, \varphi(a_n|b_n)(\omega)) \ = \psi_f(\varphi_n(\underline{a}|\underline{b})(\omega)).$$

The argument is similar for $\varphi(f(\underline{a}|\underline{b}))(\omega) = 0$ or u.

Conversely, if $\psi : \{0, u, 1\}^n \to \{0, u, 1\}$ is given, then there exists a unique Boolean-like map $(\alpha|\beta) : (R|R)^n \to R|R$ such that (*) holds. Indeed, it suffices to take

$$\alpha(\underline{a}, \underline{b}) = \bigvee_{\underline{j} \in \psi^{-1}(1)} w_{\underline{j}}(\underline{a}|\underline{b}),$$

$$\beta(\underline{a}, \underline{b}) = \bigvee_{\underline{j} \in \psi^{-1}(1) \cup \psi^{-1}(0)} w_{\underline{j}}(\underline{a}|\underline{b}).$$

Several remarks are in order.

(i) Viewing $(a|b)$ as a mathematical entity with the three possible values 0, u, or 1, the function Ψ_f uniquely associated with a map $f : (R|R)^n \to R|R$ is precisely the "truth table" of f. The function Ψ_f is completely determined once f is specified. The converse is also true: a truth table Ψ will uniquely determine a "syntactic" (mathematical) modeling of a connective on $R|R$. Moreover, (*) of Theorem 2 expresses the truth-functional property of logic, namely truth values of an n-ary connective on $R|R$

are determined from those of the components.

(ii) In the literature of three-valued logic (for example, Rescher, 1969), one usually considers a collection of sentences S in which each sentence s can be either true, false, or "undetermined" (Lukasiewicz, Bochvar, Kleene). The algebraic structure of S is rarely specified. Instead, semantically, five truth tables, one each for $\wedge$, $\vee$, $'$, $\rightarrow$, and $\leftrightarrow$ are given. Our remarks above show that, given such a system of "truth tables", one can explicitly write down their "syntactic" counter-parts, and conversely.

It is interesting to speculate about the algebraic analog of a Boolean ring as a basic space for Lukasiewicz's logic. That is, can one give a mathematical representation of a sentence s in S in such a way that as an algebraic structure, S will be equipped with the basic connectives whose truth tables are given in advance? As we shall see in Section 3.5, one such mathematical representation for S is our conditional extension $R|R$ where our logical operations introduced in Section 2.2 correspond precisely to Lukasiewicz's truth tables.

(iii) As far as we are concerned here, the easy part of Theorem 2 will serve as a way to discuss the "reasonability" of our proposed system of logical operations for conditional events. This will be carried out in two steps. First, from a given system of operations on $R|R$, one proceeds to identify their associated truth tables using normal disjunctive forms of Boolean functions and the explicit construction of Ψ_f given in the proof of Theorem 1. Next, once a system of truth tables is obtained, one looks at the names of the connectives involved (say, f = "and") and examines their truth tables. Since a truth table of a given connective (in natural language) should reflect the common sense meaning of that connective, any "unreasonable" truth table found will lead to the conclusion that its corresponding proposed operation on $R|R$ is "unreasonable". This program will be carried out in Section 3.5 with the systems of logical operations on $R|R$ proposed by Adams, Calabrese, Schay, and by us.

The other part of Theorem 2, namely that each truth table in three-valued logic corresponds uniquely to an operator on $R|R$, is useful for investigating new algebraic structures of $R|R$.

(iv) The three-valued logic viewpoint above can be used to formulate the concept of realizations of conditional events. Let R be a σ-field of subsets of a sample space Ω. The generalized indicator function φ of each $(a|b)$ is defined as $\varphi(a|b)(\omega) = 1$ on ab, u on b', and 0 on $a'b$. As in the case of ordinary events, where $a \in R$ is said to be "realized" if the "outcome" $\omega \in a$, that is, if $\varphi(a|1)(\omega) = 1$, conditional events can possess a similar concept, viewed from a three-valued logic standpoint. Recall that $(a|b) = [ab, b' \vee a]$. If $\omega \in ab$, then $\omega \in x$ for all $x \in (a|b)$, so that $(a|b)$ is "fully" realized; if $\omega \in a'b$, then $\omega \notin x$

for any $x \in (a|b)$, since $a'bx = 0$, thus $(a|b)$ is realized at "level" 0; if $\omega \in b'$, then for each $x \in (a|b)$, x may or may not occur, depending on whether $\omega \in xb'$ or not. If it is, then we can interpret the realization at some level, for example at level $P(a|b)$ for some probability measure P on R. This can be justified by the consideration of a random variable X defined on Ω having values 0 on $a'b$, 1 on ab, and $P(a|b)$ on b', and noting that $E(X) = P(a|b)$.

(v) The viewpoint of three valued logic taken here should not be confused with the three-valued logic associated with "conditional forms" of McCarthy (1967) which motivated algebraic investigations referred to in the literature as "conditional logic" (Guzman and Squier, 1990). "Conditional logic" in the literature sometimes refers to the *non-commutative (regular) extension of Boolean logic* to three truth values, the third denoted u and standing for "undefined" or "non-terminating evaluation". The non-commutativity refers to the logic connectives $\vee$ and $\wedge$ in the extended logic. This phenomenon appears in McCarthy (1967) in which it was shown that in order to define computable partial functions, it is necessary to allow undefined expressions in the recursive formulae. From a logical viewpoint, this amounts to considering a third truth value "u" for these undefined expressions.

Consider, for example, defining recursively the function $f(n) = n!$ on the domain of non-negative integers. A verbal rule is "if $n = 0$, then assign the value 1, else assign the value $n(n - 1)!$" The statement "If ..., then ..., else If ... then ..." is called a "conditional expression". In symbols, a conditional expression is denoted

$$(a_1 \to b_1, a_2 \to b_2, \ldots, a_n \to b_n) = CE(a_1, \ldots, a_n; b_1, \ldots, b_n),$$

which means "if a_1 then b_1, else if a_2 then b_2, . . . , else if a_n then b_n." Its value is defined as $CE(a_1, \ldots, a_n; b_1, \ldots, b_n) = b_j$ where j is the first i such that a_i is true. The evaluation of $CE(a_1, \ldots, a_n; b_1, \ldots, b_n)$ proceeds from left to right, and stops when the first true a_i is found. Of course, the a_i and b_i are propositions, that is, can only be true (T) or false (F). That is, we are in classical two-valued logic, where the propositions are elements of a Boolean ring R with the usual connectives. If T and F also stand for "always true" and "always false", respectively, then the usual Boolean connectives can be expressed in terms of some simple conditional expressions. Indeed, using truth tables in two-valued logic, it is readily checked that

$$\begin{aligned} a \wedge b &= (a \to b, T \to F), \\ a \vee b &= (a \to T, T \to b), \\ a' &= (a \to F, T \to T), \\ a' \vee b &= (a \to b, T \to T). \end{aligned}$$

Now consider the partial function $f(n) = n!$ on the set of integers. The recursive definition of $f(n)$ in terms of conditional expressions is

$$n! = (n = 0 \rightarrow 1, n \neq 0 \rightarrow n(n - 1)!).$$

Thus

$$2! = (2 = 0 \rightarrow 1, 2 \neq 0 \rightarrow 2(2 - 1)!) =$$

$$2(1!) = 2(1 = 1 \rightarrow 1, 1 \neq 0 \rightarrow 1(1 - 1)!) =$$

$$2\,(1)\,(0 = 0 \rightarrow 1, 0 \neq 0 \rightarrow 0(0 - 1)!) = 2(1)(1) = 2.$$

Note that $(0 - 1)!$ is undefined. To carry out the computation above, it is necessary to allow the conditional expression to be defined even if the term beyond the one that gives the value is undefined. Thus in a general $CE(a_1, \ldots, a_n; b_1, \ldots, b_n)$, one should allow the situation where a_i or b_j are undefined, which means that the range of truth values of each "proposition" is extended to $\{T, u, F\}$. In this logic, the CE are defined as follows:

$$CE(a_1, \ldots, a_n; b_1, \ldots, b_n) = b_j$$

if there is a b_j which is "defined", and a_i is false for $i < j$, and is undefined otherwise. Thus $CE(a_1, \ldots, a_n; b_1, \ldots, b_n)$ is undefined, that is, has truth value "u", when either

(i) all the a_i are false, or
(ii) a_i is false for $i < j$, a_j is true, and b_j is undefined, or
(iii) there is an undefined a_i before a true a_j.

From this, it becomes clear that the extended connectives $\vee$ and $\wedge$ are not commutative. Indeed, from $a \wedge b = (a \rightarrow b, T \rightarrow F)$, if a is F and b is u, then the value of $a \wedge b$ is F, which has truth value F, while $b \wedge a$ has value undefined (by (iii) above) with truth value u. Similarly, if a is T and b is u, then $a \vee b$ is T, but $b \vee a$ is u.

This non-commutative three-valued logic in mechanical computation theory, bearing the name of "conditional logic" because of the role played by conditional forms in recursive computations, seems not to be in the mainstream of multi-valued logic.

3.5 Comparison of various systems of logical operators.

In this section, we are going to examine three systems $(', \wedge, \vee)$ of basic connectives on $R|R$. The truth tables of these systems will be constructed, and the relative merits of these systems will be discussed. These systems have been chosen because they have been studied to some extent as algebraic systems. Indeed, the last one (Goodman and Nguyen's) is elaborated on at length in Chapter 4. Some connectives on $R|R$ arising from the truth tables of several three-valued logics will be constructed. These particular three-valued logics are chosen because of their particular interest and importance in the field. Our prinicipal tool is Theorem 2 of Section 3.4 and the comments following it with regard to making the necessary constructions.

We begin with the definitions and a bit of discussion of the three systems $(', \wedge, \vee)$ of connectives to be examined. The recent paper by Dubois and Prade (1990) is relevant here. All the systems have the same negation $'$ on $R|R$, given by $(a|b)' = (a'b|b)$. This is in agreement with the negation operator in the Boolean ring R/Rb'.

In his 1968 paper, Schay investigated the two systems which follow.

Schay's First System

$$(a|b) \wedge (c|d) = ((b' \vee a)(d' \vee c)|b \vee d),$$

$$(a|b) \vee (c|d) = (ab \vee cd|b \vee d).$$

In his original formulation of this system, Schay (1968, p. 338) wrote the operations slightly differently. Conjunction was given as

$$(a|b) \wedge (c|d) = ((abcd \vee abd' \vee cdb')|b \vee d).$$

But

$$(b' \vee a)(d' \vee c) = b'd' \vee b'c \vee ad' \vee ac,$$

and

$$(b'd' \vee b'c \vee ad' \vee ac) \wedge (b \vee d) =$$

$$b'cd \vee ad'b \vee acb \vee acd =$$

$$abcd \vee abd' \vee cdb'.$$

Disjunction was given by

$$(a|b) \vee (c|d) = ((a \vee c)bd \vee abd' \vee cdb'|b \vee d),$$

and

$$((a \vee c)bd \vee abd' \vee cdb') \wedge (b \vee d) =$$

$$abd \vee bcd \vee abd' \vee cdb' = ab \vee cd.$$

Thus the two formulations are the same. After Schay, Adams (1975) and Calabrese (1987) proposed this same system.

Schay's Second System

$$(a|b) \wedge (c|d) = (ac|bd),$$

$$(a|b) \vee (c|d) = (a \vee c|bd).$$

In their work on the foundation of the Bayesian approach to statistics, Bruno and Gilio (1985) considered connectives on conditional events corresponding to disjunction of Schay's first system and to conjunction of Schay's second system.

Goodman and Nguyen's System

$$(a|b) \wedge (c|d) = (ac|(a'b \vee c'd \vee bd)),$$

$$(a|b) \vee (c|d) = (a \vee c|(ab \vee cd \vee bd)).$$

This system arose from Goodman and Nguyen's efforts (1988) to extend operations on R (events) to those on $R|R$ (conditional events) which would be consistent with conditional probability, and resulted from realizing the elements of $R|R$ as cosets of principal ideals of R. These operations are set forth in Section 3.2.

Below is a table of these three systems of connectives in terms of conditional Boolean polynomials. Following this table is a table of several well-known three-valued logical systems. We make a slight change of notation. For $n = 2$, the Boolean polynomials involved are functions of four variables, and it is convenient to denote those variables A, C, B, and D, rather than as X_1, X_2, X_3, and X_4. We remind the reader that for $n = 2$, the evaluation of a conditional Boolean polynomial is given by

$$f|g((a|b), (c|d)) = (f(ab, cd, b, d)|g(ab, cd, b, d)).$$

Thus if

$$f|g = (AD' \vee CB' \vee AC|B \vee D)$$

then

$$f|g((a|b), (c|d)) = ((abd' \vee cdb' \vee abcd)|b \vee d).$$

System	$\wedge$	$\vee$
Schay's first	$AD' \vee CB' \vee AC|B \vee D$	$A \vee C|B \vee D$
Schay's second	$AC|BD$	$A \vee C|BD$
Goodman-Nguyen	$AC|A'B \vee C'D \vee BD$	$A \vee C|A \vee C \vee BD$

The conditional Boolean polynomials above for $\vee$ and $\wedge$ just reflect the formulas for $(a|b) \vee (c|d)$ and $(a|b) \wedge (c|d)$. The polynomials are, of course, not in disjunctive normal form, but are in much simpler forms. We have not written the relevant polynomials for $'$ since they are all $A'B|B$, using A and B for the two variables.

In truth tables below, 0, 1, and u are used for $(0|1)$, $(1|1)$, and $(0|0)$, respectively, and x and y for $(a|b)$ and $(c|d)$, respectively.

Bochvar's three-valued logic.

x	x'
0	1
1	0
u	u

$x \wedge y$

x\y	0	1	u
0	0	0	u
1	0	1	u
u	u	u	u

$x \vee y$

x\y	0	1	u
0	0	1	u
1	1	1	u
u	u	u	u

$x \rightarrow y$

x\y	0	1	u
0	1	1	u
1	0	1	u
u	u	u	u

$x \leftrightarrow y$

x\y	0	1	u
0	1	0	u
1	0	1	u
u	0	u	u

Heyting's three-valued logic

x	x'
0	1
1	0
u	0

$x \wedge y$

x\y	0	1	u
0	0	0	0
1	0	1	u
u	0	u	u

$x \vee y$

x\y	0	1	u
0	0	1	u
1	1	1	1
u	u	1	u

$x \rightarrow y$

x\y	0	1	u
0	u	u	u
1	0	1	u
u	0	u	u

$x \leftrightarrow y$

x\y	0	1	u
0	1	0	0
1	0	1	u
u	0	u	1

Kleene's three-valued logic

x	x'
0	1
1	0
u	u

$x \wedge y$

x\y	0	1	u
0	0	0	0
1	0	1	u
u	0	u	u

$x \vee y$

x\y	0	1	u
0	0	1	u
1	1	1	1
u	u	1	u

$x \rightarrow y$

x\y	0	1	u
0	1	1	1
1	0	1	u
u	u	1	u

$x \leftrightarrow y$

x\y	0	1	u
0	1	0	u
1	0	1	u
u	u	u	u

Lukasiewicz's three-valued logic

x	y'
0	1
1	0
u	u

$x \wedge y$

$x \backslash y$	0	1	u
0	0	0	0
1	0	1	u
u	0	u	u

$x \vee y$

$x \backslash y$	0	1	u
0	0	1	u
1	1	1	1
u	u	1	u

$x \rightarrow y$

$x \backslash y$	0	1	u
0	0	0	0
1	u	1	1
u	u	1	1

$x \leftrightarrow y$

$x \backslash y$	0	1	u
0	1	0	u
1	0	1	u
u	u	u	1

Sobocinski's three-valued logic

x	x'
0	1
1	0
u	u

$x \wedge y$

$x \backslash y$	0	1	u
0	0	0	0
1	0	1	1
u	0	1	u

$x \vee y$

$x \backslash y$	0	1	u
0	0	1	0
1	1	1	1
u	0	1	u

$x \rightarrow y$

$x \backslash y$	0	1	u
0	1	1	1
1	0	0	1
u	0	1	u

$x \leftrightarrow y$

$x \backslash y$	0	1	u
0	0	0	0
1	0	0	1
u	0	0	u

We will now compute the truth tables for Schay's first system. For all three systems, we have for $'$

$$(0|1)' = (0'1|1) = (1|1),$$

$$(1|1)' = (1'1|1) = (0|1),$$

$$(0|0)' = (0'0|0) = (0|0).$$

The disjunction $A \vee C|B \vee D$ for that system is simple enough so that its table can be written down easily. To get the table for $\wedge$, we make the following calculations, which is just an exercise in evaluating the conditional Boolean polynomial

$$f|g = AD' \vee CB' \vee AC|B \vee D.$$

$$(0|1) \vee (0|1) = (0 \vee 0 \vee 0|1 \vee 1) = (0|1),$$

$$(0|1) \vee (1|1) = (0 \vee 0 \vee 0|(1 \vee 1) = (0|1),$$

$$(0|1) \vee (0|0) = (0 \vee 0 \vee 0|1 \vee 0) = (0|1),$$

$$(1|1) \vee (0|1) = (0 \vee 0 \vee 0|1 \vee 1) = (0|1),$$

$$(1|1) \vee (1|1) = (0 \vee 0 \vee 1|1 \vee 1) = (1|1),$$

$$(1|1) \vee (0|0) = (1 \vee 0 \vee 0|1 \vee 0) = (1|1),$$

$$(0|0) \vee (0|1) = (0 \vee 0 \vee 0|0 \vee 1) = (0|1),$$

$$(0|0) \vee (1|1) = (0 \vee 1 \vee 0|0 \vee 1) = (1|1),$$

$$(0|0) \vee (0|0) = (0 \vee 0 \vee 0|0 \vee 0) = (0|0).$$

Thus we get the following truth tables for Schay's first system.

x	x'
0|1	1|1
1|1	0|1
0|0	0|0

$x \wedge y$

$x \backslash y$	0|1	1|1	0|0
0|1	0|1	0|1	0|1
1|1	0|1	1|1	1|1
0|0	0|1	1|1	0|0

$x \vee y$

$x \backslash y$	0|1	1|1	0|0
0|1	0|1	1|1	0|1
1|1	1|1	1|1	1|1
0|0	0|1	1|1	0|0

These tables are recognized as those of Sobocinski's three-valued logic.

Next we construct the truth tables for Schay's second system. Since the operations are particularly simple, namely $(a|b) \wedge (c|d) = (ac|bd)$ and $(a|b) \vee (c|d) = (a \vee c|bd)$, these tables can be written down easily. Here are the tables.

x	x'
0|1	1|1
1|1	0|1
0|0	0|0

$x \wedge y$

$x \backslash y$	0|1	1|1	0|0
0|1	0|1	0|1	0|0
1|1	0|1	1|1	0|0
0|0	0|0	0|0	0|0

$x \vee y$

$x \backslash y$	0|1	1|1	0|0
0|1	0|1	1|1	0|0
1|1	1|1	1|1	0|0
0|0	0|1	0|0	0|0

The tables are recongized as those of Bochvar's three valued logic.

Now to the Goodman-Nguyen system. For the connective $\vee$ for this system, we make the following calculations using the formula for $\vee$ for that system.

$$(0|1) \vee (0|1) = (0 \vee 0|0 \vee 0 \vee (1 \wedge 1)) = (0|1),$$

$$(0|1) \vee (1|1) = (0 \vee 1|0 \vee 1 \vee (1 \wedge 1)), = (1|1),$$

$$(0|1) \vee 0|0 = (0 \vee 0|0 \vee 0 \vee (1 \wedge 0)) = (0|0),$$

$$(1|1) \vee (0|1) = (1 \vee 0|1 \vee 0 \vee (1 \wedge 1)) = (1|1),$$

$$(1|1) \vee (1|1) = (1 \vee 1|1 \vee 1 \vee (1 \wedge 1)) = (1|1),$$

$$(1|1) \vee (0|0) = (1 \vee 0|1 \vee 0 \vee (1 \wedge 0)) = (1|1),$$

$$(0|0) \vee (0|1) = (0 \vee 0|0 \vee 0 \vee (0 \wedge 1)) = (0|0),$$

$$(0|0) \vee (1|1) = (0 \vee 1|0 \vee 1 \vee (0 \wedge 1)) = (1|1),$$

$$(0|0) \vee (0|0) = (0 \vee 0|0 \vee 0 \vee (0 \wedge 0)) = (0|0).$$

Making the analagous calculations for $\wedge$, and putting the results in the usual form for truth tables yields

x	x'
$0\|1$	$1\|1$
$1\|1$	$0\|1$
$0\|0$	$0\|0$

$x \wedge y$

$x \backslash y$	$0\|1$	$1\|1$	$0\|0$
$0\|1$	$0\|1$	$0\|1$	$0\|1$
$1\|1$	$0\|1$	$1\|1$	$0\|0$
$0\|0$	$0\|1$	$0\|0$	$0\|0$

$x \vee y$

$x \backslash y$	$0\|1$	$1\|1$	$0\|0$
$0\|1$	$0\|1$	$1\|1$	$0\|0$
$1\|1$	$1\|1$	$1\|1$	$1\|1$
$0\|0$	$0\|0$	$1\|1$	$0\|0$

These are recognized as truth tables for $'$, $\wedge$, *and* $\vee$, *respectively, for Lukasiewicz's and Kleene's three valued logics.* These three-valued logics are well established, and serve as a strong motivation and justification for the Goodman-Nguyen operations $'$, $\wedge$, and $\vee$ on $R|R$. *Further, the tables for* $\wedge$ *and* $\vee$ *are the truth tables for* $\wedge$ *and* $\vee$ *for Heyting's three-valued logic.* Thus, once $R|R$ is at hand, there are strong reasons from a three-valued logical perspective to define the operations $\vee$, $\wedge$, and $'$ on $R|R$ as done by Goodman and Nguyen and for making a thorough study of the resulting algebraic system.

To illustrate the method of constructing a conditional Boolean operator of $R|R$ from a truth table, we construct that operator for Lukasiewicz's $\rightarrow$ and for Sobocinski's disjunction. Following are those truth tables in terms of the elements of $V|V$, from which it is easy to make the necessary calculations.

$x \rightarrow y$

$x \backslash y$	$0\|1$	$1\|1$	$0\|0$
$0\|1$	$1\|1$	$1\|1$	$1\|1$
$1\|1$	$0\|1$	$1\|1$	$0\|0$
$0\|0$	$0\|0$	$1\|1$	$1\|1$

$x \vee y$

$x \backslash y$	$0\|1$	$1\|1$	$0\|0$
$0\|1$	$0\|1$	$1\|1$	$0\|1$
$1\|1$	$1\|1$	$1\|1$	$1\|1$
$0\|0$	$0\|1$	$1\|1$	$0\|0$

The conditional Boolean polynomial $f|g$ for $\rightarrow$ is determined by the following values for f and g, which are read off from the table for $\rightarrow$.

$f(0, 0, 1, 1) = 1,$ $\qquad g(0, 0, 1, 1) = 1,$

$f(0, 1, 1, 1) = 1,$ $\qquad g(0, 1, 1, 1) = 1,$

$f(0, 0, 1, 0) = 1,$ $\qquad g(0, 0, 1, 0) = 1,$

$f(1, 0, 1, 1) = 0,$ $\qquad g(1, 0, 1, 1) = 1,$

$f(1, 1, 1, 1) = 1,$ $\qquad g(1, 1, 1, 1) = 1,$

$f(1, 0, 1, 0) = 0,$ $\qquad g(1, 0, 1, 1) = 0,$

$f(0, 0, 0, 1) = 0,$ $\qquad g(0, 0, 0, 1) = 0,$

$f(0, 1, 0, 1) = 1,$ $\qquad g(0, 1, 0, 1) = 1,$

$f(0, 0, 0, 0) = 1,$ $\qquad g(0, 0, 0, 0) = 1.$

Thus f and g are the Boolean polynomials

$$\begin{aligned} f &= A'C'BD \vee A'CBD \vee A'C'BD' \vee ACBD \vee A'CB'D \vee A'C'B'D' \\ &= AC \vee A'B \vee B'C \vee B'D', \end{aligned}$$

and

$$\begin{aligned} g &= A'C'BD \vee A'CBD \vee A'C'BD' \vee AC'BD \vee ACBD \vee A'CB'D \vee A'C'B'D' \\ &= A'C'BD \vee A'CB \vee A'BD' \vee AC'D \vee AC \vee B'C \vee B'D' \\ &= A'B(C'D \vee C \vee D') \vee A(C'D \vee C) \vee B'C \vee B'D' \\ &= A'B \vee AD \vee B'C \vee B'D' \\ &= C \vee A'B \vee AD \vee B'D'. \end{aligned}$$

Thus Lukasiewicz's $\rightarrow$ is given by the formula

$$(a|b) \rightarrow (c|\mathrm{d}) = (ac \vee a'b \vee b'c \vee b'd' \,|\, a'b \vee \mathrm{ad} \vee b'c \vee b'd')$$

Similarly, the f and g for Sobocinski's disjunction have the values

$f(0, 0, 1, 1) = 0,$ $\qquad g(0, 0, 1, 1) = 1,$

$f(0, 1, 1, 1) = 1,$ $\qquad g(0, 1, 1, 1) = 1,$

$f(0, 0, 1, 0) = 0,$ $\qquad g(0, 0, 1, 0) = 1,$

$f(1, 0, 1, 1) = 1,$ $\qquad g(1, 0, 1, 1) = 1,$

$$f(1, 1, 1, 1) = 1, \qquad g(1, 1, 1, 1) = 1,$$

$$f(1, 0, 1, 0) = 1, \qquad g(1, 0, 1, 1) = 1,$$

$$f(0, 0, 0, 1) = 0, \qquad g(0, 0, 0, 1) = 1,$$

$$f(0, 1, 0, 1) = 1, \qquad g(0, 1, 0, 1) = 1,$$

$$f(0, 0, 0, 0) = 0, \qquad g(0, 0, 0, 0) = 0.$$

Thus f and g are the polynomials

$$\begin{aligned} f &= A'BCD \vee ABC'D \vee ABCD \vee ABC'D' \vee A'B'CD \\ &= A'BC \vee AC'D \vee AB \vee AD' \vee B'C \\ &= A(C'D \vee B \vee D') \vee C(A'B \vee B') \\ &= A \vee C \end{aligned}$$

and

$$g = (A'B'C'D')' = A \vee B \vee C \vee D = C \vee D.$$

Thus the formula for Sobocinski's disjunction is

$$(a|b) \vee (c|d) = (a \vee c|b \vee d),$$

which, of course, we already knew.

We now illustrate the use of the second proof of Theorem 2 in constructing truth tables for various three-valued logical operators.

(i) **Negation operators.** For the negation operator given by

$$(a|b)' = (a'|b) = (a'b|b),$$

the partition induced by $(a|b)$ is

$$w_0(a|b) = a'b,\ w_1(a|b) = ab,\ w_u(a|b) = b'.$$

Thus

$$(a|b)' = (\alpha(a, b)|\beta(a, b))$$

where

$$\alpha(a, b) = a'b$$

and

$$\beta(a, b) = b = ab \vee a'b.$$

Thus

$$J(\alpha) = \{0\}\ , J(\beta) = \{0, 1\},$$

and hence

for $i \in J(\alpha) \cap J(\beta) = \{0\}$, $\Psi_{\prime}(0) = 1$,

for $i \in J^c(\beta) = \{u\}$, $\psi_{\prime}(u) = u$, and

for $i \in J(\beta) \cap J^c(\alpha) = \{1\}$, $\Psi_{\prime}(1) = 0$

This is Lukasiewicz truth table for negation (see Section 3.4).

(ii) **Conjunction operators.** The conjunction operator $\wedge$ of Schay's first system is given by

$$(a|b) \wedge (c|d) = ((b' \vee a)(d' \vee c)|b \vee d) =$$

$$((b' \vee a)(d' \vee c)(b \vee d)|(b \vee d) = (abd' \vee cdb' \vee abcd|b \vee d).$$

Thus $\alpha_{\wedge}$, or α for short, is

$$w_1(a|b)w_u(c|d) \vee w_1(c|d)w_u(a|b) \vee w_1(a|b)w_1(c|d),$$

so that

$$J(\alpha) = \{(1, u), (u, 1), (1, 1)\}\ .$$

Next,

$$\beta = b \vee d = bd \vee b'd \vee bd'$$

$$= (abcd) \vee (abc'd) \vee (a'bcd) \vee (a'bc'd) \vee (abd') \vee (b'cd) \vee (a'bd') \vee (b'c'd)\ .$$

Thus

$$J(\beta) = \{(1, 1), (1, 0), (0, 1), (0, 0), (1, u), (u, 1), (0, u), (u, 0)\}$$

and

$$J(\alpha) \cap J(\beta) = \{(1, 1), (u, 1), (1, u)\},$$

$$J^c(\beta) = \{(u, u)\},$$

$$J(\beta) \cap J^c(\alpha) = \{(1, 0), (0, 1), (0, 0), (0, u), (u, 0)\}\ .$$

Therefore

$$\psi_\wedge(i, j) = \begin{cases} 1 \text{ for } (i, j) \in \{(1, 1)\,(u, 1), (1, u)\} \\ u \text{ for } (i, j) = (u, \; u) \\ 0 \text{ for } (i, j) \in \{(1, 0),(0, 1), (0, 0), (0, u), (u, 0)\} \,. \end{cases}$$

This is Sobocinski's truth table for conjunction.

For Schay's conjunction in his second system,

$$(a|b) \wedge (c|d) = (ac|bd) \,,$$

and one obtains

$$\psi_\wedge(i, j) = \begin{cases} 1 \text{ for } (i, j) = (1, 1) \\ u \text{ for } (i, j) \in \{(0, \; u), (u, 0), (u, u), \; (u, 1), (1, u)\} \,. \\ 0 \text{ for } (i, j) \in \{(0, \; 0), (0, 1), (1, 0)\} \end{cases}$$

This is Bochvar' truth function for conjunction.

For the Goodman-Nguyen conjunction,

$$(a|b) \wedge (c|d) = (ac|a'b \vee c'd \vee bd) = (abcd|a'b \vee c'd \vee bd),$$

whence $\alpha = abcd$ and so

$$J(\alpha) = \{(1, 1)\}.$$

$$\begin{aligned} \beta &= a'b \vee c'd \vee bd \\ &= a'b \vee c'd \vee (bdac \vee bd(ac)') \\ &= (abcd) \vee bda' \vee a'b \vee c'd \vee bdc' \\ &= (abcd) \vee bda' \vee (a'bd \vee a'bd') \vee (c'db \vee c'db') \vee bdc' \\ &= (abcd) \vee bda' \vee a'bd' \vee c'db \vee c'db' \\ &= (abcd) \vee ((a'b)c'd) \vee ((a'b)cd) \vee (a'bd') \vee ((abc'd) \vee (a'bc'd) \vee c'db', \end{aligned}$$

so that

$$J(\beta) = \{(1, 1), (0, 0), (0, 1), (0, u), (1, 0), (u, 0)\} \,.$$

Hence

$$J(\alpha) \cap J(\beta) = \{(1, 1)\},$$

$$J^c(\beta) = \{(u, u), (u, 1), (1, u)\},$$

and

$$J(\beta) \cap J^c(\alpha) = \{(0, 0), (0, 1), (0, u), (1, 0), (u, 0)\},$$

so that

$$\psi_\wedge(i, j) = \begin{cases} 1 \;\; for \;\; (i, j) = (1, 1) \\ u \;\; for \;\; (i, j) \in \{(u, u), (u, 1), (1, u)\} \\ 0 \;\; for \;\; (i, j) \in \{(0, 0), (0, 1), (0, u), (1, 0), (u, 0)\} \end{cases},$$

which is Lukasiewicz truth table for conjunction (see 3.4).

(iii) **Disjunction operators.** Adams and Calabrese's disjunctions are identical to the disjunction in Schay's first system, which is given by

$$(a|b) \vee (c|d) = (ab \vee cd|b \vee d).$$

We have

$$\psi(i, j) = \begin{cases} 1 \;\; for \;\; (i, j) \in \{(0, 1), (u, 1), (1, 0), (1, u), (1, 1)\} \\ u \;\; for \;\; (i, j) = (u, u) \\ 0 \;\; for \;\; (i, j) \in \{(0, 0), (0, u), (u, 0)\} \end{cases},$$

which is Sobocinski's disjunction.

The disjunction in Schay's second system is

$$(a|b) \vee (c|d) = (a \vee c|bd),$$

and

$$\psi(i, j) = \begin{cases} 1 \;\; for \;\; (i, j) \in \{(0, 1), (1, 0), (1, 1)\} \\ u \;\; for \;\; (i, j) \in \{(0, u), (u, 0), (u, u), (u, 1), (1, u)\} \\ 0 \;\; for \;\; (i, j) = (0, 0) \end{cases},$$

which is Bochvar's disjunction.

For the Goodman-Nguyen disjunction,

$$\begin{aligned}(a|b) \vee (c|d) &= (a \vee c|ab \vee cd \vee bd) \\ &= (ab \vee cd|ab \vee cd \vee bd),\end{aligned}$$

so

$$\begin{aligned}\alpha &= ab \vee cd \\ &= (abd' \vee abdc \vee abdc') \vee (cdb' \vee cdba \vee cdba') \\ &= (abcd \vee abd' \vee cdb' \vee abc'd \vee cdba').\end{aligned}$$

Thus

$$J(\alpha) = \{(1, 1), (1, u), (u, 1), (1, 0), (0, 1)\},$$

and

$$\beta = ab \vee cd \vee bd .$$

Note that

$$bd = (abcd) \vee a'bd \vee bc'd$$
$$= (abcd) \vee (a'bdc \vee a'bdc') \vee a'bc'd) ,$$

and thus

$$J(\beta) = J(\alpha) \cup \{(0, 0)\} .$$

Hence

$$\psi(i, j) = \begin{cases} 1 \text{ for } (i, j) \in \{(1, 1), (1, u), (u, 1), (1, 0), (0, 1)\} \\ u \text{ for } (i, j) \in \{(u, 0), (u, u), (0, u)\} \\ 0 \text{ for } (i, j) = (0, 0) \end{cases} ,$$

which is Lukasiewicz truth table for disjunction.

Each proposed system in three-valued logic has its own rationale. Since logical operations on conditionals correspond to truth tables in three-valued logics, the comparison of different algebras of conditional events is delicate. However, based on Rescher's discussion (Rescher, 1969, pp. 131-133), we make some comparisons below. To do that, we first complete the description of the three algebras, Schay's first and second system, and the Goodman-Nguyen system, by writing down the syntax operations corresponding to the remaining two truth tables, namely for implication ($\rightarrow$) and for equivalence ($\longleftrightarrow$). (Of course $x \longleftrightarrow y$ means $(x \rightarrow y) \wedge (y \rightarrow x)$.) Thus we will have three algebras of conditional events, corresponding respectively to the three three-valued logics of Sobocinski, Bochvar, and Lukasiewicz. In the following tables, the implication and equivalence are expressed in terms of $'$, $\wedge$, and $\vee$ within each system. As usual, we generally denote $\wedge$ by juxtaposition. Also, in R, the implication $\rightarrow$ is material implication. Here are the tables.

Schay's First System

$$(a|b) \rightarrow (c|d) = (a|b)' \vee (c|d)$$
$$(a|b) \longleftrightarrow (c|d) = ((a \longleftrightarrow c)bd|b \vee d)$$

Schay's Second System

$$(a|b) \rightarrow (c|d) = (a \rightarrow c|bd)$$
$$(a|b) \longleftrightarrow (c|d) = (a \longleftrightarrow c)|bd)$$

Goodman and Nguyen's System

$$(a|b) \rightarrow (c|d) = (b'd'|1) \vee (a|b)' \vee (c|d)$$

$$(a|b) \leftrightarrow (c|d) = ((a|b) \rightarrow (c|d)) \wedge ((c|d) \rightarrow (a|b))$$

In Heyting's three-valued logic, $\wedge$ and $\vee$ are the same as Lukasiesicz's, so in the corresponding algebra, $\wedge$ and $\vee$ are the same as those of Goodman-Nguyen. Heyting's negation is different, and is defined by

$$(a|b)' = (a'b|1).$$

Goodman and Nguyens $\wedge$ and $\vee$ make $R|R$ into a lattice, and on that lattice, Heyting's negation turns out to be a pseudo-complementation, making $R|R$ into a Stone Algebra. The details are in Chapter 4. The operations on $R|R$ corresponding to Heyting's $\rightarrow$ and $\leftrightarrow$ are

$$(a|b) \rightarrow (c|d) = b'd' \vee a'b \vee (c|d)$$

$$(a|b) \leftrightarrow (c|d) = b'd' \vee ((a \leftrightarrow c)bd|a'b \vee c'd \vee bd).$$

Now, examining the truth tables of the conjunction and disjunction operators in Schay's first and second systems, we see that they all violate plausible conditions for multi-valued logics. First, viewing u as lying between 0 and 1, any conjunction $\wedge$ should be such that $x \wedge y$ is the "falsest" of x and y. Likewise, any disjunction $\vee$ should yield the "truest" of x and y (Rescher, 1969, p. 133). Thus for Schay's first system,

$u \wedge 1$ and $1 \wedge u$ should be 0 or u, but not 1, and
$0 \vee u$ and $u \vee 0$ should be 1 or u, but not 0.

Likewise, for Schay's second system,

$u \wedge 0$ and $0 \wedge u$ should be 0 but not u, and
$u \vee 1$ and $1 \vee u$ should be 1, but not u.

Finally, one can simply require that each logical operator on $R|R$ should satisfy a list of reasonable properties. For example, let $\wedge$ denote a binary operator on $R|R$ representing "conjunction." Then the following is such a list:

P_1 $(a|1) \wedge (c|1) = (a \wedge c|1)$ ($\wedge$ extends conjunction of unconditional events),
P_2 $(a|b) \wedge (c|d) \leq (a|b), (c|d)$,

P_3 $\wedge$ is associative,
P_4 $\wedge$ is commutative,
P_5 $\wedge$ is idempotent $((a|b) \wedge (a|b)) = (a|b))$,
P_6 $(a|b) \wedge 0) = 0$,
P_7 $(a|b) \wedge 1) = (a|b)$,
P_8 $(a|b) \wedge (c|b) = (ac|b)$,
P_9 $(a|b) \wedge b = ab$ (modus ponens),
P_{10} $(a|bc) \wedge (b|c) = (ab|c)$ (a chaining property).

All of Schay, Adams, and Calabrese's conjunction operators fail to satisfy P_2. Their corresponding disjunction operators fail to satisfy the dual property $(a|b) \vee (c|d)) \geq (a|b), (c|d)$.

In terms of truth tables, $\wedge$ satisfies P_2 if and only if $i \wedge j) \leq min(i, j)$, for $i, j \in \{0, u, 1\}$.

If $\wedge$ satisfies P_1, P_2 and P_4, then the corresponding truth table must be one of the following four.

$\wedge_1$	0	u	1
0	0	0	0
u	0	0	0
1	0	0	1

$\wedge_2$	0	u	1
0	0	0	0
u	0	0	u
1	0	u	1

$\wedge_3$	0	u	1
0	0	0	0
u	0	u	0
1	0	0	1

$\wedge_4$	0	u	1
0	0	0	0
u	0	u	u
1	0	u	1

The table for $\wedge_4$ is Lukasiewicz's truth table for conjunction, which corresponds to the Goodman-Nguyen conjunction operator. Using the Theorem 2 of 3.4, we get the operations

$$(a|b) \wedge_1 (c|d) = abcd\,,$$
$$(a|b) \wedge_2 (c|d) = (abcd|abcd \vee a'b \vee c'd \vee b'd')\,,$$
$$(a|b) \wedge_3 (c|d)) = (abcd|b \vee d)\,,$$
$$(a|b) \wedge_4 (c|d) = (ac|a'b \vee c'd \vee bd)\,.$$

Now it is easily checked that

$\wedge_1$ does not satisfy P_5, P_7, P_8 and P_{10};
$\wedge_2$ does not satisfy P_5, P_8;
$\wedge_3$ does not satisfy P_7.

Only $\wedge_4$ satisfies all ten properties!

In summary, in his pioneering work on logical conditional operators, Schay (1968) proposed, at the syntax level, two systems. As Dubois and Prade (1989, 1990) have pointed out, and as we proved in Sections 3.4 and 3.5, Schay's systems correspond precisely to two well-known three-valued semantics, namely those of Sobocinski and Bochvar. The algebraic approach to logical operations on conditionals presented in Section 3.2 leads to a syntactic system corresponding to Lukasiewicz's three-valued logic. The comparisons above suggest that each choice of a logical system should be dictated by the situation at hand. This is similar to the situation in fuzzy logic (see Chapter 7). In particular, the choice between Lukasiewicz and Sobocinski's logics is a matter of debate as far as appropriate semantics for conditionals is concerned. See Chapter 6 for more details. In this book we take the viewpoint of Lukasiewicz, and investigate the mathematics of conditionals corresponding to his three-valued logic.

3.6 Connection with qualitative probability

Qualitative (or comparative subjective, or objective propensity) probability is motivated by the desire to make numerical probability measures compatible with non-numerical probability comparisons. For a general exposition, see Fine (1973, Chapter II). See also Fishburn (1983), Villegas (1967), Domotor (1969) and Suppes (1973) for further background.

In general, qualitative probability is a kind of order relation $\prec$ on a given Boolean ring R. For $a, b \in R$, the relation $a \prec b$ is interpreted as "b is at least as probable as a." Then, for $a \prec b$, one seeks probability measures P on R such that $P(a) \leq P(b)$.

More strongly, one attempts to determine a qualitative probability $\prec$ and quantitative probability measures P on R such that for all $a, b \in R$, $a \prec b$ if and only if $P(a) \leq P(b)$. In this case, P is called a representative of $\prec$. In order to achieve this, usually an *axiom of comparability* is assumed such as $a \prec b$ or $b \prec a$ for all $a, b \in R$.

In the following, we discuss Koopman's *conditional* qualitative probability system (Koopman, 1940). Interestingly, Koopman basically avoids use of any axiom of comparability -- at least initially. Koopman's axioms follow (Koopman, 1960, page 275). They are axioms for a system such as our $R|R$.

V Axiom of Verified Contingency

$(a|b) \prec (c|c)$.

I Axiom of Implication

If $(c|c) \prec (a|b)$, then $c \prec a$.

R Axiom of Reflexivity.

$(a|b) \prec (a|b)$.

T Axiom of Transitivity

If $(a|b) \prec (c|d)$, and $(c|d) \prec (e|f)$, then $(a|b) \prec (e|f)$.

A Axiom of Antisymmetry

If $(a|b) \prec (c|d)$, then $(c'|d) \prec (a'|b)$.

C Axiom of Composition

C_1 If $(a|b) \prec (c|d)$ and $(e|ab) \prec (f|cd)$, then $(ae|b) \prec (cf|d)$.

C_2 If $(a|b)$ $(f|cd)$ and $(e|ab) \prec c|d$, then $(ae|b) \prec (cf|d)$.

D Axioms of Decomposition

Suppose that $(ac|b) \prec (df|c)$. If either of $(a|b)$ or $(c|ab)$ is $\succ$ either of $(d|e)$ or $(f|de)$, then the remaining one of $(a|b)$ and $(c|ab)$ is $\prec$ the remaining one of $(d|e)$ and $(f|de)$.

P Axioms of Alternative Presumption

If $(a|bc) \prec (d|e)$ and $(a|(bc)') \prec (d|e)$, then $(a|c) \prec (d|e)$.

S Axioms of Subdivision

For any integer n, let the propositions $a_1, a_2, \ldots, a_n, b_1, b_2, \ldots, b_n$ be such that $a_ia_j = b_ib_j = 0$ for $i \neq j$; $a = a_1 \vee a_2 \vee \ldots \vee a_n \neq 0$; $b = b_1 \vee b_2 \vee \ldots \vee b_n \neq 0$; $(a_1|a) \prec (a_2|a) \prec \ldots \prec (a_n|a)$; $(b_1|b) \prec (b_2|b) \prec \ldots \prec (b_n|b)$. Then $(a_1|a) \prec (b_n|b)$.

Koopman derives many properties of his axioms. Our purpose here is to show that the relation $\leq$ on $R|R$ that we introduced in Section 3.3 satisfies all but the first and last of Koopman's axioms system.

Theorem. *Let* R *be a Boolean algebra. Then* $\leq$ *defined on* $R|R$ *by* $(a|b) \leq (c|d)$ *if and only if* $(a|b) = (a|b)(c|d)$, *satisfies Koopman's axiom I, R, T, A, C, D, P, and S. It does not satisfy his axioms V and S.*

Proof. Throughout we will use our criterion that $(a|b) \leq (c|d)$ if and only if $ab \leq cd$ and $c'd \leq a'b$. (Theorem 1, Section 3.3).

Axiom V is clearly false. Just pick a, b, c so that $ab > c$.

Axioms I and R follow almost trivially from our criterion above, and axiom T is noted immediately after the definition of $\leq$ in Section 3.3.

Axiom A is part of our criterion for $(a|b) \leq (c|d)$.

To verify the first part of C, let $(a|b) \leq (c|d)$ and $(e|ab) \prec (f|cd)$. Then

$$ab \leq cd\,,$$

$$c'd \leq a'b\,,$$

$$eab \leq fcd\,,$$

and

$$f'cd \leq e'ab.$$

To get $(ae|b) \leq (cf|d)$, we need

$$aeb \leq cfd$$

and

$$(cf)'d \leq (ae)'b.$$

The first we have, and from $f'cd \leq e'ab$, we get

$$e \vee a' \vee b' \leq f \vee c' \vee d',$$

and from $c'd \leq a'b$ we get

$$a' \vee b \leq c \vee d'.$$

Thus

$$(e \vee a' \vee b')(a' \vee b) \leq (f \vee c' \vee a')(c \vee d'),$$

or

$$e(a' \vee b) \vee a' \leq f(c \vee d') \vee d',$$

or

$$a' \vee eb \leq d' \vee fc,$$

or

$$a(e' \vee b') \geq d(c' \vee f'),$$

which we needed.

For the second part of C, let $(a|b) \le (f|cd)$ and $(e|ab) \le (c|d)$. Then

$$ab \le fcd,$$

$$f'cd \le a'b,$$

$$eab \le cd,$$

and

$$c'd \le e'ab.$$

We need

$$aeb \le cfd$$

and

$$(cf)'d \le (ae)'b.$$

Since $ab \le cfd$, certainly $aeb \le cfd$. So we need only that

$$(cf)'d \le (ae)'b,$$

or equivalently that

$$ae \vee b' \le cf \vee d'.$$

Since

$$f'cd \le a'b,$$

$$a \vee b' \le f \vee c' \vee d',$$

and since

$$c'd \le e'ab,$$

$$e \vee a' \vee b' \le c \vee d'.$$

Thus

$$(a \vee b')(e \vee a' \vee b') \le (f \vee c' \vee d')(c \vee d'),$$

and so

$$(a \vee b')e \vee b' \le f(c \vee d') \vee d',$$

or

$$ae \vee b' \le fc \vee d',$$

or finally

$$(cf)'d \le (ae)'b,$$

which we needed.

To verify D, the axioms of decomposition, suppose that $(ac|b) < (df|e)$ and $(a|b) \ge (d|e)$. We need that $(c|ab) \le (f|de)$. Thus

$$acb \leq dfe,$$

$$(df)'e \leq (ac)'b,$$

$$de \leq ab,$$

and

$$a'b \leq d'e.$$

We need

$$cab \leq fde$$

and

$$f'de \leq c'ab,$$

and the first we have. From

$$(df)'e \leq (ac)'b$$

we have

$$(d' \vee f')e \leq (a' \vee c')b,$$

and since

$$de < ab,$$

we have

$$(d' \vee f')de \leq (a' \vee c')ab,$$

or

$$f'de \leq c'ab,$$

which we needed.

Assume now that $(a|b) \geq (f|de)$, and that we have always that $(ac|b) \leq (df|e)$. We need that $(c|ab) \leq (d|e)$. So we are given

$$fde \leq ab,$$

$$a'b \leq f'de,$$

$$acb \leq dfe,$$

and

$$(df)'e \leq (ac)'b,$$

and we want

$$cab \leq de$$

and

$$d'e \leq c'ab.$$

Since $acb \leq dfe$, then $cab \leq de$. So we only need that

$$d'e \leq c'ab.$$

From

$$(df)'e \le (ac)'b,$$

and

$$a'b \le f'de,$$

we have

$$(d' \vee f')e \le (a' \vee c')b$$

and

$$f \vee d' \vee e' \le a \vee b',$$

so

$$(d' \vee f')e(f \vee d' \vee e') \le (a' \vee c')b(a \vee b').$$

Thus

$$(d' \vee f)e(f \vee d') \le (a' \vee c')ab,$$

or

$$d'e \le c'ab,$$

which is the inequality we needed. The other two parts are similar and their proofs are left to the reader.

To verify axiom P, let $(a|bc) \le (d|e)$ and $(a|(bc)') \le de$. Then

$$abc \le de,$$

$$d'e \le a'bc$$

$$a(bc)' \le de,$$

and

$$d'e \le a'(bc)'.$$

We need that

$$(a|c) \le (d|e),$$

or that

$$ac \le de$$

and

$$d'e \le a'c.$$

Now $d'e \le a'c$ since $d'e \le a'bc$. To get $ac \le de$, from $a(bc)' \le de$ we have

$$a(b' \vee c') \le de.$$

Then

$$ab' \vee ac' \le de,$$

whence

$$ab' \leq de.$$

We have from above that

$$abc \leq de.$$

Then

$$ab'c \vee abc = ac \leq de$$

and this proof is complete.

The axiom of subdivision obviously does not hold for our $\leq$. □

CHAPTER 4

ALGEBRAIC STRUCTURE OF CONDITIONAL EVENTS

This chapter is devoted to the study of the space of conditionals $R|R$ as an algebraic system. Equipped with the logical operations $\vee$, $\wedge$, and $'$ introduced in Chapter 3, it is a system generalizing Boolean algebras, or Boolean rings, and provides a vehicle for manipulating conditional events, analogous to the manipulation of events in Boolean algebras. Further, it represents a departure from classical logic, and from quantum logic. First, in Section 4.1, we examine the basic algebraic properties of $R|R$, concentrating on its similarities and its differences with those of Boolean algebra. In Section 4.2, $R|R$ is characterized as an abstract algebraic system, and a Stone Representation Theorem is established, generalizing Stone's theorem for Boolean algebras. In Section 4.3, $R|R$ is identified with a semi-simple MV algebra via the work of Belluce (1986), yielding a connection with multi-valued logic, and providing yet another Stone Representation Theorem.

4.1 Basic algebraic properties

We now turn to a detailed examination of $R|R$ as an algebraic system. Recall that $R|R$ is the set of all cosets of all principal ideals $a + Rb'$ of the Boolean ring R, and we have adopted the notation $(a|b)$ for the coset $a + Rb'$. In the Boolean ring R, there are the usual operations $\vee$, $\wedge$, $+$, and $'$, and R has a 0 and a 1. We assume as known the basic properties of Boolean rings, or Boolean algebras. Corresponding operations $\vee$, $\wedge$, $+$, and $'$ have been defined on $R|R$, and some of their properties have been noted in earlier sections. Specifically,

(1) $(a|b) \vee (c|d) = ((a \vee c)|(ab \vee cd \vee bd)) = (ab \vee cd|ab \vee cd \vee a'bc'd)$,

(2) $(a|b) \wedge (c|d) = ((a \wedge c)|(a'b \vee c'd \vee bd)) = (abcd|a'b \vee c'd \vee abcd)$,

(3) $(a|b)' = (a'|b)$,

(4) $(a|b) + (c|d) = (a + c|bd)$.

Above, if $x, y \in R$, then xy is written for $x \wedge y$. Note that the symbols $\vee$, $\wedge$, $+$ and $'$ are used both as operations in the Boolean ring R and as operations in $R|R$. The context should always make it clear what operation is meant. The most basic and elementary algebraic properties of $R|R$ are these:

Theorem 1. *The following hold in* $R|\mathrm{R}$.

(*1*) $(a|b) \vee (a|b) = (a|b)$ ($\vee$ *is idempotent*);

(2) $(a|b) \wedge (a|b) = (a|b)$ ($\wedge$ *is idempotent*);

(*3*) $(a|b) \vee (c|d) = (c|d) \vee (a|b)$ ($\vee$ *is commutative*);

(*4*) $(a|b) \wedge (c|d) = (c|d) \wedge (a|b)$ ($\wedge$ *is commutative*);

(*5*) $(a|b) + (c|d) = (c|d) + (a|b)$ (+ *is commutative*);

(*6*) $((a|b) \vee (c|d)) \vee (e|f) = (a|b) \vee ((c|d) \vee (e|f))$ ($\vee$ *is associative*);

(*7*) $((a|b) \wedge (c|d)) \wedge (e|f) = (a|b) \wedge ((c|d) \wedge (e|f))$ ($\wedge$ *is associative*);

(*8*) $((a|b) + (c|d)) + (e|f) = (a|b) + ((c|d) + (e|f))$ (+ *is associative*);

(*9*) $(a|b)'' = (a|b)$ (*′ is involutive*).

Proof. The proofs of these are routine from the definitions of the operation. However, we give proofs of (1), (4), (5), (6), (7) and (9) as illustrations of elementary manipulations in $R|R$.

(1) $(a|b) \vee (a|b) = ((a \vee a)|(ab \vee ab \vee b)) = (a|b)$.

(4) $(a|b) \wedge (c|d) = (ab|(a'b \vee c'd \vee bd)) = (ba|(c'd \vee a'b \vee db)) = (c|d) \vee (a|b)$.

(5) $(a|b) + (c|d) = ((a + c)|cd) = ((c + a)|dc) = (c|d) + (a|b)$.

(7) $((a|b) \wedge (c|d)) \wedge (e|f) = (ac|(a'b \vee c'd \vee bd)) \wedge (e|f)$

$$= (ace|((ac)'(a'b \vee c'd \vee bd) \vee e'f \vee (a'b \vee c'd \vee bd)f).$$

$$a|b \wedge (c|d \wedge e|f)) = (a|b) \wedge (ce|(c'd \vee e'f \vee df))$$

$$= (ace|(a'b \vee (ce)'(c'd \vee e'f \vee df) \vee b(c'd \vee e'f \vee df)).$$

Thus we need to show that

$$(ac)'(a'b \vee c'd \vee bd) \vee e'f \vee (a'b \vee c'd \vee bd)f =$$

$$(a'b \vee (ce)'(c'd \vee e'f \vee df) \vee b(c'd \vee e'f \vee df)).$$

The first is

$$(ac)'a'b \vee (ac)'c'd \vee (ac)'bd \vee e'f \vee a'bf \vee c'df \vee bdf=$$

$$a'b \vee c'a'b \vee a'c'd \vee c'd \vee a'bd \vee c'bd \vee e'f \vee a'bf \vee c'df \vee bdf =$$

$$a'b \vee c'd \vee e'f \vee bdf,$$

and the second is

$$a'b \vee (ce)'c'd \vee (ce)'e'f \vee (ce)'df \vee bc'd \vee be'f \vee bdf =$$

$$a'b \vee c'd \vee e'c'd \vee c'e'f \vee e'f \vee c'df \vee e'df \vee bc'd \vee be'f \vee bdf =$$

$$a'b \vee c'd \vee e'f \vee bdf.$$

(9) $(a|b)'' = (a'|b)' = (a''|b) = (a|b)$. □

None of the properties (1) - (9) above involve interactions between the various operations. We will address those properties momentarily. First, $R|R$ has some special elements, $(0|1)$ and $(1|1)$, which act as a "zero" and "one" should act. In addition, there is the "indeterminate" element $(0|0) = (1|0) = R$.

Theorem 2. *The elements* $(0|1)$, $(1|1)$ and $(0|0)$ *satisfy the following properties.*

(1) $(a|b) + (0|1) = (a|b)$ $((0|1)$ *is an additive identity);*

(2) $(a|b) \wedge (1|1) = (a|b)$ $((1|1)$ *is a multiplicative identity);*

(3) $(a|b) \wedge (0|1) = (0|1)$;

(4) $(0|1)' = (1|1)$;

(5) $(1|1)' = (0|1)$;

(6) $(a|b) = ab \vee (b' \wedge (0|0))$;

(7) *The unique element* $(a|b)$ *in* $R|R$ *such that* $(a|b)' = (a|b)$ *is* $(0|0)$.

Proof. Again, these properties are straightforward. For example,

$$(a|b) \wedge (1|1) = (a|(a'b \vee 0 \vee b)) = (a|b),$$

and

$$(a|b) \wedge (0|1) = (0|(a'b \vee 1 \vee b)) = (0|1).$$ □

Note that there are no other additive or multiplicative identities other than $(0|1)$ and $(1|1)$. If x and y were two additive identities, then $x + y = x = y$, and similarly for multiplicative identities.

The following theorem provides some connections between the various operations. They are fundamental ones.

Theorem 3. *The following hold in* $R|R$.

(1) $(a|b) \wedge ((c|d) \vee (e|f)) = ((a|b) \wedge (c|d)) \vee ((a|b) \wedge (e|f))$ ($\wedge$ *distributes over* $\vee$);

(2) $(a|b) \vee ((c|d) \wedge (e|f)) = ((a|b) \vee (c|d)) \wedge ((a|b) \vee (e|f))$ ($\vee$ *distributes over* $\wedge$);

(3) $((a|b) \vee (c|d))' = (a|b)' \wedge (c|d)'$;

(4) $((a|b) \wedge (c|d))' = (a|b)' \vee (c|d)'$ ((3) *and* (4) *are DeMorgan's laws*).

Proof. We will prove (1) and (3). Then we will see that (2) and (4) are immediate consequences of (1) and (3). For (1),

$$(a|b) \wedge ((c|d) \vee (e|f)) = (a|b) \wedge ((c \vee e)|(cd \vee ef \vee df)) =$$

$$(a(c \vee e)|(a'b \vee c'e'(cd \vee ef \vee df) \vee b(cd \vee ef \vee df))) =$$

$$a(c \vee e)|(a'b \vee c'e'df \vee bcd \vee bef \vee bdf).$$

Now

$$((a|b) \wedge (c|d)) \vee ((a|b) \wedge (e|f)) = (ac|(a'b \vee c'd \vee bd)) \vee (ae|(a'b \vee e'f \vee bf)) =$$

$$(a(c \vee e)|(acbd \vee aebf \vee a'b \vee c'de'f \vee c'dbf \vee bde'f \vee bdf)) =$$

$$(a(c \vee e)|(abcd \vee abef \vee a'b \vee c'de'f \vee bdf)).$$

Thus, we need to show that

$$a'b \vee c'e'df \vee bcd \vee bef \vee bdf = abcd \vee abef \vee a'b \vee c'de'f \vee bdf.$$

Clearly, the left side contains the right. But since $abcd \vee a'b$ contains bcd, and $abef \vee a'b$ contains bef, the right side contains the left. To prove (3), note that

$$((a|b) \vee (c|d))' = ((a \vee c)|(ab \vee cd \vee bd))' = (a'c'|(ab \vee cd \vee bd)),$$

and

$$(a|b)' \wedge (c|d)' = (a'c'|(ab \vee cd \vee bd)).$$

The equations

$$((a|b) \vee (c|d \wedge e|f)) = ((a|b) \vee (c|d \wedge e|f))''$$

$$= ((a|b)' \wedge (c|d \wedge e|f)')' = ((a|b) \vee (c|d)) \wedge ((a|b) \vee (e|f)),$$

and

$$((a|b) \wedge (c|d))' = ((a|b)'' \wedge (c|d)'')' =$$

$$((a|b)' \vee (c|d)')'' = (a|b)' \vee (c|d)',$$

establish (2) and (4). □

We now come to some negative aspects of $R|R$. These center around the operations $+$ and $'$. First, $R|R$ does not have negatives (or additive inverses). That is, given $(a|b)$ in $R|R$, there does not necessarily exist an element $(c|d)$ such that $(a|b) + (c|d) = (0|1)$. Indeed, if so, then

$$(a|b) + (c|d) = ((a + c)|bd) = (0|1),$$

whence $bd = 1$, so $b = 1$ and $d = 1$. In that case, $(a|b) = (a|1)$, and it is its own negative. Thus the elements in $R|R$ with negatives are precisely those of the form $(a|1)$. In R, $a + a = 0$. This does not carry over to $R|R$. For example $(a|b) + (a|b) = (0|b)$, which is not $(0|1)$ unless $b = 1$. That is, $R|R$ is not of characteristic 2.

Secondly, $\wedge$ does not distribute over $+$. A simple example is this:

$$(1|b)((1|d) + (1|f)) \neq (1|b)(1|d) + (1|b)(1|f),$$

the first being $(0|df)$ and the second being $(0|bdf)$.

Thirdly, $'$ is not a true complement for $R|R$. That is, $(a|b) \vee (a|b)' \neq (1|1)$. In fact,

$$(a|b) \vee (a|b)' = (a|b) \vee (a'|b) = (1|(ab \vee a'b \vee b)) = (1|(a \vee b)),$$

which is not $(1|1)$ unless $a \vee b = 1$. Also,

$$(a|b) \wedge (a|b)' = (0|(a'b \vee ab \vee b)) = (0|b) \neq (0|1)$$

unless $b = 1$.

These negative aspects of $R|R$ are summed up in the following theorem. In particular, $R|R$ is far from being a Boolean ring.

Theorem 4. *The following hold*:

(1) $R|R$ is not a group under $+$; specifically, not every element has a negative;

(2) $\wedge$ does not distribute over $+$;

(3) $'$ is not a complementation operator on $R|R$; specifically, $(a|b) \vee (a|b)'$ is not necessarily $(1|1)$, and $(a|b) \wedge (a|b)'$ is not necessarily $(0|1)$.

(4) $R|R$ is not of characteristic 2, that is, $(a|b) + (a|b)$ is not necessarily $(0|1)$.

In a Boolean ring, the four basic operations, $\vee$, $\wedge$, $+$, and $'$ are not independent.

For example, $a + b = a'b \vee ab'$, which corresponds to the "exclusive or". This relation also holds in $R|R$ as we have noted back in Section 3.2 Thus, properties of $+$ in $R|R$ are reflections of properties of $\vee$, $\wedge$, and and $'$. We have just noted some negative properties of $+$, indicating that it will be of at best limited importance and interest in $R|R$. For these reasons, and because $\vee$, $\wedge$, and $'$ are more conceptually fundamental operations in logic and probability, we will drop the operation $+$ from our considerations.

In Section 3.3, an order relation on $R|R$ was introduced. This order relation will now be examined in some detail. We first establish the appropriate language in which to discuss this topic. A good reference for the following material is Grätzer (1978).

Definition 1. *A partially ordered set is a set L with a relation $\leq$ on L such that for all $x, y, z \in L$,*

(*1*) $x \leq x$ *($\leq$ is reflexive);*
(*2*) *if $x \leq y$ and $y \leq x$, then $x = y$ ($\leq$ is anti-symmetric);*
(*3*) *if $x \leq y$ and $y \leq z$, then $x \leq z$ ($\leq$ is transitive).*

This partially ordered set is denoted $(L, \leq)$, or just L if there is no confusion as to the partial order under consideration. Let $(L, \leq)$ be a partially ordered set, and let S be a subset of L. The element x is an upper bound of S if $s \leq x$ for every $s \in S$. The element x is a least upper bound, or supremum, or simply sup, of S if x is an upper bound and $x \leq y$ for any upper bound y of S. Lower bounds, and greatest lower bounds, or infima, or inf are defined analogously.

Definition 2. *A lattice is a partially ordered set L such that every pair $\{a,b\}$ of elements of L has a sup and an inf.*

The sup of $\{a,b\}$ is usually denoted $a \vee b$ and the inf by $a \wedge b$. Note that this makes sense, namely that $\{a,b\}$ has only one sup. If x and y were both sups, then $x \leq y$ since $x \leq$ any element y which is $\geq$ all elements in $\{a,b\}$. Thus $x = y$. Similarly infs are unique. Now $\vee$ and $\wedge$ are two binary operations on L, and they satisfy the following conditions.

(1) $x \vee x = x$ and $x \wedge x = x$ ($\vee$ and $\wedge$ are idempotent);
(2) $x \vee y = y \vee x$ and $x \wedge y = y \wedge x$ ($\vee$ and $\wedge$ are commutative);
(3) $(x \vee y) \vee z = x \vee (y \vee z)$ and $(x \wedge y) \wedge z = x \wedge (y \wedge z)$ ($\vee$ and $\wedge$ are associative).
(4) $x \vee (x \wedge y) = x$ and $x \wedge (x \vee y) = x$ ($\vee$ and $\wedge$ satisfy the absorption identities).

Property (1) follows from reflexivity and anti-symmetry, and (2) and (3) directly from the definitions of sups and infs. To get $x \vee (x \wedge y) = x$ in (4), note that $x \geq x$ by reflexivity, and $x \geq x \wedge y$ by definition of $x \wedge y$, so x is an upper bound of x and $x \wedge y$. If z is another such upper bound and $z \leq$ the upper bound x, then since $z \geq x$, $z = x$. The other part of (4) follows similarly. The following two properties should also be noted.

(5) $x \geq x \wedge y$ and $x \leq x \vee y$;

(6) $x = x \vee y$ if and only if $y = x \wedge y$.

Property (5) is immediate from the definitions of upper and lower bounds, and (6) is a consequence of (4). For example, if $x = x \vee y$, then by (4), $y \wedge (x \vee y) = y = y \wedge x$. The other half of (6) follows similarly. Actually, the absorption identities imply that $\vee$ and $\wedge$ are idempotent, but we will not concern ourselves with such technical niceties here.

We provide the following theorem and its proof, since it will hold for our $R|R$, and the proof in general is as easy as for the special case of $R|R$. We have already noted its converse.

Theorem 5. *If L is a non-empty set with two binary operations $\vee$ and $\wedge$ which satisfy (1)-(4) above, then L is a lattice under the partial order given by $x \leq y$ if $x = x \wedge y$.*

Proof. First we get $\leq$ to be a partial order on L. $x \leq x$ since $x \wedge x = x$. If $x \leq y$ and $y \leq z$, then

$$x \wedge z = (x \wedge y) \wedge z = x \wedge (y \wedge z) = x \wedge y = x,$$

so $x \leq z$. If $x \leq y$ and $y \leq x$, then $x = x \wedge y$ and $y = y \wedge x$, so $x = y$. Now we show that $x \wedge y$ is the inf of $\{x,y\}$ and $x \vee y$ is the sup of $\{x,y\}$. $x \wedge (x \wedge y) = x \wedge y$, so $x \wedge y \leq x$, and similarly $x \wedge y \leq y$, so $x \wedge y$ is a lower bound of $\{x,y\}$. If $z \leq x$ and $z \leq y$, then

$$x \wedge z = z = y \wedge z = y \wedge x \wedge z$$

so $z \leq x \wedge y$. Thus $x \wedge y$ is the inf of $\{x,y\}$. By one of the absorption laws, $x \wedge (x \vee y) = x$, so $x \leq x \vee y$, and similarly $y \leq x \vee y$. If $x \leq z$ and $y \leq z$, then $x = x \wedge z$ and $y = y \wedge z$. In turn, $z = z \vee x = z \vee y$, implying

$$z = z \vee z = (z \vee x) \vee (z \vee y) = z \vee (x \vee y),$$

implying $x \vee y \leq z$, that is that $x \vee y$ is the sup of $\{x,y\}$, and $(L, \leq)$ is a lattice. □

Thus we have the following situation. If $(L, \leq)$ is a lattice, then $(L, \vee, \wedge)$ satisfies properties (1) - (4) above, where $\vee$ and $\wedge$ are defined by $a \vee b = sup\{a,b\}$ and $a \wedge b = inf\{a,b\}$. Conversely, if $L(\vee, \wedge)$ satisfies (1) - (4) above, then $(L, \leq)$ is a lattice, where $\leq$ is defined by $a \leq b$ if $a = a \wedge b$. Such an algebra $(L, \vee, \wedge)$ is also called a lattice. Thus a lattice $(L, \leq)$ yields an algebra $(L, \vee, \wedge)$ satisfying (1) - (4) above, and an algebra $(L, \vee, \wedge)$ satisfying (1) - (4) yields a lattice $(L, \leq)$. A critical fact is that these procedures are reciprocals of each other. Thus the concept of lattice, and the concept of an algebra with two binary operations satisfying (1) - (4) are the same. We refer the reader to Gratzer (1978) for details.

Now back to $R|R$. The two operations $\vee$ and $\wedge$ on $R|R$ do indeed satisfy (1) through (4) above. We have already observed that (1), (2), and (3), hold. For (4),

$$
\begin{aligned}
&(a|b) \vee ((a|b) \wedge (c|d)) \\
&\quad = (a|b) \vee (a|b)) \wedge ((a|b) \vee (c|d)) \\
&\quad = (a|b) \wedge (a \vee c|(ab \vee cd \vee bd)) \\
&\quad = (a|(a'b \vee (a \vee c)' \wedge (ab \vee cd \vee bd) \vee b(ab \vee cd \vee bd)) \\
&\quad = (a|((a'b \vee a'c'bd \vee ab \vee bcd \vee bd)) \\
&\quad = (a|((a'b \vee ab \vee bd)) = (a|b).
\end{aligned}
$$

The other absorption law follows similarly. Thus by Theorem 5, $(R|R, \leq)$ is a lattice. In considerations of $R|R$, emphasis is usually more on $\vee$ and $\wedge$ than on $\leq$, the former being the more fundamental concepts for us. Thus we prefer the following statement.

Theorem 6. $(R|R, \vee, \wedge)$ *is a lattice.*

If L is a lattice and L itself has a sup and an inf, then that sup is denoted 1 and that inf is denoted 0. In that case, L is called a lattice with 0 and 1, or a *bounded lattice.* Note now that $R|R$ is a bounded lattice. The 1 is the element $(1|1)$ and the 0 is the element $(0|1)$. To see this, recall our criteria that $(a|b) \leq (c|d)$, namely that $ac \leq cd$ and $c'd \leq a'b$. Thus $(a|b) \leq (1|1)$ since $ab \leq 1$ and $0 \leq a'b$. Thus $(1|1)$ is the 1 of the lattice $R|R$. Similarly $(0|1)$ is the 0 of it, and $R|R$ is indeed a bounded lattice.

A lattice is called *distributive* if the following conditions hold.

$$x \wedge (y \vee z) = (x \wedge y) \vee (x \wedge z);$$

$$x \vee (y \wedge z) = (x \vee y) \wedge (x \vee z).$$

We have seen that these distributive laws do hold in $R|R$, so we have the following theorem.

Theorem 7. $(R|R, \vee, \wedge)$ *is a bounded distributive lattice. The* 0 *is the element* $(0|1)$, *and the* 1 *is the element* $(1|1)$.

In a bounded lattice, an element x is a *complement* of y if $x \wedge y = 0$ and $x \vee y = 1$. Complements, if they exist, are unique in bounded distributive lattices. The complement of x is usually denoted x'. Note that $x'' = x$. A bounded lattice in which every element has a complement is called a complemented lattice. A complemented distributive lattice satisfies DeMorgan's laws:

$$(x \vee y)' = x' \wedge y';$$

$$(x \wedge y)' = x' \vee y'.$$

The details may be found in Grätzer (1978), and we will not pursue them there, mainly because $R|R$, our lattice of interest is not complemented. Were it complemented, then the complement $(c|d)$ of $(0|b)$ would have the property that

$$(c|d) \vee (0|b) = (cd|d) \vee (0|b) = (1|1) = (cd|cd \vee bd),$$

whence $cd = 1 = c = d$. But the complement of $(1|1)$ must then be $(0|b)$, but is $(0|1)$ instead. Thus no element of the form $(0|b)$ can have a complement unless $b = 1$. In particular, our operator $'$ on $R|R$ is not a complementation operator. There does not exist a complementation operator on $R|R$ with respect to $\vee$ and $\wedge$.

There is a weaker notion than complement. In a bounded lattice, an element x^* is a *pseudocomplement* of x if $x \vee y^* = 0$, and if $x \wedge y = 0$ implies that $y \leq x^*$. An element can have at most one pseudocomplement; if a and b are pseudocomplements of x, then $a \leq b$ and $b \leq a$, so $a = b$. Thus a pseudocomplement of an element x is that unique largest element whose intersection with x is 0. A *pseudocomplemented lattice* is one in which every element has a pseudocomplement.

Definition 3. *A Stone algebra L is a distributive pseudocomplemented (bounded) lattice which satisfies Stone's identity: for all* $a \in L$,

$$a^* \vee a^{**} = 1.$$

It is a fact that in any Stone algebra, the pseudocomplementation operator * satisfies DeMorgan's laws (Grätzer, pages 113, 119). A crucial fact is that $R|R$ is a Stone algebra, and this is not entirely obvious.

Theorem 8. $(R|R, \vee, \wedge)$ *is a Stone algebra. The pseudocomplement* $(a|b)^*$ *of an element* $(a|b)$ *is* $(a'b|1)$ that is $a'b$. *DeMorgan's laws hold for* *:

$$((a|b) \vee (c|d))^* = (a|b)^* \wedge (c|d)^*;$$

$$((a|b) \wedge (c|d))^* = (a|b)^* \vee (c|d)^*;$$

Proof. First, we show that $(a|b)^* = (a'b|1)$.

$$(a|b) \wedge (a'b|1) = (0|a'b \vee a\vee b' \vee b) = (0|1).$$

If $(c|d) \wedge (a|b) = (0|1)$, then $(ca|(c'd \vee a'b \vee bd)) = (0|1)$. Now, from the characterization of $\leq$ on $R|R$, $(c|d) \leq (a'b|1)$ if and only if $cd \leq a'b$ and $a\vee b' \leq c'd$. We have $c'd \vee a'b \vee bd = 1$ and $ac = 0$. From the equation

$$c'd \vee a'b \vee bd = 1,$$

by first conjoining with b' and then separately with d' we get $b' \leq c'd$, and $d' \leq a'b$. Thus $b' \leq c'$, $a\vee b' \leq d$, and since $ac = 0$, also $a \leq c'$. We have then that $a\vee b' \leq c'd$. It remains to get $cd \leq a'b$: But, from $ac = 0$ we have $c \leq a'$ and from $b' \leq c'$ we have $c \leq b$. Thus $cd \leq a'b$, and so $(a|b)^* = (a'b|1)$.

DeMorgan's laws can be verified easily now that we have an explicit formula for *.

$$((a|b) \vee (c|d))^* = ((a\vee c|(ab \vee cd \vee bd))^* = ((a'c')(ab \vee cd \vee bd)|1) = (a'c'bd|1),$$

and

$$(a|b)^* \wedge (c|d)^* = (a'b|1) \wedge (c'd|1) = (a'c'bd|1).$$

The other part of DeMorgan's laws follows similarly. □

Remark. Thus we have that $R|R$ has a rather rich algebraic structure, being a pseudo-complemented distributive lattice, in fact, a Stone algebra. However, it is not complemented, that is, does not have an operator $^\circ$ on it such that $a^\circ \wedge a = 0$ and $a^\circ \vee a = 1$ for all a. This situation is somewhat different from that of quantum logic. Indeed, a space of quantum events is a collection of closed subspaces of some complex Hilbert space and its algebraic structure is also that of a lattice, but is a non-distributive yet complemented one. (See Gudder, 1988.) As pointed out in Section 3.5, the truth table of the pseudocomplementation operator * on $R|R$ is the truth table of Heyting's negation operator in his three-valued logic.

The structure $R|R$ is one generalization of Boolean algebras. It is a special kind of Stone algebra, and will be so characterized in the next section. But it can be viewed other ways, depending on which operations on $R|R$ to investigate. For example, looking at other operations on $R|R$ in conjunction with our $\vee$ and $\wedge$ makes $R|R$ into a semi-simple

MV algebra, and this will be discussed with Section 4.3, with an attendant Stone Representation Theorem.

4.2 An abstraction of the space of conditional events

Section 4.1 culminated with the theorem that $R|R$ is a Stone algebra. One way to give an algebraic characterization of $R|R$ is to identify it among Stone algebras. That is what we will do. Thus we need to determine just what conditions on Stone algebras make them precisely of the form $R|R$. For such a characterization to be a good one, the conditions added should be succinct and conceptionally pleasing, involving fundamental entities associated with Stone algebras. Two such entities are in the following definition.

Definition 1. *Let L be a Stone algebra and* * *its pseudo-complementation operator. The skeleton of L is the image of* * *, that is, the set* $L^* = \{a^* : a \varepsilon L\}$. *The* ***dense set*** *of L is the kernel of* * *, that is, the set* $D(L) = \{a : a \in L, a^* = 0\}$.

We need a number of properties of L^* and $D = D(L)$. A more complete discussion may bc found in Grätzer (1978).

Theorem 1. *Let L be a Stone algebra. The following hold:*

(1) $a \leq a^{**}$;

(2) $a \leq b$ *implies that* $a^* \geq b^*$;

(3) $a^* = a^{***}$;

(4) $a \varepsilon L^*$ *if and only if* $a = a^{**}$;

(5) $(a \wedge b)^* = a^* \vee b^*$;

(6) $(a \vee b)^* = a^* \wedge b^*$;

Proof. (1) and (2) follow immediately from the definition of pseudocomplement. (1) and (2) imply that $a^* \geq a^{***}$, and (1) applied to a^* yields $a^* \leq a^{***}$. Thus (3) holds. If $a \in L^*$, then $a = b^*$, so $a^{**} = b^{***} = b^* = a$. If $a = a^{**}$, then $a = (a^*)^*$, whence $a \in L^*$, so (4) holds.

To prove (5), we have

$$(a \wedge b) \wedge (a^* \vee b^*) = (a \wedge b \wedge a^*) \vee (a \wedge b \wedge b^*) = 0 \vee 0 = 0.$$

If $(a \wedge b) \wedge x = 0$, then $(b \wedge x) \wedge a = 0$, so that $(b \wedge x) \leq a^*$. Thus

$$(b \wedge x) \wedge a^{**} \leq a^{*} \wedge a^{**} = 0,$$

so $(x \wedge a^{**}) \wedge b = 0$, implying that $x \wedge a^{**} \leq b^{*}$. By the Stone identity, $a^{*} \vee a^{**} = 1$, and thus

$$x = x \wedge 1 = x \wedge (a^{*} \vee a^{**}) = (x \wedge a^{*}) \vee (x \wedge a^{**}) \leq a^{*} \vee b^{*}.$$

Thus (5) is proved.

To prove (6),

$$(a \vee b) \wedge (a^{*} \wedge b^{*}) = (a \wedge a^{*} \wedge b^{*}) \vee (b \wedge a^{*} \wedge b^{*}) = 0 \vee 0 = 0.$$

If $x \wedge (a \vee b) = 0$, then $x \leq (a \vee b)^{*}$. But $a \vee b \geq a$ implies that $(a \vee b)^{*} \leq a^{*}$, so $x \leq a^{*}$, and similarly $x \leq b^{*}$, whence $x \leq a^{*} \wedge b^{*}$. Thus $(a \vee b)^{*} = a^{*} \wedge b^{*}$, and (6) is proved. □

The properties in the theorem yield the following fundamental facts about L^{*} and D.

Theorem 2. *Let L be a Stone algebra. Then*

(1) L^{*} is a Boolean algebra whose 0 and 1 are those of L;

(2) D is a filter (dual ideal) and $1 \in D$. In particular, D is a distributive lattice with 1.

Proof. Clearly $0^{*} = 1$ and $1^{*} = 0$, so that 0 and 1 are in L^{*}. From Theorem 1, $(a^{*} \wedge b^{*}) = (a \vee b)^{*}$ and $(a^{*} \vee b^{*}) = (a \wedge b)^{*}$, so that L^{*} is a sublattice of L. Since $a^{*} \vee a^{**} = 1$, * is a complementation operator on L^{*}. Thus L^{*} is a Boolean algebra.

If $a, b \in D$, then $(a \vee b)^{*} = a^{*} \wedge b^{*} = 0 \wedge 0 = 0$ and $(a \wedge b)^{*} = a^{*} \vee b^{*} = 0$, so D is a sublattice. If $a \in D$, then for all $x \in L$, $(a \vee x)^{*} = a^{*} \wedge x^{*} = 0 \wedge x^{*} = 0$, whence D is a filter. Since $1^{*} = 0$, $1 \in D$. □

Now we turn to $R|R$, identify its skeleton and dense set, and note some of their special properties. Recall that the pseudocomplementation operator * on $R|R$ is given by $(a|b)^{*} = (a'b|1)$. Thus it is clear that $(R|R)^{*} = \{a|1 : a \in R\}$, which we denote by $R|1$. For $(a|b)^{*}$ to be $(0|1)$, we must have $(a'b|1) = (0|1)$, so $a'b = 0$. Thus $b \leq a$, so $(a|b) = (b|b) = (1|b)$. It follows that $D(R|R) = \{1|b : b \in R\}$, which we denote by $1|R$. Thus we have the following theorem.

Theorem 3. *The skeleton of $R|R$ is $(R|R)^{*} = \{(a|1) : a \in R\} = R|1 = R$, and the dense set of $R|R$ is $D(R|R) = \{(1|a) : a \in R\} = \{(a|a) : a \in R\} = 1|R$.*

Both $R|1$ and $1|R$ are copies of R. In fact, the elements of $R|1$ are identified with the (unconditional) events of R. The mapping $R|1 \to 1|R : (a|1) \to (1|a)$ is clearly a bijection. Since

$$(1|a) \vee (1|b) = (1|(a \vee b)) \text{ and } (1|a) \wedge (1|b) = (1|(a \wedge b)),$$

that mapping preserves $\vee$ and $\wedge$. Further, $(0|1)$ and $(1|1)$ of $R|1$ go to $(1|0)$ and $(1|1)$, respectively of $1|R$, and $(a|1)' = (a'|1)$ goes to $(1|a')$. Thus $1|R$ is a Boolean algebra with its 0 and 1 the elements $(1|0)$ and $(1|1)$, respectively, and with $(1|a)' = (1|a')$. Thus the dense set of $R|R$ is also a Boolean algebra, and is isomorphic to the skeleton of $R|R$. Since $1|R$ is a Boolean algebra, it has an operation + given by $x + y = x'y \vee xy'$ making it into a Boolean ring. This + is not the + inherited from $R|R$ since the complementation operation $'$ on $1|R$ is not the restriction of the complementation $'$ on $R|R$.

Now suppose that L is a Stone algebra, and it is known that its dense set D is a Boolean algebra isomorphic to its skeleton L^*. There is no obvious way to effect this isomorphism. However, since D is a filter, $a \vee x$ is in D for any $a \in L$ and any $x \in D$. The mapping $a \to a \vee x$ is a homomorphism from L into D, and in particular from L^* into D. Just observe that $(a \vee x)(b \vee x) = ab \vee x$ so that the mapping preserves $\vee$, and similarly it preserves $\wedge$. If D is Boolean, or more generally, if D is a lattice and thus has a 0, say $\underline{0}$, then that is a natural element to pick in hopes of yielding an isomorphism between L^* and D. In $R|R$, the element $\underline{0}$ is $(1|0) = (0|0)$ as noted above, and indeed the mapping $(a|1) \to (a|1) \vee (1|0) = (a|a) = (1|a)$ effects the isomorphism already noted between $R|1$ and $1|R$.

We sum up.

Theorem 4. *In $R|R$, the skeleton $R|1$ and the dense set $1|R$ are Boolean algebras, and the mapping $(a|1) \to (a|1) \vee (1|0)$ is an isomorphism between them.*

It turns out that the conditions expressed in Theorem 4, namely that the skeleton and the dense set are Boolean algebras, and the mapping $a \to a \vee \underline{0}$ is an isomorphism between these two Boolean algebras, characterize $R|R$ among Stone algebras. This is made precise in the following theorem.

Theorem 5. *Let L be a Stone algebra, L^* its skeleton, and D its dense set. Suppose that D is a Boolean algebra, and that the mapping $a \to a \vee \underline{0}$ is an isomorphism from L^* to D, where $\underline{0}$ is the 0 of D. Then, the mapping $\varphi : L \to D|D : a \to ((a \vee \underline{0})|(a \vee a^*))$ is an isomorphism.*

Proof. First, note that φ is indeed a mapping from L into $D|D$. $a \vee \underline{Q}$ is in D since D is a dual ideal and $\underline{Q}$ is in D. $(a \vee a^*)^* = a^* \wedge a^{**} = 0$, so $a \vee a^*$ is also in D. We now break the proof up into several steps.

(1) φ is one-to-one.

Suppose that $(a \vee \underline{Q})|(a \vee a^*) = (b \vee \underline{Q})|(b \vee b^*)$. Then

$$(a \vee \underline{Q}) \wedge (a \vee a^*) = (b \vee \underline{Q}) \wedge (b \vee b^*) = a \vee (\underline{Q} \wedge (a \vee a^*)) = a \vee \underline{Q} = b \vee \underline{Q}.$$

We also have $a \vee a^* = b \vee b^*$, so that $a \vee a^* = a \vee a^* \vee \underline{Q} = b \vee \underline{Q} \vee a^*$. Multiplying through by a, we get $a \wedge (a \vee a^*) = a \wedge (b \vee a^*) = a = ab$, and by symmetry, $b = ab$, so $a = b$. Thus φ is one-to-one.

(2) φ preserves $\vee$.

$$\varphi(a) \vee \varphi(b) = (a \vee \underline{Q})|(a \vee a^*) \vee (b \vee \underline{Q})|(b \vee b^*) =$$

$$(a \vee b \vee \underline{Q})|[((a \vee \underline{Q}) \wedge (a \vee a^*)) \vee ((b \vee \underline{Q}) \wedge (b \vee b^*)) \vee ((a \vee a^*) \wedge (b \vee b^*))] =$$

$$(a \vee b \vee \underline{Q})|(a \vee \underline{Q} \vee b \vee \underline{Q} \vee (a \wedge b) \vee (a^* \wedge b) \vee (a \wedge b^*) \vee (a^* \wedge b^*)) =$$

$$(a \vee b \vee \underline{Q})|(a \vee b \vee (a^* \wedge b^*)) =$$

$$(a \vee b \vee \underline{Q})|(a \vee b \vee (a \vee b)^*) = \varphi(a \vee b).$$

Some preliminaries are needed before showing that φ preserves $\wedge$. Since $\wedge$ in $D|D$ involves the complement in the Boolean algebra D, we need to figure out what it is. We have the isomorphism $a \to a \vee \underline{Q}$ from L^* to D, and the complement operator on L^* is * itself. For $a \in L$, let x_a be the (unique) element in L^* such that $x_a \vee \underline{Q} = a \vee \underline{Q}$. Thus the complement in D, which we will denote by $'$, is given by $(a \vee \underline{Q})' = x_a^* \vee \underline{Q}$. This is simply because the mapping $a \to a \vee \underline{Q}$ is an isomorphism between L^* and D. If a itself is in D, then $a' = x_a^* \vee \underline{Q}$.

For $a \in L$, it turns out that a pertinent question for us is the relation between a^* and x_a^*. Note that for $a \in L$, $a = a^{**} \wedge (a \vee a^*)$. This is because $a^{**} \geq a$ and $a^{**} \wedge a = 0$. Thus $a \vee \underline{Q} = (a^{**} \vee \underline{Q}) \wedge (a \vee a^*)$ since $\underline{Q} \wedge (a \vee a^*) = \underline{Q}$, so that

$$a \vee \underline{Q} = x_a \vee \underline{Q} = (a^{**} \vee \underline{Q}) \wedge (a \vee a^*) =$$

$$(a^{**} \vee \underline{Q}) \wedge (a \vee \underline{Q} \vee a^* \vee \underline{Q}) =$$

$$(a^{**} \vee \underline{Q}) \wedge (x_a \vee \underline{Q} \vee a^* \vee \underline{Q}) =$$

$$(a^{**} \wedge x_a) \vee \underline{Q}.$$

Now from $x_a \vee \underline{Q} = (a^{**} \wedge x_a) \vee \underline{Q}$, we get $a^{**} \wedge x_a = x_a$, so that $a^{**} \geq x_a$. In particular, $a^{***} = a^* \leq x_a^*$. We sum up these facts.

(3) For $a \in L$, let x_a be the element in L^* such that $x_a \vee Q = a \vee Q$. Then the complement operator $'$ in D is given by $(a \vee Q)' = x_a^* \vee Q$, and $a^* \leq x_a^*$.

(4) φ preserves $\wedge$.

$$\varphi(a) \wedge \varphi(b) = ((a \vee Q)|(a \vee a^*)) \wedge ((b \vee Q)|(b \vee b^*)) =$$

$$((a \vee Q) \wedge (b \vee Q))|[((a \vee Q)' \wedge (a \vee a^*)) \vee ((b \vee Q)' \wedge (b \vee b^*)) \vee ((a \vee a^*) \wedge (b \vee b^*))] =$$

$$((a \vee Q)\wedge(b \vee Q))|[((x_a^* \vee Q) \wedge (a \vee a^*)) \vee ((x_b^* \vee Q) \wedge (b \vee b^*)) \vee ((a \vee a^*) \wedge (b \vee b^*))].$$

Since

$$(x_a^* \vee Q) \wedge (a \vee a^*) = (x_a^* \vee Q) \wedge (a \vee Q \vee a^* \vee Q) =$$

$$((x_a^* \vee Q) \wedge (a \vee Q)) \vee ((x_a^* \vee Q) \wedge a^* \vee Q)) =$$

$$Q \vee x_a^* a^*,$$

and since $a^* \leq x_a^*$, we have $Q \vee x_a^* a^* = a^* \vee Q$. Thus

$$\varphi(a) \wedge \varphi(b) =$$

$$((a \vee Q)\wedge(b \vee Q))|[(a^* \vee Q \vee b^* \vee Q \vee (a \wedge b) \vee (a^* \wedge b) \vee (a \wedge b^*) \vee (a^* \wedge b^*)] =$$

$$((a \vee Q)\wedge(b \vee Q))|(a^* \vee b^* \vee (a \wedge b))$$

$$((a \wedge b) \vee Q)|(a \wedge b)^* \vee (a \wedge b)) =$$

$$\varphi(a \wedge b).$$

To complete the proof, we need that φ is onto. Given an element $(a|b)$ in $D|D$, it is not obvious just what element φ takes onto it. How would we find this element in the case L were $R|R$ itself? In that case, we are given an element $(1|a)|(1|b) = (1|ab)|(1|b)$ in $D(R|R)|D(R|R)$, recalling the fact that $1|R$ is a Boolean ring, and need the element $(a|b) = (ab|b)$, which φ does indeed take to $(1|a)|(1|b)$. There is one key observation to be made. First note that for $(a|b)$ in $R|R$, say, $(a|b) = (ab|b) = (ab|1) \vee (0|b)$. Now consider our map φ as applied to $R|R$. Then

$$(a|b) =$$

$$((a|b \vee 1|0)|(a|b \vee (a|b)^*)) =$$

$$((ab|1 \vee 0|b \vee 1|0)|(ab|1 \vee 0|b \vee (ab|1 \vee 0|b)^*)) =$$

$$((1\,|\,ab)\,|\,(1\,|\,b)).$$

Since

$$(ab\,|\,1) \vee (0\,|\,b) \vee (1\,|\,0) = (1\,|\,ab) = (ab\,|\,1) \vee (1\,|\,0),$$

and

$$(ab\,|\,1) \vee (0\,|\,b) \vee (ab\,|\,1 \vee 0\,|\,b)^* = (1\,|\,b) = (0\,|\,b) \vee (0\,|\,b)^*,$$

we will be in business if from $(1\,|\,ab)\,|\,(1\,|\,b)$ we can construct, in a *Stone algebra satisfying our hypotheses*, elements corresponding to $(ab\,|\,1)$ and $(0\,|\,b)$. So we know the elements $(1\,|\,ab)$ and $(1\,|\,b)$, and so know ab and b. The element corresponding to $(ab\,|\,1)$ is, in the notation above, $x_{a\wedge b}$. The element corresponding to $0\,|\,b$ is the element $x_b^* \wedge \underline{0}$, since in $R\,|\,R$, $(0\,|\,b) = (b\,|\,1)^* \wedge (1\,|\,0)$. Now this dictates that given the element $(a\,|\,b)$ in $D\,|\,D$, φ should take $x_{a\wedge b} \vee (x_b^* \wedge \underline{0})$ onto it. We check:

$$\varphi(x_{a\wedge b} \vee (x_b^* \wedge \underline{0})) =$$

$$(x_{a\wedge b} \vee (x_b^* \wedge \underline{0}) \vee \underline{0})\,|\,(x_{a\wedge b} \vee (x_b^* \wedge \underline{0}) \vee (x_{a\wedge b} \vee (x_b^* \wedge \underline{0}))^* =$$

$$(x_{a\wedge b} \vee \underline{0})\,|\,(x_{a\wedge b} \vee \underline{0} \vee ((x_{a\wedge b})^* \wedge x_b) =$$

$$(a \wedge b)\,|\,(a \wedge b \vee ((x_{a\wedge b}^* \vee \underline{0}) \wedge (x_b \vee \underline{0})) =$$

$$(a \wedge b)\,|\,(a \wedge b) \vee ((a \wedge b)' \wedge b) =$$

$$(a \wedge b)\,|\,((a \wedge b) \vee (a' \vee b') \wedge b) =$$

$$(a \wedge b)\,|\,((a \wedge b) \vee (a' \wedge b)) =$$

$$((a \wedge b)\,|\,b).$$

This completes the proof. □

Several comments are in order. First, since L^* is isomorphic to D, L is isomorphic to $L^*\,|\,L^*$. The theorem was stated using $D\,|\,D$ since the isomorphism from L into $D\,|\,D$ is more simply and elegantly defined than the one from L into $L^*\,|\,L^*$.

Second, in the statement of the theorem, one need only assume that D is a bounded lattice and that $a \rightarrow a \vee \underline{0}$ is a one-to-one mapping from L^* onto D. That mapping is then automatically an isomorphism since $\vee$ and $\wedge$ are preserved in any case.

In $R\,|\,R$, one has the "complementation" $'$ given by $(a\,|\,b)' = (a'\,|\,b)$. No mention or use of it has been made in our theorem. In the Boolean algebra L^*, * is the complementation operator. In a Stone algebra satisfying the hypothesis of Theorem 5, the Boolean algebra D must itself also have a complementation operator, and we identified the complement of an element a in D as an element of the Boolean algebra D to be the element $x_a^* \vee \underline{0}$. In a Stone algebra satisfying the hypothesis of Theorem 5, what is the operator corresponding to $'$ in $R|R$? The element a in L corresponds to $(a \vee \underline{0})|(a \vee a^*)$ in $D|D$, and

$$((a \vee \underline{0})|(a \vee a^*))' = (a \vee \underline{0})'|(a \vee a^*) = (x_a^* \vee \underline{0})|(a \vee a^*).$$

The preimage of this element under our isomorphism is the element

$$x_{(x_a^* \vee 0)\wedge(a\vee a^*)} \vee ((x_{a\vee a^*})^* \wedge \underline{0}),$$

which by a routine calculation is $a^* \vee (x_a^* \wedge a^{**} \wedge \underline{0})$. In other words, for a in L,

$$a' = a^* \vee (x_a^* \wedge a^{**} \wedge \underline{0}) = x_a^* \wedge (a^* \vee \underline{0}) = a^* \vee (x_a^* \wedge \underline{0}).$$

Definition 2. *An abstract conditional space is a Stone algebra L such that*

(1) its dense set D is a bounded lattice, and

(2) the mapping from its skeleton L^* to D given by $a \to a \vee \underline{0}$, where $\underline{0}$ is the 0 of D, is a bijection.

Of course, D is also a Boolean algebra. An alternate way to phrase this definition is to require that D is a Boolean algebra and that the mapping is an isomorphism. That is the phraseology in Theorem 5. The conditions in Definition 2 are not really weaker although they appear to be. In any case, an abstract conditional space is just $R|R$ for some Boolean ring R.

There are other versions of Theorem 5 available to us. In $R|R$, the operator $'$ plays a significant role, as does the special element $(1|0)$. One can arrive at a representation theorem by postulating these two entities on a Stone algebra and requiring certain properties of them. The following is an example along this line.

Theorem 6. *Let L be a Stone algebra, L^* its skeleton and D its dense set. Suppose that there is an element $\underline{0} \in L$ such that $D = L^* \vee \underline{0}$, and that there is a unary operator $'$ on L that coincides with * on L^*, and satisfies $(x \vee \underline{0})' = x' \wedge \underline{0}'$ for all $x \in L^*$, and $\underline{0}'^* = 0$. Then L is isomorphic to $L^*|L^*$.*

Proof. If $x \in L^*$, then $(x \vee \underline{0})'^* = (x' \wedge \underline{0}')^* = x'^* \vee \underline{0}'^* = x \vee 0 = x$, so the map

$L^* \to D : x \to x \vee \underline{0}$ is a bijection. Since $0 \vee \underline{0} = \underline{0}$ is in its image, $\underline{0} \in D$. The map $L \to D : x \to x \vee d$ for any $d \in D$ preserves $\vee$ and $\wedge$, so L^* and D are isomorphic Boolean algebras, $\underline{0}$ is the zero of the Boolean algebra D, and Theorem 5 applies. □

4.3 Semi-simple MV-algebras

This section consists of proving that the algebraic system $R|R$ can be enriched in a simple way to obtain a (Chang) MV-algebra, so that a formal relationship with fuzzy logics is established. This latter fact follows from Belluce (1986).

First, as stated earlier, in view of three-valued logic connection, $R|R$, equipped with any given system of basic operators on it, is an algebraic structure generalizing Boolean ring structure. This generalization can be viewed in various different ways, depending upon the given system of operators. In Section 4.1 we have seen that when $R|R$ is equipped with our operators $(\wedge, \vee, ')$, then $R|R$ is a special type of a Stone algebra where the associated pseudo-complementation $*$ is

$$(a|b)^* = a' \wedge b \quad (= a' \cdot b) .$$

In a (independent) pioneering work, Schay (1968) took the equivalent viewpoint by modeling conditional events as generalized three-valued indicator functions. By doing so, he considered $R|R$ as an algebraic structure with a system of five operators $(\cap, \cup, \wedge, \vee, (\cdot)')$ (where his $\wedge, \vee, (\cdot)'$ are different from ours).

Abstracting this algebraic structure, he spent almost half of his work on establishing a Stone's Representation Theorem for his new structure (Schay, 1968, p. 338-342). While the mathematics involved is interesting, his axioms for the abstract structure are quite complicated.

In another direction, motivated by the desire of establishing a three-way relationship among formal systems, MV-algebras and fuzzy sets in the context of multi-valued logics (as an analog to the case of classical two-valued logic, where there is such a relationship among formal systems, Boolean rings and set theory), Belluce (1986) considered a generalized structure known as Chang MV-algebras. This algebraic structure is known in multi-valued logics (Chang, 1958, 1959). Roughly speaking, such a structure is obtained when the idempotency and the distributive law in a Boolean ring $R(+, \cdot)$ are both dropped.

Specifically, following Belluce, an MV-algebra is a non-empty set A with two binary operators $+$, $\cdot$, and one unary operator $^-$ with $0, 1$ satisfying the following conditions.

(i) $\langle A, +, 0\rangle$ and $\langle A, \cdot, 1\rangle$ are commutative semi-groups with identity.

(ii) For all $x, y \in A$,

$$x + \bar{x} = 1, \; x \cdot \bar{x} = 0, \; \bar{x} = 0, \; \bar{0} = 1 .$$

(iii) For all $x, y \in A$,

$$\overline{(x + y)} = \bar{x} \cdot \bar{y}, \; \overline{x \cdot y} = \bar{x} + \bar{y}, \; \bar{\bar{x}} = x$$

(that is, $\bar{}$ is involutive, and of "De-Morgan" type with respect to $+$ and $\cdot$).

(iv) $+$ and $\cdot$ are such that, if one defines two "Boolean-like" operators

$$x \vee y = x + \bar{x}y, \; x \wedge y = (x + \bar{y})y ,$$

then $\langle A, \vee, 0\rangle$, $\langle A, \wedge, 1\rangle$ are also commutative semi-group with identity.

(v) For all $x, y, z \in A$,

$$x \cdot (y \vee z) = x \cdot y \vee x \cdot z, \; (x + y) \wedge (x + z) = x + (y \wedge z) .$$

Notice that $\langle A, \wedge, \vee, 1, 0\rangle$ is a bounded commutative lattice where the associated order relation $\leq$ is $x \leq y$ if and only if $x \wedge y = x$.

Definition. *An MV-algebra A is said to be archimedean when for each $x, y \in A$, if $(x + \ldots + x) = nx \leq y$ for all $n \geq 0$, then $x \cdot y = x$.*

A result in Belluce (1986) stated that archimedean MV-algebras and semi-simple MV-algebras are the same.

With analogous algebraic concepts for MV-algebras, a MV-algebra A is said to be *semi-simple* if its radical is zero. (See Belluce, 1986, for details.) The point is this: semi-simple (or equivalently, archimedean) MV-algebras are precisely "bold" algebras of fuzzy sets (Belluce, 1986, Theorem 4), where by a "bold" algebras of fuzzy sets, one means a subalgebra of the MV-algebra (under induced operations) of all fuzzy subsets of some space Ω, that is, the collection of all functions $f: \Omega \to [0, 1]$. Specifically, $[0, 1]^{\Omega}$ becomes an MV-algebra with:

$$(f + g)(\omega) = Min(1, f(\omega) + g(\omega))$$

$$(f \cdot g)(\omega) = Max(0, f(\omega) + g(\omega) - 1)$$

$$\bar{f}(\omega) = 1 - f(\omega)$$

$$(f \vee g)(\omega) = Max(f(\omega), g(\omega))$$

$$(f \wedge g)(\omega) = Min(f(\omega), g(\omega)) .$$

We proceed now to show that $R|R$ can be viewed as an archimedean MV-algebra, so that algebraically speaking, conditional logic in a sense is a form of fuzzy logic. The search for operations on $R|R$ making $R|R$ an MV-algebra is dictated by the operations $+$, $\cdot$ of fuzzy sets or of corresponding operations on $[0, 1]$, and this using Theorem 1 of Section 3.4.

In the three-valued logic setting, viewing u as "lying" between 0 and 1, say $u = 1/2$, we can treat u as a real number in $[0, 1]$. In this vein, consider $\oplus$ and $\circ$ defined on $[0, 1]$ by

$$x \oplus y = Min(1, x + y),$$

$$x \circ y = Max(0, x + y - 1) .$$

The restrictions of $\oplus$ and $\circ$ to $\{0, u, 1\}^2$ yield values in $\{0, u, 1\}$, and hence correspond to truth functions of operations on $R|R$. So let $\psi : \{0, 1/2, 1\}^2 \to \{0, 1/2, 1\}$ be defined by

$$\psi(i, j) = Max(0, i + j - 1) .$$

We have

$$\psi^{-1}(1) = \{(1, 1)\},$$

$$\psi^{-1}(0) = \{(0, 0), (0, 1/2), (1/2, 0), (1/2, 1/2), (0, 1), (1, 0)\} .$$

Recall that, for $a, b, c, d \in R$ with $a \le b$, $c \le d$, the pair (i, j) corresponds to $w_i(a|b)w_j(c|d)$, where

$$w_i(a|b) = \begin{cases} a'b & \text{if } i = 0 \\ b' & \text{if } i = 1/2 \\ ab(=a) & \text{if } i = 1 \end{cases}$$

and

$$f_\psi : (R|R)^2 \to R|R$$

is determined by

$$f_\psi((a|b), (c|d)) = \left[\bigvee_{(i,j) \in \psi^{-1}(1)} w_i(a|b)w_j(c|d) \,\Big|\, \bigvee_{(i,j) \in \psi^{-1}(1) \vee \psi^{-1}(0)} w_i(a|b)w_j(c|d) \right].$$

Thus, here

$$\bigvee_{(i,j) \in \psi^{-1}(1)} w_i(a|b)w_j(c|d) = ac ,$$

and

$$\bigvee_{(i,j)\in\psi^{-1}(1)\vee\psi^{-1}(0)} w_i(a|b)w_j(c|d)$$

$$= ac \vee a'bc'd \vee a'bd' \vee b'c'd \vee b'd' \vee a'bc \vee ac'd$$

$$= ac \vee b'd' \vee a'b(c'd \vee d' \vee c) \vee c'd(b' \vee a)$$

$$= ac \vee b'd' \vee a'b \vee c'd(a'b)',$$

$$= ac \vee b'd' \vee a'b \vee c'd,$$

noting that since $c \leq d$, $c'd \vee d' \vee c = 1$, and that $a'b \vee (c'd)(a'b)' = a'b \vee c'd$). Hence

$$f_\psi((a|b), (c|d)) = (ac|ac \vee a'b \vee c'd \vee b'd')$$

$$= (a|b)(c|d)(b \vee d|1).$$

Similarly, for $\psi(i, j) = Min(1, i + j)$, we have

$$\psi^{-1}(1) = \{(1/2, 1/2), (1/2, 1), (1, 1/2), (1, 1), (0, 1), (1, 0)\},$$

$$\psi^{-1}(0) = \{(0, 0)\},$$

$$\bigvee_{(i,j)\in\psi^{-1}(1)} w_i(a|b)w_j(c|d) = b'd' \vee b'c \vee ad' \vee ac \vee a'bc \vee ac'd$$

$$= b'd' \vee a(d' \vee c'd \vee c) \vee c(b' \vee a'b)$$

$$= b'd' \vee a \vee c(a')$$

$$= a \vee c \vee b'd',$$

and

$$\bigvee_{(i,j)\in\psi^{-1}(1)\vee\psi^{-1}(0)} w_i(a|b)w_j(c|d) = a \vee c \vee b'd' \vee a'bc'd$$

$$= a \vee c \vee b'd' \vee bd(a \vee c)'$$

$$= a \vee c \vee b'd' \vee bd$$

$$= (a|b) \vee_{.} (c|d) \vee (b'd'|1).$$

This suggests the following new operations on $R|R$:

$$(a|b) \oplus (c|d) = (a|b) \vee (c|d) \vee (b'd'|1),$$

$$(a|b) \circ (c|d) = (a|b)(c|d)(b \vee d|1) .$$

Theorem 1. *$(R|R, \oplus, \circ, ', (0|1), (1|1))$ is a semi-simple MV-algebra.*

Proof. We verify the axioms of an MV-algebra. $<R|R, \oplus, (0|1)>$ is a commutative semi-group with identity $(0|1)$. Indeed, the commutativity follows from the symmetry in the definition of $\oplus$ above; when $c = 0$ and $d = 1$,

$$(a|b) \vee (0|1) \vee (0|1) = (a|b) .$$

Similarly, $<R|R, \circ, (1|1)>$ is a commutative semi-group with identity $(1|1)$. Next, the operation $^-$ is taken to be $'$, and we have

$$(a|b) \oplus (a|b)' = (a|b) \oplus (a'|b) = (a|b) \vee (a'|b) \vee (b'|1) = (1|b) \vee (b'|1) = (1|1);$$

$$(a|b) \circ (a'|b) = (a|b)(a'|b)(b|1) = (0|b)(b|1) = (0|1),$$

$$(0|1)' = (1|1).$$

Next, always assuming that $a \leq b, c \leq d$,

$$\begin{aligned}((a|b) \oplus (c|d))' &= (a \vee c \vee b'd' | a \vee c \vee bd \vee b'd')' \\ &= (a'c'b \vee a'c'd | a \vee c \vee bd \vee b'd') \\ &= ((a'c'b \vee a'c'd)(a \vee c \vee bd \vee b'd') | a \vee c \vee bd \vee b'd') \\ &= (a'c'bd | a \vee c \vee bd \vee b'd') \\ &= (a'|b) \circ (c'd) = (a|b)' \circ (c|d)';\end{aligned}$$

$$\begin{aligned}((a|b) \circ (c|d))' &= (ac | a'b \vee c'd \vee bd \vee b'd') \\ &= (a' \vee c' | a'b \vee c'd \vee c'd \vee bd \vee b'd') \\ &= ((a' \vee c')(a'b \vee c'd \vee bd \vee b'd') | a'b \vee c'd \vee bd \vee b'd') \\ &= (a'b \vee c'd \vee b'd' | a'b \vee c'd \vee bd \vee b'd') \\ &= (a'|b) \oplus (c'|d) = (a|b)' \oplus (c|d)' ,\end{aligned}$$

noting that $a \leq b$ and $c \leq d$ imply that $a'b' = b'$ and $c'd' = d'$.

It is easy that $(a|b)'' = (a|b)$, and is is readily checked that

$$(a|b) \oplus (a|b)' \circ (c|d) = (a|b) \vee (c|d)$$

and

$$((a|b) \oplus (c'|d)) \circ (c|d) = (a|b) \wedge (c|d) .$$

Thus, $\vee$ and $\wedge$ on $R|R$ are precisely the derived operations. Also, $\langle R|R, \vee, (0|1)\rangle$ and $\langle R|R, \wedge, (1|1)\rangle$ are commutative semi-groups with identity. Finally, it can be checked that

$$(a|b) \circ [(c|d) \vee (e|f)] = [[(a|b) \circ (c|d)] \vee [(a|b) \circ (e|f)]$$

$$(a|b) \oplus ((c|d) \wedge (e|f)) = [(a|b) \oplus (c|d)] \wedge [(a|b) \oplus (e|f)].$$

For $n \geq 2$,

$$\underbrace{(a|b) \oplus \ldots \oplus (a|b)}_{n \ times} = (b' \vee a|1)$$

Thus if

$$(c|d) \geq (b' \vee a|1),$$

then for all $n \geq 0$,

$$\underbrace{(a|b) \oplus \ldots \oplus (a|b)}_{n \ times} \leq (c|d) ,$$

and for $a \leq b$,

$$(a|b) \circ (b' \vee a|1) = (a|b).$$

Hence

$$(a|b) \circ (c|d) = (a|b),$$

that is, $R|R$ is archimedean. Indeed,

$$(a|b) \circ (c|d) = (a|b)(c|d)(b \vee d|1)$$

$$= (a|b)(b \vee d|1) = (a|b),$$

since $(a|b) \leq (b' \vee a|1) \leq (c|d)$, using the criterion that since $(a|b) \leq (c|d)$ if and only if $ab \leq cd$ and $c'd \leq a'b$. □

Remarks. There are some similarities with fuzzy logic.

(i) The two additional operations $\oplus$ and $\circ$ on $R|R$ are defined in terms of the original logical operations $\vee$, $\wedge$. The algebraic structure

$$(R|R, \oplus, \circ, ', (0|1), (1|1), \leq)$$

is somewhat similar to a quantum logic, since with respect to $\oplus$ and $\circ$, the operator $'$ is an *ortho-complementation*, so that the law of excluded middle holds, and $\circ$ is not distributive over $\oplus$. However, $\circ$ is not idempotent.

(ii) In fuzzy logic (for example, Zadeh, 1983), the basic connectives are defined in terms of operations on the unit interval $[0, 1] : \vee = max, \wedge = min, \; ' = 1 \; ^{-}$. As in Belluce (1986), $[0, 1]$ becomes an MV-algebra when one introduces new operations $\oplus$, $\circ$, and $^{-}$ given by

$$^{-} = ',$$

$$x \oplus y = 1 \wedge (x + y),$$

$$x \circ y = 0 \vee (x + y - 1).$$

for $x, y \in [0, 1]$. In turn, $\wedge$ and $\vee$ are expressed in terms of $\oplus$ and $\circ$ by

$$x \wedge y = (x \oplus \bar{y}) \circ y,$$

and

$$x \vee y = x \oplus \bar{x} \circ y.$$

(iii) For $u = 1/2$, Lukasiewicz's three-valued logic is a subalgebra of the MV-algebra $[0, 1]$, that is, is a "bold" algebra of fuzzy sets. An alternative proof of Theorem 1 is obtained by using Theorem 2 of Section 3.4, and making the easy verification of the above fact.

Let $A = \{0, 1/2, 1\}$. Define, for $x, y \in A$,

$$x \oplus y = min(1, x + y),$$

$$x \circ y = max(0, x + y - 1),$$

$$\bar{x} = 1 - x.$$

Then

$$x \vee y = max(x, y),$$

$$x \wedge y = min(x, y),$$

and the order relation $\leq$ is the ordinary order relation on real numbers. A is a MV-algebra. Moreover, it is archimedean. Indeed, for $x \in \{0, 1/2, 1\}$, if $x = 0$, then for $n \geq 0$, we have

$$\underbrace{0 \oplus \cdots \oplus 0}_{n \ times} = 0 \leq 0, 1/2, 1 ,$$

with

$$0 \circ 0 = 0 \circ 1/2 = 0 \circ 1 = 0.$$

If $x = 1/2$, and $n \geq 2$, then

$$\underbrace{1/2 \oplus \cdots \oplus 1/2}_{n \ times} = 1,$$

and

$$1/2 \circ 1 = 1/2.$$

If $x = 1$, then

$$\underbrace{1 \oplus \cdots \oplus 1}_{n \ times} = 1$$

with $1 \circ 1 = 1$. □

CHAPTER 5

CONDITIONAL EVENTS AND PROBABILITY

The connection between logic and probability is apparent in automated reasoning processes under uncertainty. A systematic study of the extension of probability logic to the conditional case will be presented in Chapter 6. In this Chapter 5, we establish various basic properties of probability measures extended to the algebra of conditional events as well as the justification of assigning conditional probabilities to conditional events. We discuss the association of randomness to conditional events (such as random sets, random conditional events, random conditional variables). Finally, a general concept of qualitative (or measure-free) conditional independence is introduced.

5.1 Uncertainty measures on conditionals

It is an accepted thesis that uncertainty is essentially conditional, that is, the uncertainty of an event is always conditioned upon some other events. At the numerical level, that is, when uncertainty is taken in a quantitative way, a natural domain for uncertainty measures is a conditional space $R|R$. For example, in order to rigorize Lindley's discussions on the inadmissability of uncertainty measures in expert systems, via the scoring rule approach (Lindley, 1982), it is necessary to evoke conditional events (Goodman, Nguyen and Rogers, 1990).

By an uncertainty measure μ on $R|R$, we mean a map $\mu : R|R \to \mathbb{R}$, say, where $\mathbb{R}$ denotes the set of real numbers. Now, for $(a|\mathrm{b}) \in R|R$, we have $(a|b) = [ab, b' \vee a]$, an interval in R (see Section 2.3). Thus, an uncertainty measure μ on $R|R$ can be derived from a map $\nu : R \to \mathbb{R}$ as follows.

$$\mu(a|b) = F(\nu(ab), \nu(b' \vee a)) ,$$

where

$$F : \mathbb{R} \times \mathbb{R} \to \mathbb{R}$$

is some given function. For example, if $\nu = P$, a probability measure on R, and

$$F(x, y) = \frac{x}{1 + x - y},$$

we have $\mu(a|b) = P(a|b)$, provided $P(b) > 0$. (See also Dubois and Prade, 1991.) It is obvious that, in this way, various uncertainty measures on $R|R$ can be considered. In his book, however, we are concerned only with uncertainty measures derived from probability measures. For other types of uncertainty measures, see Dubois and Prade, (1991). Specifically, we will proceed to justify conditional probability as a means to assign uncertainty to conditional events. Note that, in the development of the theory of measure-free conditioning (Chapter 2), the condition of compatibility with probability was used in an essential way, so that it is possible to assign conditional probabilities to conditional events in a consistent manner. We emphasize, however, the appearance of conditional events as cosets of Boolean rings in our present work. The problem has been: define mathematically objects $(a|b)$ representing implicative propositions of the form "if b, then a " (implicit or explicit) or " a on condition b " or "a given b" in such a way that it is possible to quantify the strengths of these propositions by conditional probabilities. Of course, if $(a|b)$ is modeled as material implication $b \rightarrow a = b' \vee a$, then one can quantify it by unconditional probability $P(b \rightarrow a)$. The general problem in reasoning under uncertainty in artificial intelligence is this. Given a knowledge base consisting of uncertain conditional information, how does one combine these conditional propositions and do inference? At the syntax level, one first needs to define or model "if b, then a" by $b \Rightarrow a$, say. Next, define appropriate connectives among such objects so that one can combine $b \Rightarrow a$ with $d \Rightarrow c$ through the use of these connectives. For example, $(b \Rightarrow a) \wedge (d \Rightarrow c)$. At the numerical quantification level, one chooses an uncertainty measure μ which can operate on the $(b \Rightarrow a)$ and proceeds to compute, for example, $\mu((b \Rightarrow a) \wedge (d \Rightarrow c))$. When $\mu(b \Rightarrow a)$ is chosen to be $P(a|b)$, then $b \Rightarrow a$ has to be a coset. The logical operations among cosets developed in Chapter 3 provide connectives for conditional propositions. One combines several conditional propositions at the syntax level, obtaining another coset, and then evaluates its conditional probability which is considered as the measure of uncertainty of the combined evidence. Furthermore, it will be shown in Chapter 6 that an entailment relation among conditional propositions can be established so that deduction or inference can be carried out at the numerical level. If $b \Rightarrow a$ is modeled differently, for example, as in a "first-order conditional logic" of Delgrande (1987), then the quantification measure μ should be different than a conditional probability operator. As an example, one can model $b \Rightarrow a$ as $b \rightarrow a$ (material implication) and use some appropriate *non-additive* "measure" μ on the ring R so that $\mu(b \rightarrow a) = \mu(a|b)$, where $\mu(\cdot|b)$ is defined to be a "conditional measure". A typical situation is when μ is chosen to be a Dempster-Shafer belief function (see Sombé, 1990, p. 405-406, or Pearl, 1990, p. 371-373). For example, let R be the power set of a finite set Ω. Let $m : R \rightarrow [0, 1]$ be such that

$$m(\phi) = 0 , \sum_{a \leq \Omega} m(a) = 1$$

and

$$\mu : R \to [0, 1] \text{ with } \mu(b) = \sum_{a \le b} m(a) .$$

For fixed b , define

$$m_b(a) = \sum m(x)$$

where the summation is over all x in R such that $bx = a$. Then the conditional belief function $\mu(\cdot \mid b)$ is defined by

$$\mu(a \mid b) = \sum_{x \le a} m_b(x) .$$

It is easy to verify that

$$\{x : x \le b' \vee a\} = \{x : xb = y \le a\} .$$

Thus, $\mu(a \mid b) = \mu(b \to a)$.

It is relevant here to describe the works of Rényi (1970) and Cox (1961). First, let $(\Omega, \mathcal{A})$ be a measureable space. If P is a probability measure on $\mathcal{A}$, then the associated *conditional probability "operator"* $\hat{P}$ is defined as follows. Let

$$w_P = \{a : a \in \mathcal{A}, P(a) > 0\}.$$

Then define

$$\hat{P} : \mathcal{A} \times w_P \to [0, 1]$$

by

$$\hat{P}(a, b) = P(a \mid b) = P(ab)/P(b) .$$

Here, $\hat{P}$ is viewed as a "global" map, that is, with domain $\mathcal{A} \times w_P$, rather than "locally," that is, rather than a collection of maps $P(\cdot \mid b)$, one for each $b \in w_P$. This is in line with Rényi's concept of *conditional probability spaces* (Rényi, 1970). See later for details. The map $\hat{P}$ has the following basic properties:

(i) For each $b \in w_P$, $\hat{P}(\cdot, b) : \mathcal{A} \to [0, 1]$ is a non-negative and σ-additive set function (that is, a measure).

(ii) For every $b \in w_P$, $\hat{P}(b, b) = 1$.

(iii) For $b, c \in w_P$ with $b \le c$, one has $\hat{P}(b, c) > 0$, and if $a \in \mathcal{A}$, then

$$\hat{P}(a, b) = \hat{P}(ab, c)/\hat{P}(b, c) .$$

The subset w_P of $\mathcal{A}$ has the following basic properties:

(iv) If $b_1, b_2 \in w_P$, then $b_1 \vee b_2 \in w_P$.

(v) There exists a sequence $b_n \in W_P$, $n \geq 1$, such that $\bigvee_{n=1}^{+\infty} b_n = \Omega$.

(vi) $\emptyset \notin w_P$.

Following Rényi, a subset $\mathscr{B} \subseteq \mathscr{A}$, satisfying (iv), (v) and (vi), is called a *bunch*.

The abstraction of the above is clear: an *abstract conditional probability operator* (or conditional probability operator or CPO for short) on $(\Omega, \mathscr{A}, \mathscr{B})$, where $\mathscr{B} \subseteq \mathscr{A}$ is a bunch, is a map $\hat{P}$ defined on $\mathscr{A} \times \mathscr{B}$, satisfying (i), (ii) and (iii).

Note that, from (iii) and (ii), with $b = c$, we get

$$\hat{P}(a, b) = \hat{P}(ab, b) .$$

By (i), $\hat{P}(\cdot, b)$ is non-decreasing, so that

$$\hat{P}(a, b) = \hat{P}(ab, b) \leq \hat{P}(b, b) = 1.$$

Also, $\hat{P}(\emptyset, b) = 0$, since $\hat{P}(\cdot, b)$ is a measure by (i). Thus, the range of $\hat{P}$ is $[0, 1]$.

The main result of Rényi (Rényi, 1970, p. 40) is this. If $\hat{P}$ is a CPO on $(\Omega, \mathscr{A}, \mathscr{B})$, then there exists a σ-finite measure μ on $\mathscr{A}$, unique up to a positive constant factor, such that:

$$\mathscr{B} \subseteq \{a : a \in \mathscr{A}, 0 < \mu(a) < +\infty\},$$

and for all $a \in \mathscr{A}$ and $b \in \mathscr{B}$,

$$\hat{P}(a, b) = \mu(ab)/\mu(b).$$

For μ to characterize $\hat{P}$, we need to extend $\hat{P}$ so that

$$\mathscr{B} = \{a : a \in \mathscr{A}, 0 < \mu(a) < +\infty\}.$$

This can be done as follows (Rényi, 1970, p. 43). Clearly,

$$\mathscr{B}^* = \{a : a \in \mathscr{A}, 0 < \mu(a) < +\infty\}$$

is a bunch. Note that $\mathscr{B}^*$ is the same for all measures μ in Rényi's theorem above. We have $\mathscr{B} \subseteq \mathscr{B}^*$. If $\mathscr{B} \neq \mathscr{B}^*$, we extend $\hat{P}$ to $\mathscr{A} \times \mathscr{B}^*$ by

$$\hat{P}(a, b) = \mu(ab)/\mu(b) ,$$

for $b \in \mathscr{B}^* - \mathscr{B}$ and $a \in \mathscr{A}$. This extended operator $\hat{P}$ is a CPO on $(\Omega, \mathscr{A}, \mathscr{B}^*)$. Therefore, there is no loss of generality to assume that any CPO $\hat{P}$ on $(\Omega, \mathscr{A}, \mathscr{B})$ is characterized by μ with $\mathcal{B} = \{a : a \in \mathcal{A}, 0 < \mu(a) < +\infty\}$. Thus, each CPO $\hat{P}$ on $(\Omega, \mathcal{A}, \mathcal{B})$ is derived from some (σ-finite) measure μ on $(\Omega, \mathcal{A})$. In particular, if μ is finite, that is, $\mu(\Omega) < +\infty$, then $\Omega \in \mathcal{B}$, so that $\hat{P}(\cdot, \Omega)$ is an ordinary probability measure on $(\Omega, \mathcal{A})$, where for $a \in \mathcal{A}$, $\hat{P}(a, \Omega)$ is interpreted as the unconditional probability of the event a.

As a final note on Rényi's work, recall that Rényi's concept of conditional probability spaces $(\Omega, \mathcal{A}, \mathcal{B}, \hat{P})$ was motivated by the thesis mentioned at the beginning of this section that "every probability is in reality a conditional probability." Thus, it is intuitive to define CPO first, and then derive ordinary probability measures as special cases. Conditional probability spaces are consistent with Kolmogorov's model of probability spaces in the sense that they generalize Kolmogorov's probability spaces. Note, however, that Kolmogorov defined probability measures first and then derived conditional probability measures.

Next, we outline Cox's work (Cox, 1961) concerning a class of uncertainty measures which can be transformed into conditional probability measures. In passing, we will mention the analogy with Lindley's message on the inevitability of probability (Lindley, 1982).

Let R be a Boolean ring of propositions. Taking the same thesis that numerical uncertainty is conditional in nature, Cox proceeded to derive a calculus of uncertainty as follows.

Let μ be a map on an appropriate domain in $R \times R$. Cox (1961, pp. 18-22) proved that if

(1) $\mu(a, b) = f(\mu(a', b))$ with f differentiable, and

(2) $\mu(ab, c) = \mu(a, c)\mu(b, ac)$,

then $\mu(\cdot, b)$ is finitely additive and $f(x) = 1 - x$.

More generally, Cox replaced (2) by

(3) $\mu(ab, c) = F(\mu(a, c), \mu(b, ac))$ with $F(x, y)$ differentiable.

Then he showed that there exist functions g of one variable such that (1) and (2) are satisfied when μ is replaced by $g \circ \mu$. As a consequence, $g \circ \mu(\cdot, b)$ is finitely additive. In other words, the uncertainty measures μ satisfying (1) and (3) can be transformed into (conditional) probabilities. Cox argued that there is no difference between μ and $g \circ \mu$ since "if $\mu(a|b)$ measures probability, so also does an arbitrary function of $\mu(a|b)$" (Cox, 1961, p. 16). This is precisely what we should understand years later when Lindley declared that "one cannot avoid probability" (Lindley, 1982).

Now, suppose P is a probability measure on $(\Omega, \mathcal{A})$, and let $g(x) = x^r$ for some $r \geq 1$. Then $g \circ P = P^r$ is no longer a probability measure. In fact, P^r is a *belief function* in the sense of Dempster-Shafer (Shafer, 1976). See also Section 5.3. Of course, P^r can be computed from P, but as a set function on $\mathcal{A}$, P^r satisfies a weaker set of axioms than that of a probability measure. In an abstract setting, that is, when belief functions are defined from axioms, not all belief functions are functions of probability measures (see Goodman, Nguyen and Rogers, 1990). In Lindley's sense, belief functions which cannot be transformed into probability measures are "inadmissible." More generally, if $\mu : \mathcal{A} \to \mathbb{R}^+$, say, is a set-function representing a quantification of uncertainty, then μ is "admissible" if there is some function g such that $g \circ \mu$ is a (finitely) additive set-function. It is clear that μ need not be a probability measure. Thus, in the view of Lindley, whenever an uncertainty measure μ is considered, one should find some function g such that $g \circ \mu$ is a probability, and then inferences should be based upon $g \circ \mu$ and not upon μ. As we have seen, a sufficient condition for the existence of such g is the set of conditions (i) and (iii) in Cox's program. Note that the work of Lindley is "conditional" in nature. Any μ which cannot be transformed into probabilities should be ruled out! Because of this important view on decision making in uncertain systems, we present below an outline of Lindley's paper. For more details, see Goodman, Nguyen and Rogers (1990).

Let R be a Boolean ring, viewed as a field of subsets of some set Ω. Roughly speaking, an uncertainty measure $\mu : (R|R) \to \mathbb{R}$ is said to be "admissible" if there is a function g such that $g \circ \mu$ is finitely additive. To make this statement precise, we need to explain the concept of admissibility and the sense in which $g \circ \mu$ is finitely additive. The most general framework in which admissibility can be addressed is game theory.

Consider the following special class of games called *uncertainty games*. These are triples $(\Lambda_1, \Lambda_2, L)$ of the following form. Λ_1 is regarded as a space

$$\Lambda_1 = \{(((a_1|b_1), \ldots, (a_n|b_n)), \omega) : a_i, b_i \in R,\ i = 1, \ldots, n;\ \omega \in \Omega,\ n \geq 1\}$$

of all possible "moves" or "pure strategies" of player I. Fix, once and for all, two real numbers $\alpha_0 < \alpha_1$, and let

$$\Lambda_2 = \{\mu : (R|R) \to [\alpha_0, \alpha_1]\}.$$

Each element of Λ_2 is a map assigning a number (describing the uncertainty) to each conditional event. Λ_2 is regarded as the space of "moves" of player II. Consider now the choice of loss function L. As in Lindley's paper, a function

$$f : [\alpha_0, \alpha_1] \times \{0, u, 1\} \to \mathbb{R}$$

is called *score function* if

(i) for each $j \in \{0, 1\}$, $f(\cdot, j) : [\alpha_0 , \alpha_1] \to \mathbb{R}$ is continuously differentiable, with a unique global minimum in $[\alpha_0 , \alpha_1]$ at α_j, and

(ii) $f(x, u) = 0$ for all $x \in [\alpha_0 , \alpha_1]$.

We extend f to $[\alpha_0, \alpha_1]^n \times \{0, u, 1\}^n$, $n \geq 1$, as usual. For

$$\hat{x}_n = (x_1, ..., x_n) \in [\alpha_0, \alpha_1]^n,$$

$$\hat{t}_n = (t_1, ..., t_n) \in \{0, u, 1\}^n,$$

$$f(\hat{x}_n, \hat{t}_n) = (f(x_1, t_1), ..., f(x_n, t_n)) \in \{(y_1, ..., y_n) \in \mathbb{R}^n , n \geq 1\} .$$

Similarly, μ is extended to $(R|R)^n$ componentwise. For

$$(\underline{a}|\underline{b})_n = ((a_1|b_1), ..., (a_n|b_n)) \in (R|R)^n,$$

$$\mu(\underline{a}|\underline{b})_n = (\mu(a_1|b_1), ..., \mu(a_n|b_n)) \in [\alpha_0 , \alpha_1]^n .$$

Let $\varphi(a|b)$ denote the generalized indicator function of $(a|b)$. A natural way to combine individual "scores"

$$f(\mu(a_i|b_i) , \varphi(a_i|b_i)(\omega)) , \ i = 1, 2, ..., n,$$

to obtain the total score is using addition on $\mathbb{R}$. That is, take

$$L_{f,+}((\underline{a}|\underline{b})_n, \omega, \mu) = \sum_{i=1}^{n} f(\mu(a_i|b_i) , \varphi(a_i|b_i))(\omega) .$$

The loss function $L_{f,+}$ depends on two functions, the score function f and the *additive aggregation* function $+$.

In general, by an *aggregation function*, we mean a function

$$\psi : \{(y_1, ..., y_n) \in \mathbb{R}^n, n \geq 1\} \to \mathbb{R}$$

such that

a) ψ is continuous differentiable in all of its arguments,

b) ψ is increasing in each of its arguments, and

c) $\psi(0_n) = 0$, $\forall\, n \geq 1$, where 0_n denotes the zero vector in $\mathbb{R}^n$.

The additive aggregation function is generated by ordinary addition on $\mathbb{R}$. Taking $\psi = +$

is equivalent to the sequence of functions g_n , $n \geq 1$, where

$$g_n : \mathbb{R}^n \to \mathbb{R}, \ g_n(x_1, ..., x_n) = \sum_{i=1}^{n} x_i.$$

Similarly, an aggregation function ψ can be identified with the sequence ψ_n, $n \geq 1$, where ψ_n is the restriction of ψ to $\mathbb{R}^n$. While the additive aggregation function is symmetric, there is no a priori reason to impose such a condition on arbitrary aggregation functions.

The game $(\Lambda_1, \Lambda_2, L_{f,+})$ will be denoted by $G_{f,+}$. It is simpler to formulate the concept of admissibility of uncertainty measures using an equivalent reduced form of $G_{f,+}$. In the expression of $L_{f,+}$, the value of $L_{f,+}((\underline{a}|\underline{b})_n , \omega, \mu)$, for each fixed μ, at $((\underline{a}|\underline{b})_n , \omega) \in \Lambda_1$, depends on the "configuration"

$$\varphi(\underline{a}|\underline{b})_n(\omega) = (\varphi(a_1|b_1)(\omega) , ..., \varphi(a_n|b_n)(\omega)) \in \{0, u, 1\}^n.$$

Thus, $L_{f,+}((\underline{a}|\underline{b})_n, \cdot, \mu)$ is constant on each element of the *canonical partition* $\pi(\underline{a}|\underline{b})_n$ of Ω generated by $(\underline{a}|\underline{b})_n$. Specifically,

$$\pi(\underline{a}|\underline{b})_n = \{B_j , j = 1, 2, ..., 2^{3n}\} ,$$

where each B_j is of the form $\bigwedge_{k=1}^{3n} D_k^{\varepsilon_k}$, for

$$D_k \in \{a_ib_i , a_i'b_i , b_i' , i = 1, ..., n\} ,$$

$$\varepsilon_k = 1 \text{ or } 0 , a^1 = a , a^0 = a'.$$

(See Rényi, 1970, p. 12-15.) Thus, we can replace Λ_1 by

$$\Lambda_1^* = \{((\underline{a}|\underline{b})_n, B) : (\underline{a}|\underline{b})_n \in (R|R)^n, B \in \pi(\underline{a}|\underline{b})_n , n \geq 1\} .$$

$L_{f,+}$ is modified to

$$L^*_{f,+} : \Lambda_1^* \times \Lambda_2 \to \mathbb{R},$$

$$L^*_{f,+}((\underline{a}|\underline{b})_n , B, \mu)) = \sum_{i=1}^{n} f(\mu(a_i|b_i) , (a_i|b_i)(B))$$

where $(a_i|b_i)(B) = \varphi(a_i|b_i)(\omega)$ for $\omega \in B$. The equivalent reduced form of $G_{f,+}$ is

$$G^*_{f,+} = (\Lambda^*_1, \Lambda_2, L^*_{f,+}).$$

The development above, as well as various concepts of admissibility with respect to $G^*_{f,+}$ which will be formulated below, are extended in a straightforward manner to an arbitrary aggregation function ψ, replacing $+$ by ψ.

First, $\mu \in \Lambda_2$ is (*ordinary*) *admissible* with respect to $G^*_{f,+}$ if there is no $\nu \in \Lambda_2$ such that

$$L^*_{f,+}((\underline{a}|\underline{b})_n, B, \nu) \leq L^*_{f,+}((\underline{a}|\underline{b})_n, B, \mu)$$

for all $((\underline{a}|\underline{b})_n, B) \in \Lambda^*_1$, with strict inequality holding for at least one $((\underline{a}|\underline{b})_n, B)$.

For each fixed $(\underline{a}|\underline{b})_n$, a *subgame* of $G^*_{f,+}$ is $G^*_{f,+}(\underline{a}|\underline{b})_n$ where Λ_2 is replaced by

$$\Lambda_2(\underline{a}|\underline{b})_n = \{\mu : \{((a_i|b_i) : i = 1, ..., n) \to \mathbb{R}\}.$$

With respect to $G^*_{f,+}(\underline{a}|\underline{b})_n$, $\mu \in \Lambda_2(\underline{a}|\underline{b})_n$ is admissible if there is no $\nu \in \Lambda_2(\underline{a}|\underline{b})_n$ such that

$$L^*_{f,+}((\underline{a}|\underline{b})_n, B, \nu) \leq L^*_{f,+}((\underline{a}|\underline{b})_n, B, \mu)$$

for all $B \in \pi(\underline{a}|\underline{b})_n$, with strict inequality holding for at least one B. $\mu \in \Lambda^*_2$ is said to be *uniformly admissible* with respect to $G^*_{f,+}$ if it is admissible with respect to $G^*_{f,+}(\underline{a}|\underline{b})_n$ for all $(\underline{a}|\underline{b})_n \in (R|R)^n$. It is clear that uniform admissibility implies ordinary admissibility. It turns out that under mild conditions, uniform admissibility of μ with respect to $G^*_{f,+}$ is equivalent to the existence of a function g such that the restriction of $g\circ\mu$ to R is a finitely additive probability measure. (R is considered as a subset of $R|R$, by identifying $(a|\Omega)$ with a.)

As in Lindley (1982), let $f'(x, j)$, $j = 0, 1$, denote the derivative of $f(x, j)$ with respect to x; the function g above is

$$P_f(x) = \frac{f'(x,0)}{f'(x,0) - f'(x,1)}, \quad x \in [\alpha_0, \alpha_1].$$

The following result was proved in Goodman, Nguyen and Rogers (1990): With respect to the game $G^*_{f,+}$ with score function f such that P_f is increasing, μ is uniformly admissible if and only if the restriction of $P_f\circ\mu$ to R is a finitely additive probability measure. But if f is not a proper score function, that is if $P_f(x) \neq x$ for some x, then it can be shown that no non-atomic probability μ on R can be $G^*_{f,+}$- uniformly admissible, so that we have to consider the concept of admissibility in a wide sense. Specifically, μ is said to be *generally admissible* if there is a game $G^*_{f,+}$ such that μ is uniformly admissible with respect to that game. In this sense, any probability measure is "admissible" by taking the score function f to be a proper score function, that is by taking f such that $P_f(x) = x$, for all x.

Consider Dempster-Shafer belief functions (Shafer, 1976). For simplicity, consider the case where Ω is a finite set (see Section 5.3 for the general case). A belief function Bel on the power set of Ω, denoted as $\mathcal{P}(\Omega)$, can be defined as follows.

Let $m : \mathcal{P}(\Omega) \to [0, 1]$ be such that

$$m(\emptyset) = 0\,, \qquad \sum_{a \in \mathcal{P}(\Omega)} m(a) = 1\,;$$

$$Bel(b) = \sum_{a \leq b} m(a)\,.$$

Note that if $\mu : \mathcal{P}(\Omega) \to [0, 1]$ is such that $\mu(\Omega) = 1$ and for all $a \leq \Omega$,

$$\sum_{b \leq a} (-1)^{|a-b|} \mu(b) \geq 0\,,$$

then μ is a belief function. ($|b|$ denotes the cardinality of the set b)

If we think of "sets" as "points", then m plays the role of a probability mass function, and Bel is the "cumulative distribution function" of some *random set*. See Section 5.3 . Since Ω is finite, we have

$$Bel(a) = P(X \in \mathcal{P}(a)),\ a \leq \Omega\,,$$

where X is a *random set*, defined on some probability space $(E, \mathcal{E}, P)$, and taking values in $\mathcal{P}(\Omega)$ with "density" m, that is

$$P(X = a) = m(a).$$

Note that $Bel(a) + Bel(a') \leq 1$.

We extend Bel from $\mathcal{P}(\Omega)$ to the conditional space $\mathcal{P}(\Omega) | \mathcal{P}(\Omega)$ as follows. For $a, b \in \mathcal{P}(\Omega)$, such that $P(X \leq b) > 0$, define $Bel(a|b) = P(X \leq a | X \leq b)$. By the nature of belief functions, we take $[\alpha_0, \alpha_1] = [0, 1]$.

It is easy to construct Bel such that there exist $a, b \in \mathcal{P}(\Omega)$ with $Bel(a) = Bel(b)$

but $Bel(a') \neq Bel(b')$. It can be shown that such a belief function is inadmissible. This is due essentially to the fact that, for such belief functions, $Bel(a'|b)$ is not a function of $Bel(a|b)$.

Now observe that

$$\begin{aligned} Bel(ac|b) &= P(X \leq ac|X \leq b) \\ &= P(X \leq a, X \leq c|X \leq b) \\ &= P(X \leq a|X \leq b)P(X \leq c|X \leq a, X \leq b) \\ &= Bel(a|b)Bel(c|ab). \end{aligned}$$

Thus, there is a differentiable function $h : [0, 1] \to [0, 1]$ such that for all $a, b \leq \Omega$ with $Bel(b) > 0$, we have

$$Bel(a'|b) = h(Bel(a|b)).$$

Then by Cox's result, Bel is admissible. For example, if $Bel = P^r$ where $r \geq 1$, and P is a probability measure, then Bel is admissible.

As another example of non-additive uncertainty measures which are admissible, we turn to *fuzzy logics*. For background, see Chapter 7. A t-conorm T is said to be *archimedean* if T is continuous and for $x \in (0, 1)$, $x < T(x, x)$. (See, for example, Schweizer and Sklar, 1983). $T(x, y) = min(x + y \ , 1)$ is archimedean, while $T(x, y) = max(x, y)$ is not. T is an archimedean t-conorm if and only if there exists an increasing, continuous function g (called the *additive generator*, or *generator* of T) which maps $[0, 1] \to [0, +\infty]$ with $g(0) = 0$ and such that for $x, y \in [0, 1]$,

$$T(x, y) = g^*(g(x) + g(y)).$$

The *pseudo-inverse* g^* of g is a function $g^* : [0, +\infty] \to [0, 1]$ defined by

$$g^*(x) = \begin{cases} g^{-1}(x) & \text{if } x \in [0, g(1)] \\ 1 & \text{if } x \geq g(1) \end{cases}$$

(See Ling, 1965).

For example, for $p \geq 1$, $T_p(x, y) = [min(x^p + y^p, 1)]^{1/p}$ has generator $g_p(x) = x^p$ and

$$g_p^*(x) = \begin{cases} x^{1/p} & \text{if } x \in [0, 1] \\ 1 & \text{if } x > 1 \end{cases}$$

(noting that $g_p(1) = 1$ here).

Since each *t*-conorm T is associative and commutative, we can extend T to any *n*-tuples, $n \geq 1$, as follows. $T(x_1) = x_1$, by convention, and for $n \geq 2$,

$$T(x_1, x_2, ..., x_n = T(x_1, T(x_2, x_3, ..., x_n)).$$

The representation of an archimedean *t*-conorm T becomes

$$T(x_1, x_2, ..., x_n) = g^*\left(\sum_{i=1}^{n} g(x_i)\right), \quad n \geq 1.$$

For a be a *t*-conorm T, a *T-possibility measure* is a map μ from $\mathscr{P}(\Omega)$ to $[0, 1]$ such that for $a, b \leq \Omega$ with $ab = \emptyset$, $\mu(a \vee b) = T(\mu(a), \mu(b))$. Zadeh's possibility measure corresponds to $T(x, y) = max(x, y)$. The following result is from Goodman, Nguyen and Rogers (1990).

Let Ω be finite and $\mu : \mathscr{P}(\Omega) \to [0, 1]$. Then μ is admissible if and only if μ is T-possibility measure with T being an archimedean t-conorm with generator g such that $g(1) = 1$ and $\sum_{\Omega} g(\mu(\{\omega\})) \leq 1$.

A T-possibility measure with $T(x, y) = max(x, y)$ is not admissible, but it can be approximated by admissible ones. Indeed, if μ is such that $\sum_{\Omega} \mu(\omega) \leq 1$, then for $p \geq 1$, $\nu_p(a) = T_p(\mu(\omega), \omega \in a)$ is admissible since

$$\sum_{\Omega} g_p(\nu_p(\omega)) = \sum_{\Omega} (\mu(\omega))^p \leq 1.$$

On the other hand, for each fixed n,

$$T_p(x_1, x_2, ..., x_n) \to max(x_1, x_2, ..., x_n)$$

as $p \to \infty$, uniformly in $(x_1, x_2, ..., x_n)$.

Thus, if $\mu : \mathscr{P}(\Omega) \to [0, 1]$ is defined by

$$\mu(a) = \max_{\omega \in a} \mu(\omega),$$

for $a \leq \Omega$, then μ is a T-possibility measure with $T(x, y) = max(x, y)$, and

$$\mu(a) = \lim_{p \to +\infty} \nu_p(a) ,$$

uniformly in a.

Finally, note that while the concept of admissibility of uncertainty measures with respect to a game is general, its equivalent form, namely that admissible uncertainty measures have to be transforms of finitely additive probability measures, is valid only in games with *additive* aggregation functions. Specifically, there are non-additive aggregation functions ψ such that with respect to games $G_{f,\psi}$, admissible uncertainty measures need not be transformable to finitely additive probability measures. (For this analysis, see again Goodman, Nguyen and Rogers, 1990).

We turn now to the justification of assigning conditional probabilities to conditional events. From the standard viewpoint of conditional probabilities, not via conditional events, the assignment of $P(a|b) = P(ab)/P(b)$ to the conditional event $(a|b)$ can be justified through a functional equation approach of Aczel (1966, p. 319-324). See also the discussions concerning Cox and Rényi's works presented earlier in this Section.

We present another justification based upon conditional event considerations (Goodman, 1991). Let P be a probability measure on a Boolean ring R. If $A \subseteq R$, then $P(A)$ is the image of A under P, that is

$$P(A) = \{P(a) : a \in A\}.$$

In particular, for $(a|b) \in R|R$, we have $(a|b) \subseteq R$, so that, formally,

$$P(a|b) = \{P(x) : x \in (a|b)\} .$$

But $(a|b) = [ab, b' \vee a]$, a closed interval in R (with the partial order relation $\leq$ on R). Thus,

$$P(a|b) = \{P(x) : ab \leq x \leq b' \vee a\} \subseteq [P(ab), P(b' \vee a)],$$

a closed interval in the unit interval $[0, 1]$. If $P(ab) = P(b' \vee a)$, then $P(b) = 1$ and

$$P(a|b) = \{P(ab)\} = P(ab) = P(ab)/P(b) .$$

If $P(b' \vee a) - P(ab) = 1$, then $P(b) = 0$, and conditional probability $P(a|b)$ is not defined. Thus, consider the case $[s, t] \subseteq [0, 1]$ with $s < t$ and $t - s \neq 1$. Let $h_1 : [0, 1] \to [0, 1]$ be given by

$$h_1(\lambda) = \lambda t + (1 - \lambda)s$$

and for $n \geq 2$,

$$h_n(\lambda) = h_1(h_{n-1}(\lambda)) .$$

Then, for $\lambda \in [0, 1]$, $\lim_{n\to\infty} h_n(\lambda)$ exists and is equal to

$$\lambda_{s,t} = \frac{s}{1 + s - t} .$$

The proof goes as follows. Since $h_1(\cdot)$ is non-decreasing and $\lambda_{s,t}$ is the (unique) fixed point of $h_1(\cdot)$, we have that $\lambda_{s,t} \le \lambda$ implies $\lambda_{s,t} \le h_n(\lambda)$, for $n \ge 1$. But if $\lambda_{s,t} \le h_n(\lambda)$, then $h_{n+1}(\lambda) \le h_n(\lambda)$. Thus, the sequence $(h_n(\lambda), n \ge 1)$ is decreasing and bounded from below by $\lambda_{s,t}$ (and from above by λ). Hence $\lim_{n\to\infty} h_n(\lambda) = h_\infty(\lambda)$ exists. Similarly, if $\lambda \le \lambda_{s,t}$, then $h_n(\lambda) \le \lambda_{s,t}$ for $n \ge 1$, and hence $h_n(\lambda) \le h_{n+1}(\lambda)$. Thus,

$$\lambda \le h_1(\lambda) \le h_2(\lambda) \le \ldots \le \lambda_{s,t},$$

and hence $\lim_{n\to\infty} h_n(\lambda)$ exists.

In any case, for $\lambda \in [0, 1]$, we have

$$h_\infty(\lambda) = \lim_{n\to\infty} h_n(\lambda) = h_1(\lim_{n\to\infty} h_{n-1}(\lambda)) = h_1(h_\infty(\lambda)) .$$

Therefore, $h_\infty(\lambda) = \lambda_{s,t}$ for $\lambda \in [0, 1]$. □

The above procedure of assigning the value $\lambda_{s,t}$ to the sub-interval $[s, t] \subseteq [0, 1]$ can be extended to an arbitrary subset A of $[0, 1]$ by considering $[inf(A), sup(A)]$, that is, if $s = inf(A)$ and $t = sup(A)$, then one assigns to A the value $s/(1 + s - t)$.

Now, back to the case

$$A = \{P(x), ab \le x \le b' \vee a\} \subseteq [0, 1],$$

with

$$inf(A) = P(ab) , \ sup(A) = P(b' \vee a).$$

It is natural to assign to the conditional event $(a|b)$ the value

$$\frac{P(ab)}{1 + P(ab) - P(b' \vee a)}$$

when $P(ab) < P(b' \vee a)$, which is the conditional probability $P(ab)/P(b)$.

5.2 Conditional probability evaluations

Let P be a probability measure on a Boolean ring R. Unlike the traditional approach to conditional probability measures, namely, viewing a quantity like $P(a|b)$ as a probability $P_b(\cdot)$ on R, for each fixed $b \in R$, we are going to extend P globally to the algebra of conditional events $R|R$, so that, if we denote this extension of P as $\tilde{P}$, then

$$\tilde{P} : R|R \longrightarrow [0, 1] .$$

Note that for fixed $b \in R$ with $P(b) > 0$, the probability measure $P_b(\cdot)$ on R defined by

$$P_b(a) = P(ab)/P(b) , \ \forall a \in R,$$

is "equivalent" to the probability measure $\tilde{P}_b(\cdot)$ on the Boolean (quotient) ring $R|Rb'$ where $\tilde{P}_b(a|b) = P_b(a)$. Indeed, first $\tilde{P}_b(\cdot)$ is well-defined on $R|Rb'$; next, with respect to Boolean operations on $R|Rb'$ (that is, coset operations), $\tilde{P}_b(\cdot)$ is a probability measure. Conversely, let $\tilde{P}$ be a probability measure on $R|Rb'$. If we define $P_b(\cdot) : R \to [0, 1]$ by $P_b(a) = \tilde{P}(a|b)$, then obviously $P_b(\cdot)$ is a probability measure. In $\tilde{P}(a|b)$, $(a|b)$ is an argument of the map $\tilde{P}(\cdot)$. Note that, although, the extended value $\tilde{P}(a|b)$ is taken to be $P(a|b) = P(ab)/P(b)$, for $P(b) > 0$, in the usual sense, care should be exercised upon $\tilde{P}(\cdot)$ as an extended map. In particular, with algebraic domain $(R|R, \cdot, \vee, ')$, $\tilde{P}$ is not a probability measure. As we have seen, $R|R$ is not a Boolean ring. Moreover $\tilde{P}$ is not additive. The situation is somewhat different from the axiomatic setting for quantum probability theory (for example, Gudder, 1988) where the domain is a σ-additive class (generalizing the usual concept of a σ-field): there, a form of σ-additivity is reasonable to retain. This is possible because not only the physical reality supports such a mathematical modeling, but because quantum probability measures are not derived from classical probability measures the way $\tilde{P}$ is derived from P.

Obviously, the advantage of viewing $\tilde{P}$ as a global map on $R|R$ is the fact that, when the uncertainty is handled in a more quantitative way, one can combine conditional evidence with different antecedents. From a pure mathematical viewpoint, one can view $R|R$ as an algebraic structure generalizing Boolean rings, say, a Stone algebra which does contain an underlying Boolean ring, and consider maps on $R|R$ such that their restrictions to the underlying Boolean ring are probability measures. However, here we are simply content with examining properties of $\tilde{P}$ for probabilistic inference purposes. First of all, extending the concept of disjointness of events, that is, elements of R, to $R|R$, we say that $(a|b)$ and $(c|d)$ are disjoint, if

$$(a|b) \vee (c|d) = (a|b) + (c|d) .$$

In this case, we have

$$P((a|b) \vee (c|d)) = P((a|b) + (c|d)) = P((a + c|bd)) = P(a|bd) + P(c|bd) .$$

Now

$$(a|b) \vee (c|d) = (a|b) + (c|d)$$

implies

$$ab \vee cd \vee bd = bd,$$

that is $ab \vee cd \leq bd$, so that $ab \leq d, cd \leq b$, hence, $abd = ab, bcd = cd$. Thus

$$P(a|bd) + P(c|bd) = \frac{P(a|b)}{P(a|b)} + \frac{P(c|d)}{P(b|d)} \geq P(a|b) + P(c|d) .$$

Thus, $\tilde{P}$ is not additive on $R|R$. However,

Theorem 1. *$\tilde{P}$ is monotone increasing on $R|R$.*

Proof. Suppose $(a|b) \leq (c|d)$. Then

$$(c|d) = (a|b) \vee (c|d) =$$

$$(ab|b) \vee (cd|d) = (ab \vee cd|ab \vee cd \vee bd) .$$

Since $ab \vee cd \vee bd \leq b \vee cd$, we have

$$P(c|d) = P(ab \vee cd)/P(ab \vee cd \vee bd) \geq P(ab \vee cd)/P(b \vee cd) .$$

Now,

$$ab \vee cd = ab \vee (ab)'cd = ab + (ab)'cd \geq ab + b'cd$$

and

$$b \vee cd = b + b'cd.$$

We have

$$P(ab \vee cd) \geq P(ab) + P(b'cd), P(b \vee cd) = P(b) + P(b'cd) .$$

Putting these together yields

$$P(c|d) \geq \frac{P(ab) + P(b'cd)}{P(b) + P(b'cd)}.$$

It is easy to see that $\frac{P(ab) + t}{P(b) + t}$ is monotone increasing in $t \geq 0$, so that

$$P(c|d) \geq P(ab)/P(b) = P(a|b). \qquad \square$$

Remarks.

(i) Since $ab \leq (a|b) \leq (b \to a)$, we have

$$P(ab) \leq P(a|b) \leq P(b \to a) = P(b' \vee a).$$

(ii) It is easy to check that

$$(ac|b \vee d) \leq (a|b) \cdot (c|d) \leq (ac|bd),$$

and

$$a \vee c|bd) \leq (a|b) \vee (c|d) \leq (a \vee c|bd),$$

so that

$$P(ac|b \vee d) \leq P((a|b) \cdot (c|d)) \leq P(ac|bd),$$

and

$$P(a \vee c|bd) \leq P((a|b) \vee (c|d)) \leq P(a \vee c|bd).$$

For combining conditional evidence, from a quantitative viewpoint, we present an extension of Fréchet's bounds to the conditional case. First, we recall the unconditional case. Let P be a probability measure on R. Then for any $a, b \in R$,

$$P(ab) \leq min\{P(a), P(b)\}.$$

In fact, for $a \leq b$,

$$P(ab) = min\{P(a), P(b)\},$$

so that $min\{P(a), P(b)\}$ is the best possible upper bound for $P(ab)$. Similarly,

$$P(a \vee b) \leq min\{1, P(a) + P(b)\},$$

and equality is achieved when $ab = 0$. Now

$$P(a' \vee b') \leq min\{1, P(a') + P(b') = min\{1, 2 - P(a) - P(b)\},$$

so that

$$P(ab) = 1 - P(a' \vee b') \geq 1 - \min\{1, 2 - P(a) - P(b)\} = \max\{0, P(a) + P(b) - 1\},$$

which is the best possible lower bound for $P(ab)$. By the same token, the best possible lower bound for $P(a \vee b)$ is

$$1 - \min\{P(a'), P(b')\} = \max\{P(a), P(b)\}.$$

More generally, for $n \geq 2$, we have

$$\max\{0, \sum_{i=1}^{n} P(a_i) - (n-1)\} \leq P(\bigwedge_{i=1}^{n} a_i) \leq \min\{P(a_i), i = 1, ..., n\},$$

and

$$\max\{P(a_i), i = 1, ..., n\} \leq P(\bigvee_{i=1}^{n} a_i) \leq \min\{1, \sum_{i=1}^{n} P(a_i)\}.$$

Now, let $\varphi(a_1, ..., a_n)$ be a Boolean function of n variables. Write φ in its normal disjunctive form

$$\varphi(a_1, a_2, ..., a_n) = \bigvee_{i_1=0,1} \cdots \bigvee_{i_n=0,1} \varphi(i_1, i_2, ..., i_n) a_1^{i_1} a_2^{i_2} \cdots a_n^{i_n},$$

with the usual convention $a^0 = a'$, $a^1 = a$. It is easy to see that one can determine two functions

$$U_\varphi, L_\varphi : [0, 1]^n \to [0, 1]$$

such that

$$L_\varphi(P(a_1), ..., P(a_n)) \leq P[\varphi(a_1, ..., a_n)] \leq U_\varphi(P(a_1), ..., P(a_n)),$$

where $L_\varphi = 1 - U_{\varphi'}$, with $\varphi'(a_1, ..., a_n) = [\varphi(a_1, ..., a_n)]'$.

These results were also obtained by Hailperin (1965, 1984) using the technique of linear programming. This latter technique provides a feasible procedure for computing the bounds L_φ and U_φ of $P[\varphi]$, and can be adapted to computational procedures in probability logic (Nilsson, 1986). To find U_φ, let $\alpha_i = P(a_i)$, $i = 1, 2, ..., n$. Then from the normal disjunctive form of $\varphi(a_1, ..., a_n)$, we have

$$P[\varphi(a_1, ..., a_n)] = \sum_{i_1=0}^{1} \cdots \sum_{i_n=0}^{1} \varphi(i_1, ..., i_n)\beta(i_1, ..., i_n),$$

where

$$\beta(i_1, ..., i_n) = P(a_1^{i_1} a_2^{i_2} \dots a_n^{i_n}) .$$

Next, for each $j = 1, ..., n$,

$$a_j = \bigvee_{i_1=0,1} \cdots \bigvee_{i_{j-1}=0,1} \bigvee_{i_{j+1}=0,1} \cdots \bigvee_{i_n=0,1} a_1^{i_1} \dots a_{j-1}^{i_{j-1}} a_j a_{j+1}^{i_{j+1}} \dots a_n^{i_n}$$

so that

$$(*) \qquad \alpha_j = \sum_{i_1=0}^{1} \cdots \sum_{i_{j-1}=0}^{1} \sum_{i_{j+1}=0}^{1} \cdots \sum_{i_n=0}^{1} \beta(i_1, ..., i_{j-1}, 1, i_{j+1}, ..., i_n) ;$$

also, since the $a_1^{i_1} \dots a_n^{i_n}$ form a partition of 1 (the greatest element of R), we have

$$(**) \qquad \sum_{i_1=0}^{1} \cdots \sum_{i_n=0}^{1} \beta(i_1, ..., i_n) = 1 .$$

Thus the least upper bound of $P[\varphi(a_1, ..., a_n)]$ is obtained by maximizing $P[\varphi(a_1, ..., a_n)]$, as a function of the variables $\beta(i_1, ..., i_n)$, subject to the constraints (*), (**), and $\beta(i_1, ..., i_n) \geq 0$ (the α_i's are constants). Note that since the $\varphi(i_1, ..., i_n)$'s are either 0 or 1 (elements of R), the linear constraints (*) and (**) can be put in a matrix form with a "design matrix" whose entries are 0's and 1's.

The linear programming technique above for actually determining lower and upper bounds L_φ, U_φ of $P[\varphi]$ for any boolean expression φ can be adapted to a similar situation in probability logic (Nilsson, 1986). Since Chapter 6 will deal with conditional probability logic, it is relevant here to say a few words about basic aspects of probability logic (see also, Rescher, 1969 and Hailperin, 1984). We follow Nilsson (1986).

Although the collection of sentences of interest forms a Boolean ring R, and hence, one can talk about probability measures P on it in an abstract setting, in practice, only a small set of sentences is to be considered, for example, evidence in an expert system. The problem of probabilistic entailment is this. Suppose we have a set of sentences s_i, $i = 1, ..., n$ with known probabilities $P(s_i)$, $i = 1, ..., n$; compute $P(r)$, for some sentence r of interest, in terms of the $P(s_i)$'s. First, sentences are taken to be "propositions," that is, each sentence is either true or false only. However, the uncertainty emerges since we do not know whether a given sentence is true or false, based on available information. By Stone's Representation Theorem, one can view each sentence s as a subset of a "sample space" or universe Ω on which probability measures are defined. In this setting, a "possible world" is simply an element of Ω. Thus, for each $s \subseteq \Omega$, there are two sets of possble worlds s and s': in s, s is true; and in s', s is false. One can also consider the indicator function of $s : 1_s : \Omega \to \{0, 1\}$; and as in statistical theory, before performing an experiment, it is meaningful to consider the chance that s will be "realized." In expert systems, for example, a sentence (evidence) s is to be considered, and it is desired to know its probability of being true. This is the common interpretation for probabilities of sentences. On the other hand, inference mechanisms in, say, expert systems, require some form of logical deduction to reach decisions. In the presence of uncertainty (about the trueness and falseness of sentences), it is reasonable to invent a multi-valued logic in which the (probabilistic) truth value of a sentence s is taken to be its probability $P(s)$ of being true. This logic is termed probability logic. Its base space remains a Boolean ring as in classical two-valued logic, while its truth evaluations range over the unit interval $[0, 1]$. Its difference with the simplest form of fuzzy logic lies in its non-truth functional calculus (derived from axioms of probability measures) as well as in the interpretation of the meaning of degrees of beliefs.

Consider a finite collection of sentences, that is n subsets $a_1, ..., a_n$ of Ω (or equivalently, n elements of a Boolean ring R). These sets generate a finite partition of Ω, namely $\{a_1^{i_1} \dots a_n^{i_n}\}$ where $i_j \in \{0, 1\}$, $j = 1, .., n$ (with the usual convention $a^0 = a'$, $a^1 = a$, as before).

In logical terms, these sentences generate m ($\leq 2^n$) sets of possible worlds. In each of these sets of possible worlds, one can specify the true/false values of any Boolean expression of the variables a_i (that is, component sentences) using its normal disjunctive form as usual. For example, two sentences a, b generate four sets of possible worlds, namely $ab, ab', a'b, a'b'$. A possible world is a state of nature, or, in the "sample space" setting an element $\omega \in \Omega$. However, unlike statistics, one cannot "perform an experiment" to get the "outcome" ω. Consider three Boolean expressions $f_1(a, b) = a$, $f_2(a, b) = a \to b = a' \vee b$, $f_3(a, b) = b$. A "truth matrix" for these expressions is obtained by specifying their truth values on each of the above sets of possible worlds (in the order written)

$$M = \begin{bmatrix} 1 & 1 & 0 & 0 \\ 1 & 0 & 1 & 1 \\ 1 & 0 & 1 & 0 \end{bmatrix} \begin{matrix} f_1 \\ f_2 \\ f_3 . \end{matrix}$$
$$\quad ab \;\; ab' \;\; a'b \;\; a'b'$$

In terms of normal disjunctive forms,

$$a = ab \vee ab',$$

$$a' \vee b = ab \vee a'b \vee a'b',$$

$$b = ab \vee a'b,$$

so that

$$P(a) = P(ab) + P(a'b),$$

$$P(a' \vee b) = P(ab) + P(a'b) + P(a'b'),$$

$$P(b) = P(ab) + P(a'b) .$$

If we set

$$x_1 = P(ab), \quad x_2 = P(ab'), \quad x_3 = P(a'b), \quad x_4 = P(a'b')$$

and

$$\pi_1 = P(a), \quad \pi_2 = P(a' \vee b), \quad \pi_3 = P(b),$$

then $\pi = MX$, where

$$\pi = \begin{bmatrix} \pi_1 \\ \pi_2 \\ \pi_3 \end{bmatrix}, \quad X = \begin{bmatrix} x_1 \\ x_2 \\ x_3 \\ x_4 \end{bmatrix} .$$

The equation $\pi = MX$ represents a "consistent" condition for the assignments x_i's. To include the condition $x_1 + x_2 + x_3 + x_4 = 1$ (besides $x_i \geq 0$), one usually adds the row $(1, 1, 1, 1)$ to the top of M and modify π to

$$\pi = \begin{bmatrix} 1 \\ \pi_1 \\ \pi_2 \\ \pi_3 \end{bmatrix} .$$

The general probabilistic entailment problem is this. Given a_i and $\pi_i = P(a_i)$, $i = 1, ..., n$ and a sentence of interest b. In view of the above procedure detailed in the example, one first needs to include b into the collection of the a_i's, so that a partition of

Ω is formed by the $a_1^{i_1} \ldots a_n^{i_n} b^k$, $i_j, k \in \{0, 1\}$. Label these sets in some order, say c_j, $j = 1, \ldots, m$ $(\leq 2^{n+1})$, and set $x_j = P(c_j)$. Let M be the $(n + 2)$ by m matrix (with first row consisting of all 1's) representing the true/false values of the a_i's and b in the c_j's, let

$$\pi = \begin{bmatrix} 1 \\ \pi_1 \\ \vdots \\ \pi_n \\ P(b) \end{bmatrix}$$

and let

$$X = \begin{bmatrix} x_1 \\ \vdots \\ x_m \end{bmatrix} .$$

Formally, to solve for $P(b)$, delete the last row of M (corresponding to true/false values $M_{n+2,j}$, $j = 1, \ldots, m$ of b in the c_j's), and $P(b)$ in π. Then solve for X in the equation $MX = \pi$. If a solution X is found, then

$$P(b) = \sum_{j=1}^{m} M_{n+2,j}\, x_j .$$

To obtain bounds for $P(b)$, a similar procedure as in Hailperin's work is used.

In the following, we will first determine best upper and lower bounds for basic connective $\wedge$ and $\vee$ in the conditional case, then proceed to outline a generalization of Nilsson's computational procedure to a conditional setting. Specifically, we are seeking best lower and upper bounds for $P((a|b)\cdot(c|d))$ and $P((a|b) \vee (c|d))$. First, observe that

$$(a|b)\cdot(c|d) \leq (a|b), (c|d).$$

Thus

$$P((a|b)\cdot(c|d)) \leq min(P(a|b), P(c|d)) ,$$

by Theorem 1, and equality is achieved, say, when $(a|b) \leq (c|d)$. Next,

$$(a|b) \vee (c|d) = (ab|b) \vee (cd|d) = (ab \vee cd|ab \vee cd \vee bd) ,$$

so that

$$P((a|b)\vee(c|d)) = \frac{P(ab\vee cd)}{P(ab\vee cd\vee bd)} \leq$$

$$\frac{P(ab\vee cd)}{P(bd)} \leq \frac{P(ab)+P(cd)}{P(bd)} = \frac{P(a|b)}{P(d|b)} + \frac{P(c|d)}{P(b|d)}.$$

Hence

$$P((a|b) \vee (c|d)) \leq min\{1, \frac{P(a|b)}{P(d|b)} + \frac{P(c|d)}{P(b|d)}\}.$$

The fact that this is the best upper bound follows from $abcd = 0$ and $ab \vee cd \leq bd$. Note that, since $\tilde{P}$ is not additive on $R|R$, this upper bound is not a function of $P(a|b)$ and $P(c|d)$ alone. Obviously, when $b = d = 1$, it reduces to the bound in the unconditional case. However, as in the unconditional case, we still have

$$((a|b) \vee (c|d))' = (a|b)' \cdot (c|d)' = (a'|b) \cdot (c'|d),$$

$$((a|b) \cdot (c|d))' = (a'|b) \vee (c'|d)$$

and

$$P(a|b)' = P(a'|b) = 1 - P(a|b),$$

so that the lower bounds for $P((a|b) \cdot (c|d))$ and $P((a|b) \vee (c|d))$ can be obtained from the upper bounds of $P((a'|b) \vee (c'|d))$ and $P((a'|b) \cdot (c'|d))$, respectively as

$$P((a|b) \cdot (c|d)) \geq 1 - min\{1, \frac{P(a'|b)}{P(b|b)} + \frac{P(c'|d)}{P(b|d)}\} = max\{0, s + t - 1\},$$

where

$$s = [P(a|b) + P(d|b) - 1]/P(d|b),$$

$$t = [P(c|d) + P(b|d) - 1]/P(b|d),$$

and

$$P((a|b) \vee (c|d)) \geq 1 - min\{P(a'|b), P(c'|d)\} = \max\{P(a|b), P(c|d)\}.$$

Turning to computational procedures in the conditional case, we first observe that a conditional event $(a|b)$ (with $a \leq b$) generates a partition of 1 consisting of the three sets ab, $a'b$, and b'. (since $a \leq b$ implies $ab' = 0$ and $a'b' = b'$). Also,

$$(*) \qquad P(a|b) = P(ab) + P(a|b)P(b').$$

To put it differently,

$$P(a|b) = [1 \;\; 0 \;\; P(a|b)] \begin{bmatrix} P(ab) \\ P(a'b) \\ P(b') \end{bmatrix}.$$

Thus, consistent with three-valued logic framework, for $(a|b)$, we assign the truth value 1 on the set of possible worlds ab, the value 0 on $a'b$ and the value $P(a|b)$ on b'. See Chapter 7. More generally, consider n conditional events $(a_i|b_i)$, $i = 1, 2, ..., n$. The associated partition of 1 consists of m sets of possible worlds $a_1^{i_1}...a_n^{i_n}b_1^{j_1}...b_n^{j_n}$ with $m < 2^{2n}$. Label these possible worlds as c_j, $j = 1, 2, ..., m$. The "truth values" of each $(a_i|b_i)$ in these c_j are determined as follows.

$$t(a_i|b_j) = \begin{cases} 0 & \text{if } \; c_j \le a_i' b_i \\ 1 & \text{if } \; c_j \le a_i b_i \\ P(a_i|b_i) & \text{if } \; c_j \le b_i' \end{cases}.$$

Thus, if we let the "truth values" matrix $M = [t(a_i|b_j)]$, $\pi_i = P(a_i|b_i)$, $x_j = P(c_j)$,

$$\pi = \begin{bmatrix} \pi_1 \\ \vdots \\ \pi_n \end{bmatrix},$$

and

$$X = \begin{bmatrix} x_1 \\ \vdots \\ x_m \end{bmatrix},$$

then $\pi = MX$.

Now if $(c|d)$ is a conditional event of interest, and it is desired to compute or approximate $P(c|d)$ in terms of the π_i's (conditional beliefs), that is, to see how strong $(c|d)$ is entailed probabilistically by these conditional beliefs, one proceeds exactly as in the unconditional case. Specifically, add $(c|d)$ to the $(a_i|b_i)$'s and consider the collection of sets of possible worlds $a_1^{i_1}...a_n^{i_n}b_1^{j_1}...b_n^{j_n}c^k d^\ell$ (of m elements, $m < 2^{2(n+1)}$). Label these elements as c_j, $j = 1, ..., m$. Add the top row consisting of all 1's to M, and 1 to the top row of π. Solve for X (where $x_j = P(c_j)$) in

$$\pi = MX.$$

Then $P(c|d) = NX$ where N is a $1 \times m$ row vector whose entries are truth values of $(c|d)$ on the c_j's.

5.3 Random conditional objects

I. Conditional random variables

Conceptually, random variables are well-behaved numerical (scalar or vector-valued) functions whose domain is some initial conceptualized space reflecting the problem of interest. Mathematically, one begins with a probability space $(\Omega, \mathcal{A}, P)$ and two random variables say $X : \Omega \to \mathbb{R}^m$, $Y : \Omega \to \mathbb{R}^n$. The relevant joint random variable here is $(X, Y) : \Omega \to \mathbb{R}^{m+n}$ where, for any $\omega \in \Omega$,

$$(X, Y)(\omega) = (X(\omega), Y(\omega)) .$$

Then, X and Y can be considered marginal random variables relative to (X, Y), with all three inducing probability spaces

$$(\mathbb{R}^m, B^m, P\circ X^{-1}), (\mathbb{R}^n, B^n, P\circ Y^{-1}), (\mathbb{R}^{m+n}, B^{m+n}, P\circ(X, Y)^{-1}),$$

where B^k is the real Borel field of subsets over k-dimensional Euclidean space $\mathbb{R}^k$. For any sets $a_1 \in B^m$, $a_2 \in B^n$, and

$$X^{-1} : B^m \to \mathcal{A},$$
$$Y^{-1} : B^n \to \mathcal{A},$$

and

$$(X, Y)^{-1} : B^{m+n} \to \mathcal{A},$$

we have

$$(P\circ(X, Y)^{-1})(a_1 \times a_2) = P(X^{-1}(a_1) \cap Y(a_2)) ,$$
$$(P\circ X^{-1})(a_1) = P(X^{-1}(a_1)) = P\circ(X, Y)^{-1})(a_1 \times \mathbb{R}^n) ,$$

and

$$(P\circ Y^{-1})(a_2) = P(Y^{-1}(a_2)) = (P\circ(X, Y))^{-1}(\mathbb{R}^m \times a_2) .$$

Note, also, using the notation $\underline{b} = (b_1, ..., b_r) \in (B^{m+n})^r$ and obvious notation to indicate arbitrary combinations of basic operators, that any Boolean operator over B^{m+n}

is preserved by $(X, Y)^{-1}$ with corresponding evaluation

$$(P\circ(X, Y)^{-1})(comb(\cap, [, '; \underline{b})) = P(comb(\cap, [, '; (X, Y)^{-1}(\underline{b}))),$$

where

$$(X, Y)^{-1}(\underline{b}) = ((X, Y)^{-1}(b_1), \ldots, (X, Y)^{-1}(b_r)) \in \Omega^r .$$

Consider again $a_2 \in \boldsymbol{B}^n$, with $P(Y^{-1}(a_2)) > 0$. Then one defines the *single conditional random variable*

$$(X|Y^{-1}(a_2)) : Y^{-1}(a_2) \to \mathbb{R}^m$$

by

$$(X|Y^{-1}(a_2))(\omega) = X(\omega), \quad \omega \in Y^{-1}(a_2) .$$

That is, $(X|Y^{-1}(a_2))$ is the restriction of X to $Y^{-1}(a_2)$. In turn, $(X|Y^{-1}(a_2))$ induces the (conditional) probability space $(X(Y^{-1}(a_2)), \mathcal{A}_{X|Y;a_2}, P_{X|Y;a_2})$ on its range, where one assumes

$$X(Y^{-1}(a_2)) = \{X(\omega) : \omega \in Y^{-1}(a_2)\} = \{X(\omega) : Y(\omega) \in a_2\} \in \boldsymbol{B}^m$$

and

$$\mathcal{A}_{X|Y;a_2} = X(Y^{-1}(a_2)) \wedge \boldsymbol{B}^m = \{X(Y^{-1}(a_2)) \cap b : b \in \boldsymbol{B}^m\} .$$

Now, for any $a_1 \in \boldsymbol{B}^m$, and hence for any $b = X(Y^{-1}(a_2)) \cap a_1 \in \mathcal{A}_{X|y;a_2}$,

$$(X|Y^{-1}(a_2))^{-1}(b) = X^{-1}(b) \cap Y^{-1}(a_2)$$

$$= X^{-1}(a_1) \cap Y^{-1}(a_2) = (X, Y)^{-1}(a_1 \times a_2) .$$

Hence,

$$P_{X|Y;a_2}(b) = P\circ(X|Y^{-1}(a_2))^{-1}(b)/P(Y^{-1}(a_2))$$

$$= P(X^{-1}(a_1) \cap Y^{-1}(a_2))/P(Y^{-1}(a_2))$$

$$= P\circ(X, Y)^{-1}(a_1 \times a_2)/P(Y^{-1}(a_2))$$

$$= P(X^{-1}(a_1)|Y^{-1}(a_2)) . \tag{1}$$

Our approach here of viewing a single-conditional random variable $(X|Y^{-1}(a))$ as the restriction of X to the fixed event $Y^{-1}(a)$ is the same as Rényi's (1970, p. 72) random variable (defined on a conditional probability space) with respect to the condition $Y^{-1}(a)$. The *double conditional random variable* is given by $(X|Y^{-1}(\cdot)) : \Omega_{X,Y} \to \mathbb{R}^m$, where

$$\Omega_{X,Y} = \bigcup_{\{a_2 : a_2 \in B^n, P(Y^{-1}(a_2))>0\}} (Y^{-1}(a_2) \times \{a_2\}) ,$$

and for any $a_2 \in B^n$ with $P(Y^{-1}(a_2)) > 0;\ \omega \in Y^{-1}(a_2)$,

$$(X|Y^{-1}(\cdot))(\omega, \{a_2\}) = (X|Y^{-1}(a_2))(\omega).$$

Hence, for any $a_2 \in B^n$ with $P(Y^{-1}(a_2)) > 0$ and for any $a_1 \in B^m$,

$$(X|Y^{-1}(\cdot))^{-1}(a_1, \{a_2\}) = (X|Y^{-1}(a_2))^{-1}(a_1) . \tag{2}$$

The extension of single and double conditional random variables to include sets a_2 of probability measure zero in B^n can be accomplished through use of the Radon-Nikodym Theorem. Given all of the above standard development of conditional random variables, it is natural to inquire whether a direct connection can be established between these entities and an appropriately constructed random mechanism over the class of all conditional events. For any $a \in B^m$, $b \in B^n$, define the conditional event

$$(a|b) = (a \times b|\mathbb{R}^m \times b),$$

and define the Krönecker form $\delta : \Omega^2 \to \Omega\ [\{\emptyset\}$

$$\delta(\omega_1, \omega_2) = \begin{cases} \emptyset & \text{if } \omega_1 \neq \omega_2 \\ \omega_1 & \text{if } \omega_1 = \omega_2 . \end{cases}$$

Similarly, for any $\omega_1, \omega_2 \in \Omega$ and $s \in \mathbb{R}^m$, $t \in \mathbb{R}^n$,

$$(\omega_1|\omega_2) = (\{(\omega_1, \omega_2)\}|\Omega \times \{\omega_2\}) ,$$

$$(s|t) = (\{(s, t)\}|\mathbb{R}^m \times \{t\}) ,$$

with

$$(\mathbb{R}^m|\mathbb{R}^n) = \{(s|t) : s \in \mathbb{R}^m, t \in \mathbb{R}^n\} ,$$

$$(B^m|B^n) = \{(a|b) : a \in B^m, b \in B^n\} ,$$

$$(\Omega|\Omega) = \{(\omega_1|\omega_2) : \omega_1, \omega_2 \in \Omega\} ,$$

$$(\mathcal{B}|\mathcal{B}) = \{(c|d) : c, d \in \mathcal{B}\} .$$

Use also the convention for equal exponents

$$(B^m \times B^n|B^m \times B^n) = \{(a|b) : a, b \in B^m \times B^n\} .$$

Then define the *random conditional event mapping* $(X|Y) : (\Omega|\Omega) \to (\mathbb{R}^m|\mathbb{R}^n)$ by

$$(X|Y)((\omega_1|\omega_2)) = ((X, Y)(\delta(\omega_1, \omega_2))|\mathbb{R}^m \times Y(\omega_2)) .$$

It follows, by a slight abuse of notation, omitting the $\{\emptyset\}$ term, that, extending via functional images, $(X|Y)$ to $(X|Y) : (\mathcal{B}|\mathcal{B}) \to (B^m|B^n)$, one obtains for any $c, d \in \mathcal{B}$,

$$(X|Y)((c|d)) = (\{(X, Y)(\delta(\omega_1, \omega_2)) : \omega_1 \in c, \omega_2 \in d\}|\{(\mathbb{R}^m \times Y(\omega_2)) : \omega_2 \in d\})$$
$$= ((X, Y)(c \cap d)|\mathbb{R}^m \times Y(d)) ,$$

where

$$(X, Y)(c \cap d) = \{(X(\omega), Y(\omega)) : \omega \in c \cap d\} ,$$

and

$$Y(d) = \{Y(\omega) : Y(\omega) : \omega \in d\} .$$

Next, consider the inverse mapping for $(X|Y)$, at any $(a_1|a_2) \in (B^m|B^n)$.

$$(X|Y)^{-1}(a_1|a_2) = \{ (c|d) : c, d \in \mathcal{B} \text{ and } (X|Y)(c|d) = (X, Y)((X, Y)^{-1}(a_1|a_2)) \}$$
$$= \{(c|d) : c, d \in \mathcal{B} \text{ and } cd = (X, Y)^{-1}(a_1, a_2) , d = Y^{-1}(a_2)\}$$
$$= (\{c : c \in \mathcal{B} \text{ and } c \cdot Y^{-1}(a_2) = X^{-1}(a_1) \cap Y^{-1}(a_2)\}|Y^{-1}(a_2))$$
$$= ((X^{-1}(a_1)|Y^{-1}(a_2))|Y^{-1}(a_2))$$
$$= (X^{-1}(a_1)|Y^{-1}(a_2)) . \tag{3}$$

Comparing (1)-(3) shows that

$$P((X|Y)^{-1}(a_1|a_2)) = P(X^{-1}(a_1)|Y^{-1}(a_2))$$

$$= (P\circ(X|Y^{-1}(a_2))^{-1})(a_1)$$

$$= P\circ(X|Y^{-1}(\cdot))^{-1}(a_1, \{a_2\}) , \qquad (4),$$

so that in a natural sense $(X|Y)$ and $(X|Y)^{-1}$ can be considered the equivalent conditional event random mechanism corresponding to the double conditional random variable $(X|Y^{-1}(\cdot))$. Moreover, using (3) and the operation preserving property of X^{-1} and Y^{-1}, it readily follows that if $f : (B^m|B^n)^r \to (B^m|B^n)$ is any r-ary extended Boolean function, then $(X|Y)^{-1}$ preserves f, that is,

$$(X|Y)^{-1}\circ f = f\circ(X|Y)^{-1} , \qquad (5)$$

analogous to the preserving properties of $X, Y, (X, Y)$ relative to unconditional operators.

In summary, analogous to how the double conditional random variable

$$(X|Y^{-1}(\cdot)) : Y^{-1}(\cdot) \to \mathbb{R}^m$$

and its operator inverse $(X|Y^{-1}(\cdot))^{-1} : B^m \to \mathcal{A}$ determine from the probability space $(\Omega, \mathcal{A}, P)$ the induced probability spaces

$$(X(Y^{-1}(\cdot)), \mathcal{A}_{X|Y;}, P_{X,Y;}),$$

one gets that the random conditional event mapping

$$(X|Y) : (\Omega|\Omega) \to (\mathbb{R}^m|\mathbb{R}^n) \to (\mathcal{A}|\mathcal{A})$$

determines from the probability space $(\Omega, \mathcal{A}, P)$ the induced "conditional probability" space

$$((\mathbb{R}^{m+n}|\mathbb{R}^{m+n}), (B^m \times B^n|B^m \times B^n), \tilde{P}\circ(X|Y)^{-1}),$$

where $\tilde{P}$ is the conditional probability extension of P. That is, $\tilde{P} : (\mathcal{A}|\mathcal{A}) \to [0, 1]$ is defined for any $c, d \in \mathcal{A}$, and hence $(c|d) \in \mathcal{A}$ by

$$\tilde{P}((c|d)) = P(c|d) = P(c \cap d)/P(d) ,$$

provided $P(d) > 0$. The chief relations between, and evaluations for, double conditional random variables and random conditional events are given in equations (3) and (4).

II. Random sets

It now becomes clear in the literature of uncertainty in AI that the mathematical concept of a random set is the cornerstone for evidential reasoning (for example, Hestir et al, 1990). For background on random sets see for example Matheron (1975), Goodman and Nguyen (1985). Below, we will illustrate, through an example, the use of (measure-free) conditional events and their algebraic structure in a problem of combining conditional evidence. For more details, see Nguyen and Rogers (1990).

In a given problem, we make the basic assumption that our knowledge at each state is expressed by a probability measure. When new evidence is obtained, this is to be "updated" by some "combining." We interpret a piece of evidence as a realized event supplied by some "test" which might be merely the opinion of an expert. This lack of precision in evidence suggests looking at a less precise formulation of randomness, namely random sets. Roughly speaking, a random set S is a measurable function from $(\Omega, \mathcal{A})$ to the power set of some set Θ, equipped with an appropriate σ-field.

For ease of reference, we present below basic aspects of random sets in the context of the Dempster-Shafer theory of belief functions. For more details see Hestir, Nguyen and Rogers (1990). A random set S on a space Θ is described as follows.

Let $\mathscr{C}$ be a subset of the power set $\mathscr{P}(\Theta)$, $\sigma(\mathscr{C})$ a σ-field on $\mathscr{C}$, and $(\Omega, \mathcal{A}, P)$ a probability space. A random set with values in $\mathscr{C}$ is a map S from Ω to $\mathscr{C}$ which is $\mathcal{A}$-$\sigma(\mathscr{C})$-measurable. Briefly, a random set S on Θ is a triple $(\mathscr{C}, \sigma(\mathscr{C}), Q)$, where $Q = PS^{-1}$.

For a given Θ, there are two general ways to specify the objects making up a random set.

(i) If $\mathscr{C} = \mathscr{P}(\Theta)$, then $\sigma(\mathscr{C})$ is constructed as follows. Let $\mathscr{I}$ be the collection of all finite subsets of Θ. For $i, j \in \mathscr{I}$, $[i, j] = \{x \in \mathscr{P}(\Theta) : i \leq x \leq j\}$, where $\leq$ denotes set inclusion. Let $\mathscr{M} = \{ [i, j'] : i, j \in \mathscr{I}\}$. Then $\sigma(\mathscr{C})$ is taken to be the σ-field generated by $\mathscr{M}$, denoted as $\sigma(\mathscr{M})$. Each probability measure Q on $\sigma(\mathscr{M})$ determines a random set with values in $\mathscr{P}(\Theta)$.

(ii) If Θ is a topological space, then the topological structure of Θ can be taken into account. For example, consider the case $\Theta = \mathbb{R}$, the real line, or more generally, Θ a locally compact space. Let $\mathscr{F}$, $\mathscr{K}$, $\mathscr{G}$ be the classes of closed, compact, and open subsets of $\mathbb{R}$, respectively. If $\mathscr{C} = \mathscr{F}$, that is, if we are concerned with *closed random sets*, then $\mathscr{F}$ can be given a topology τ using the open subbase

$$\{F \in \mathscr{F} : F \cap K = \emptyset \text{ for } K \in \mathscr{K}\}$$

and

$$\{F \in \mathscr{F} : F \cap G \neq \emptyset, \text{ for } G \in \mathscr{G}\}.$$

Then $\sigma(\mathscr{C}) = \sigma(\tau)$ is the Borel σ-field on $\mathscr{F}$ in this topology. Each probability measure on $(\mathcal{F}, \sigma(\tau))$ determines a closed random set.

As in the case of random vectors, one can associate with each random set S a *generalized distribution function* (GDF) characterizing S. In case (i), given by the two spaces $(\Omega, \mathcal{A}, P)$ and $(\mathcal{P}(\Theta), \sigma(\mathcal{M}), Q)$, and the map $S: \Omega \to \mathcal{P}(\Theta)$, let $\mathcal{J}' = \{ j' : j \in \mathcal{J}\}$. Define

$$F_S : \mathcal{J}' \to [0, 1]$$

by

$$F_S(j') = P(S \leq j') = Q[\emptyset, j'] .$$

It can be shown that

$$F_S(\Theta) = F_S(\emptyset') = Q([\emptyset, \emptyset']) = Q(P(\Theta)) = 1,$$

and for $i, j \in \mathcal{J}$

$$Q([i, j']) = \sum_{\alpha=0}^{|i|} \sum_{t \in i_\alpha} (-1)^\alpha F_S((i \vee t)'),$$

where $|i|$ denotes the cardinality of i, and

$$i_\alpha = \{t : t \leq i, |t| = \alpha\} .$$

Thus, in this case, a function $F : \mathcal{J}' \to [0, 1]$ uniquely determines a GDF if and only if $F(\Theta) = 1$ and for all $i, j \in \mathcal{J}$

$$\sum_{\alpha=0}^{|i|} (-1)^\alpha \sum_{t \in i_\alpha} F((i \vee t)') \geq 0 .$$

For example, if Θ is finite, then $\mathcal{J} = \mathcal{J}' = \mathcal{P}(\Theta) \to [0, 1]$. Let

$$f(a) = \sum_{b \leq a} (-1)^{|a-b|} F(b) \geq 0.$$

By the Möbius inversion formula, we get

$$F(a) = \sum_{b \leq a} f(b),$$

so that f is a "density function" on $\mathcal{P}(\Theta)$. Define Q on $\mathcal{P}\mathcal{P}(\Theta)$ by

$$Q(\{a_1, \ldots, a_n\}) = \sum_{i=1}^{n} f(a_i) .$$

The probability space $(\mathcal{P}(\Theta), \mathcal{P}\mathcal{P}(\Theta), Q)$ specifies a random set S on Θ.

In case (ii), given by the spaces $(\Omega, \mathcal{A}, P)$ and $(\mathcal{F}, \sigma(\tau), Q)$, and the map $S : \Omega \to \mathcal{F}$, the domain of a GDF F will be $\mathcal{K}' = \{K' : K \in \mathcal{K}\}$. Let $T: \mathcal{K} \to [0, 1]$, and $T(K) = 1 - F(K')$. As an application of Choquet's theorem (Matheron, 1975, p. 30-35), a function F on K' uniquely determines a GDF if and only if

(1) $T(\emptyset) = 0$,

(2) if the sequence K_n in $\mathcal{K}$ decreases to K in $\mathcal{K}$, then $T(K_n) \to T(K)$,

(3) for all $n \geq 1$, all $K, K_1, \ldots, K_n$ in $\mathcal{K}$, the following functions are non-negative:

$$\varphi_1(K; K_1) = T(K \vee K_1) - T(K),$$

$$\varphi_2(K; K_1, K_2) = \varphi_1(K; K_1) - \varphi_1(K \vee K_2; K_1) ,$$

$$\ldots$$

$$\varphi_n(K; K_1, \ldots, K_n) = \varphi_{n-1}(K; K_1, \ldots, K_{n-1}) - \varphi_{n-1}(K \vee K_n; K_1, \ldots, K_{n-1}) .$$

Such an F uniquely determines a probability measure Q on $(\mathcal{F}, \sigma(\tau))$ such that for all $K \in \mathcal{K}$, $F(K') = Q([\emptyset, K'])$.

When Θ is finite, a belief function Bel on Θ is a map $Bel : \mathcal{P}(\Theta) \to [0, 1]$ defined by

$$Bel(a) = \sum_{b \leq a} m(a),$$

where the basic probability assignment function m satisfies: $m(\emptyset) = 0$ and

$$\sum_{b \in \mathcal{P}(\Theta)} m(b) = 1.$$

Thus, a *belief function* is nothing more than a GDF of a random set S such that $P(S = \emptyset) = 0$. (See Shafer, 1990.) Belief functions on finite sets can be characterized by various set functions. Indeed, let S be a non-empty random set on a finite set Θ. If

$$Q_S(a) = P(a \leq S) ,$$

then

$$Q_S(a) = \sum_{a \le b} m_S(b) .$$

Also, by the Möbius inversion formulae (See Rota, 1964, or Aigner, 1979.),

$$m_S(a) = \sum_{b \le a} (-1)^{|a-b|} Bel_S(b) = \sum_{a \le b} (-1)^{|b-a|} Q_S(b) .$$

Finally,

$$PL_S(a) = P(S \cap a \neq \emptyset) = 1 - P(S \cap a = \emptyset)$$
$$= 1 - P(S \le a') = 1 - Bel_S(a').$$

It is interesting to note that the *commonality* Q_S can be viewed as the Fourier transform of m_S (see Thoma, 1989, 1991). In other words, Q_S is the "characteristic function" of S. Of course, the harmonic analysis involved is over a semi-group structure.

The interpretation of belief functions in terms of random sets allows the rigorous formulation of the problem of combining evidence, where each piece of evidence is assumed to be represented by a belief function. Specifically, using the concept of conditional events, two (non-empty) random sets S_1 and S_2 can be combined into one non-empty random set $(S_1 \cap S_2 | S_1 \cap S_2 \neq \emptyset)$.

In the following, the range of S will be simply a finite Boolean ring R or a finite subset of an arbitrary Boolean ring R. In this case, S is completely characterized by its generalized distribution function (GDF) F_S, called in the literature a belief function (Shafer, 1976). $F_S(a) = P(S \le a)$, where P is a probability on $(\Omega, \mathcal{A})$.

A typical situation in the problem of updating of knowledge is the following. The measure P_0 on the range of the variable of interest is postulated, but only partial information about P_0 is available. This is the Bayesian case of incomplete prior information. Specifically, consider the case where P_o is unknown, but we are given (say, by an expert) that $a, b, c \in R$, a and c are P_0-independent given c, that $P_0(a|b) = \alpha$ and $P_0(c|b) = \beta$. The question is: what can be said about the values $P_0(r|b)$ for the other $r \in R$? Consider the (Boolean) quotient ring $R|Rb'$. We extract the prior information as follows.

Let X, Y be random sets with values in the power set of R/Rb', with ranges $\{(a|b), (a'|b)\}$, $\{(c|b), (c'|b)\}$, respectively. Also, (note that P is on $(\Omega, \mathcal{A})$),

$$P(X = (a|b)) = \alpha = 1 - P(X = (a'|b)),$$

and

$$P(Y = (c|b)) = \beta = 1 - P(Y = (c'|b)) .$$

We combine X and Y through the random set Z with range

$$\{(ac|b), (ac'|b), (a'c|b), (a'c'|b)\},$$

with probabilities (in view of conditional independence assumption)

$$\alpha\beta, \alpha(1 - \beta), (1 - \alpha)\beta, (1 - \alpha)(1 - \beta),$$

respectively. See Section 5.4 for the concept of qualitative conditional independence. The GDF of Z is the map $F_Z : R/Rb' \longrightarrow [0, 1]$ defined by

$$F_Z(r|b) = \sum_{(a_ic_i|b)\leq(r|b)} P_Z(a_ic_i|b),$$

where $P_Z = PZ^{-1}$ as usual, and where a_i is a or a', and c_i is c or c'. In terms of the order relation $\leq$ among conditional events, $(a_ic_i|b) \leq (r|b)$ if and only if $a_ic_ib \leq r$. For all $(r|b) \in R|Rb'$,

$$F_Z(r|b) \leq P_0(R|b) .$$

Replacing $(r|b)$ by $(r|b)' = (r'|b)$ in this inequality, yields

$$P_0(r|b) \leq 1 - F_Z(R'|b) .$$

Based on the available evidence, for $r \in R$, an interval approximation for $P_0(r|b)$ is $[F_Z(r|b), 1- F_Z(r'|b)]$.

5.4 Qualitative conditional independence

Qualitative independence (or Q-independence for short), or measure-free or algebraic independence of ordinary events is a well-known concept (for example, Rényi, 1970). It is not simply for academic interest that the above concept should be extended to conditional events. In fact, our motivation for considering Q-independence comes from the problem of fast computations in inference networks of expert systems. For example, in some models of medical diagnosis, the variables of interest are represented as nodes in a graph, the causal relationships among these variables are represented by (directed) edges of the graph, and the strengths of such relationships are usually quantified by an uncertainty measure (Bayesian probability, Dempster-Shafer belief functions, or Zadeh possibility measure). In AI activities, there is no general agreement on the choice of such uncertainty measures (see for example, Henkind and Harrison, 1988). Thus the design of inference networks (or influence diagrams) should be done without reference to the uncertainty measure used. Not only some sort of "independence" assumption generally simplifies the calculations within the knowledge representation, but by the very nature of many application domains, neighboring interactions among variables exhibit some form of conditional independence. This is typically the case of Markov random fields (for example, Lauritzen and Spiegelhalter, 1988).

First, we give a brief historical background. The concept of Q-independence of events was treated in some detail in Rényi (1970). With our notation concerning a Boolean ring R, two elements a and b of R are said to be Q-independent if and only if $ab, ab', a'b, a'b'$ are all not 0 (implying also that a, a', b, b' are all not 0). One possible interpretation is clear: viewing elements of R as events, if, for example, $a'b = 0$, then $a \leq b$ so that when b is "realized," a is also realized, it follows that a and b cannot be "independent." It is easy to check that P-independence implies Q-independence: $P(ab) = P(a)P(b) > 0$ implies $ab \neq 0$, and the rest follow by the use of complements. This concept of Q-independence of non-zero a and b can be reformulated as follows.

Let $\pi(a) = \{a, a'\}$, $\pi(b) = \{b, b'\}$ be partitions of 1. Then a and b are Q-independent if and only if for all $\alpha \in \pi(a)$, and all $\beta \in \pi(b)$, one has $\alpha\beta \neq 0$. If we fix a and b through their indicator functions 1_a and 1_b, then the following equivalent definition can be used to extend the concept of Q-independence to variables. The σ-field generated by 1_a is $\sigma(1_a) = \{0, 1, a, a'\}$; similarly, $\sigma(a_b) = \{0, 1, b, b'\}$. Then a and b are Q-independent if and only if for all $\alpha \in \sigma(1_a)\backslash\{0\}$, and all $\beta \in \sigma(1_b)\backslash\{0\}$, one has $\alpha\beta \neq 0$.

Next, to be concrete, let R be a σ-field of subsets of some set Ω. Let X and Y be measurable functions, defined on (Ω, R), with countable ranges in the real line $\mathbb{R}$. Let the countable partitions (with no empty subsets) generated by X and Y be correspondingly,

$$\pi(X) = \{a_n\}\ ,\ \pi(Y) = \{b_m\}\ .$$

Then X and Y are said to be Q-independent if and only if $\pi(X)$ and $\pi(Y)$ are Q-independent in the sense that for all $\alpha \in \pi(X)$, $\beta \in \pi(Y)$, one has $\alpha\beta \neq \emptyset$. Note that the σ-field generated by X is

$$\sigma(X) = \{\bigcup_{i \in I} a_i : I \subseteq X(\Omega)\},$$

where $a_i = X^{-1}(i)$ for $i \in X(\Omega)$, and I could be $\emptyset$. From this, X and Y are Q-independent if and only if for all $\alpha \in \sigma(X)\backslash\{\emptyset\}$, $\beta \in \sigma(Y)\backslash\{\emptyset\}$, one has $\alpha\beta \neq \emptyset$.

Recently, Shafer et al (1987), also motivated by the study of inference networks in expert systems, defined Q-conditional independence for finite partitions, or equivalently, for variables with finite ranges. They did not consider the concept of Q-conditional independence for events. As far as we know, Q-independence of "continuous" variables has not been discussed in the literature; also, Q-conditional independence was not addressed in Rényi (1970). Below, we follow the recent work of Nguyen and Rogers (1990) to present a comprehensive discussion of all the notions mentioned above.

Back to the abstract setting, let $R(\vee, \wedge, ', 0, 1)$ be a Boolean ring.

Definition 1.

(*i*) *Let A and B be two subsets of R consisting of non-zero elements. Then*

$$A \underset{R}{\perp} B \quad \textit{if and only if for } a \in A \textit{ and } b \in B, \textit{we have } ab \neq 0.$$

(*ii*) *Let $a, b \in R$. Then $a \underset{R}{\perp} b$ if and only if*

$$\pi(a) = \{a, a'\} \underset{R}{\perp} \{b, b'\} = \pi(b).$$

(*iii*) *Let X and Y be discrete variables. Then $X \underset{R}{\perp} Y$ if and only if $\pi(X) \underset{R}{\perp} \pi(Y)$ if and only if for $a \in \sigma(X)\backslash\{0\}$* and *$b \in \sigma(Y)\backslash\{0\}$, we have $ab \neq 0$.*

For the concept of Q-conditional independence of events, we observe that P (probabilistic)-conditional independence of a and b given c is expressed by

$$P(ab|c) = P(a|c)P(b|c),$$

which can be rewritten as

$$\tilde{P}((ab|c)) = \tilde{P}((a|c))\tilde{P}((b|c)),$$

where we use $\tilde{P}$ as a function on $R|R$ with arguments $(ab|c)$, $(a|c)$, $(b|c)$, ..., viewed as conditional events. This suggests that one could define $(a|b)$ and $(c|d)$ to be independent with respect to $\tilde{P}$ if and only if

$$\tilde{P}((a|b)\cdot(c|d)) = \tilde{P}((a|b))\tilde{P}((c|d)) .$$

Definition 2.

(*i*) *Let* $a, b, c \in R$. *Then* $a \underset{P}{\perp} b|c$ *if and only if*

$$P(ab|c) = P(a|c)P(b|c)$$

if and only if

$$\tilde{P}((a|c)(b|c)) = \tilde{P}(a|c)\tilde{P}(b|c) .$$

(*ii*) *Let* $a, b, c \in R$. *Then* a *and* b *are* Q*-independent given* c, *in symbols,* $a \underset{Q}{\perp} b|c$, *if and only if* $(a|c) \underset{R/Rc'}{\perp} (b|c)$, *if and only if for* $(\alpha|c) \in \pi(a|c) = \{(a|c), (a'|c)\}$, *and for* $(\beta|c) \in \pi(b|c) = \{(b|c), (b'|c)\}$, *we have*

$$(\alpha|c)\cdot(\beta|c) \neq (0|c) .$$

Remark. $(\alpha|c)\cdot(\beta|c) \neq (0|c)$ is equivalent to $(\alpha\beta|c) \neq (0|c)$ or to $\alpha\beta c \neq 0$. Also, in considering $\pi(a|c)$, it is implicitly assumed that $(a|c)$ and $(a'|c)$ are not $(0|c)$ which is equivalent to $ac \neq 0$ and $a'c \neq 0$. Moreover, it can be checked that the Q-conditional independence in Definition 2 (ii) is strictly weaker than that for finite partitions in Shafer et al (1987). Indeed, in our notation, their definition is expressed as follows.

Let X, Y and Z be discrete variables. Let

$$A(c, \pi(X)) = \{(a|c) : a \in \pi(X), ac \neq 0)\}.$$

Then, $X \underset{Q}{\perp} Y|Z$ if and only if $\pi(X) \underset{Q}{\perp} \pi(Y)|\pi(Z)$ if and only if for $c \in \pi(Z)$,

$$A(c, \pi(X)) \underset{R/Rc'}{\perp} A(c, \pi(Y)) .$$

We see immediately that if $1_a \underset{Q}{\perp} 1_b|1_c$ according to this last definition, then $a \underset{Q}{\perp} b|c$ according to Definition 2 (ii), but that the converse does not hold. Thus, unlike the unconditional case, Q-conditional independence of events cannot be defined in terms of variables. In order to define a Q-independence which will be compatible with stochastic independence for "continuous" variables, it is necessary to pay attention to "small sets." In probability theory, these are P-null sets which form a σ-ideal of subsets. This structure is abstracted to σ-ideals (for example, Halmos, 1963), a notion dual to that of the "bunch" in Rényi (1970). See Section 5.1 .

In the discrete case, one needs to consider only the trivial σ-ideal $\{0\}$. If P is a probability measure on R, then $\mathcal{M}_P = \{a \in R : P(a) = 0\}$ is clearly a σ-ideal. Now $X \underset{P}{\perp} Y$ if and only if $\sigma(X) \underset{P}{\perp} \sigma(Y)$ in the sense that for $a \in \sigma(X)$ and $b \in \sigma(Y)$, $P(ab) = P(a)P(b)$. If $a \in \sigma(X) \backslash \mathcal{M}_P$ and $b \in \sigma(Y) \backslash \mathcal{M}_P$, then $P(ab) > 0$, which implies that $ab \notin \mathcal{M}_P$ so that $ab \neq 0$. If we were to require only that $a \in \sigma(X) \backslash \{0\}$, it could happen that either $P(a) = 0$ or $ab = 0$. Thus we are led to

Definition 3. *Let X, Y, Z be real-valued measurable functions (defined, say, on (Ω, R)).*

(*i*) $X \underset{Q}{\perp} Y$ *if and only if there is a σ-ideal $\mathcal{M}$ such that* for $a \in \sigma(X) \backslash \mathcal{M}$ and $b \in \sigma(Y) \backslash \mathcal{M}$, *we have* $ab \neq 0$.

(*ii*) $X \underset{Q}{\perp} Y | Z$ *if there is a σ-ideal $\mathcal{M}$ such that* for $a \in \sigma(X) \backslash \mathcal{M}$, $b \in \sigma(Y) \backslash \mathcal{M}$, and $c \in \sigma(Z) \backslash \mathcal{M}$, *we have* $a \underset{Q}{\perp} b | c$.

CHAPTER 6

CONDITIONAL PROBABILITY LOGIC

Unlike Adams' approach to a logic of conditionals (Adams, 1975), we will take advantage of the rich algebraic structure of the space of conditional events $R|R$ to develop a conditional probability logic (CPL). The concrete syntactic component of this logic is especially useful for the purpose of automation. The problem of modeling defaults and production rules in expert systems using measure-free conditionals as well as aspects of non-monotonic deduction will be discussed in Chapter 8.

6.1 Essentials of probability logic

In a sense, logic is about the study of knowledge representation languages in which the basic notion of entailment (for inference) can be captured. We are concerned here with the situation in which the uncertainty in our knowledge is taken in a quantitative way. See for example, Bibel (1986) for general methods of automated reasoning.

However, because of the relevancy to the treatment of conditional events, we address only the probability logic approach to managing quantitative uncertainty in expert systems. See, for example, Bibel (1986) and Pearl (1988) for both Bayesian and non-Bayesian formalisms. The so-called *probabilistic logic* (Nilsson, 1986) in AI has been discussed in Chapter 5, together with an extension to the conditional case. In this chapter, we are concerned with probability logic and its extension to conditional probability logic from the viewpoint of mathematical logic. Since the CPL developed in this chapter is a direct extension of probability logic (PL), we will first review the basics of the latter. We start with a review of classical two-valued logic (C_2).

In C_2, the base space is a Boolean ring R (representing propositions) with its usual operators and relations. Taking the concept of truth as the (only) primitive notion, one proceeds to derive the concept of logical entailment. Each element of R is either true (T or 1) or false (F or 0), that is the truth-space of R is $\{0, 1\}$. To emphasize the fact that elements of R are true or false on different "possible worlds" one introduces the concept of *models*. Roughly speaking, a model (or semantic valuation) of R is an assignment of truth values to elements of R. However, such an assignment should be logical (or consistent), that is, it should be such that no element of R could be simultaneously true and false in the same assignment. Further two elements a, b are both true if and only if their conjunction ab is true. The mathematical translation of the concept of consistent assignments is that of a Boolean homomorphism. The truth-space $\{0, 1\}$ is viewed as a 2-element Boolean algebra. That is, for $x, y \in \{0, 1\}$,

$$xy = min\{x, y\} ,$$

$$x \vee y = max\{x, y\} ,$$

$$0' = 1, \ 1' = 0 .$$

We use the same notation $'$, $\wedge$ (or $\cdot$), and $\vee$ on both the spaces R and $\{0, 1\}$. A map $h : R \to \{0, 1\}$ is a (Boolean) homomorphism if for $a, b \in R$

$$h(a') = [h(a)]',$$

$$h(ab) = h(a)h(b),$$

and

$$h(a \vee b) = h(a) \vee h(b).$$

The first condition may be replaced by $h(a) \neq h(a')$.

A *model* is defined to be a homomorphism $R \to \{0, 1\}$, and we denote the set of all models of R by H. Thus *an element* $a \in R$ *is true in the model* $h \in H$ *if and only if* $h(a) = 1$.

For further syntactic development, and for concreteness, we look at an alternative way of formalizing the concept of models. For elementary background on ideals and filters, as well as some algebraic logic, see for example, Mendelson, (1970), or Halmos, (1962, 1963). Since each $h : R \to \{0, 1\}$ can be identified with a subset of R, namely $h^{-1}(1)$, we can consider the space $\Omega = \{h^{-1}(1) : h \in H\}$ as that of all models of R. We describe now the elements of Ω. Let $\omega = h^{-1}(1)$. Then first, $\omega \subseteq R$, is a *filter* of the ring R. That is,

(1) $1 \in \omega$ (1 is the greatest element of R),

(2) If $a, b \in \omega$ then $ab \in \omega$, and

(3) If $a \in \omega$ and $b \in R$, then $a \vee b \in \omega$.

Let $a \in \omega$. Then $a = a \cdot 1$ and $h(a) = h(a)h(1)$, implying that $h(1) = 1$, that is, that $1 \in \omega = h^{-1}(1)$. If $a, b \in \omega$, then $h(ab) = h(a)h(b) = 1$, so $ab \in \omega$. For (3), $a = a(a \vee b)$, so that $1 = h(a) = h(a)h(a \vee b) = h(a \vee b)$. Thus $= h^{-1}(1)$ is a filter.

Moreover, each $\omega = h^{-1}(1)$ is *maximal*, that is, ω is a proper filter, meaning that $\omega \neq R$, or equivalently, that $0 \notin \omega$, and if $\gamma \subseteq R$ is a filter such that $\omega \subseteq \gamma$, then either $\gamma = \omega$ or $\gamma = R$. Since $h(1) = 1 \in \omega$, $h(0) = h(1') = 1' = 0$, whence $0 \notin \omega$, and ω is proper. Let γ be a filter such that $\omega \subseteq \gamma$. If $\omega \neq \gamma$, then there exists $b \in \gamma$ with $b \notin \omega$. Then $h(b) = 0$, so $h(b') = 1$ and $b' \in \omega$ and hence $b' \in \gamma$. But then, since γ is a filter, $bb' = 0 \in \gamma$, and $\gamma = R$.

Thus, elements of Ω are maximal filters of R. In fact, all maximal filters of R can be described by homomorphisms, that is, Ω is the set of all maximal filters of R. To see this, it suffices to show that if γ is a maximal filter of R, then its indicator function $1_\gamma : R \to \{0, 1\}$, defined by $1_\gamma(a) = 1$ or 0 according as to whether $a \in \gamma$ or $a \notin \gamma$, is a homomorphsim. The condition that $h(a) \neq h(a')$ turns out to be a characterization of maximality for filters.

Lemma 1. *A filter γ of R is maximal if and only if for $a \in R$, either $a \in \gamma$ or $a' \in \gamma$ (but not both).*

Proof. Suppose that γ is a maximal filter and that $b \notin \gamma$. Then

$$\beta = \{xy : x \in \gamma, b \leq y\}$$

is a filter). Taking $y = 1$ gets $\gamma \subseteq \beta$. Taking $x = 1$ and $y = b$ gets $b \in \beta$. Thus β strictly contains γ, and thus $\beta = R$. Hence $0 = xy$ for some $x \in \gamma$ and $y \geq b$, and so $x \leq y' \leq b'$. Thus $b' \in \gamma$

The proof of the converse parallels the proof above that $\omega = h^{-1}(1)$ is maximal.

From the lemma above, it is easy to check that indicator functions of maximal filters are homomorphisms. Indeed, by Lemma 1, $1_\gamma(a') = [1_\gamma(a)]'$. For $a, b \in R$, we have $1_\gamma(ab) = 1$ if and only if $ab \in \gamma$ if and only if $a, b \in \gamma$, $1_\gamma(ab) = 1_\gamma(a)1_\gamma(b)$. Similarly, $1_\gamma(a \vee b) = 1_\gamma(a) \vee 1_\gamma(b)$, and 1_γ is a homomorphism.

Regarding the set Ω of maximal filters of R as the set of models of R, an element $a \in R$ is true in a model $\omega \in \Omega$ if $a \in \omega$.

Remarks

1. Since filters and ideals are dual in the sense that if α is a filter of R, then $\alpha' = \{x' : x \in \alpha\}$ is an ideal, and if γ is an ideal, then $\gamma' = \{x' : x \in \gamma\}$ is a filter, the Stone Representation Theorem for Boolean rings can be also stated in terms of maximal filters (that is, models). Specifically, define $\psi : R \to \mathcal{P}(\Omega)$, the power set of Ω, by

$$\psi(a) = \{\omega \in \Omega : a \in \omega\} .$$

Then $\psi(0) = \emptyset$, $\psi(1) = \Omega$, and for $a \neq 0$, $\psi(a) \neq \emptyset$. (The third property is not a trivial one. See the second remark below.) By maximality, for $\omega \in \Omega$ and $a \in R$, if $a \in \omega$ then $a' \notin \omega$, so that

$$\psi(a') = [\psi(a)]^c.$$

(We use $(\cdot)^c$, $\cap$, and $\cup$ for set operations on $\mathscr{P}(\Omega)$). Also, $a, b \in \omega$ if and only if $ab \in \omega$, implying

$$\psi(ab) = \psi(a) \cap \psi(b) .$$

This, and DeMorgan's laws, readily yield

$$\psi(a \vee b) = \psi(a) \cup \psi(b).$$

Hence ψ is a homomorphism from R into $\mathscr{P}(\Omega)$. ψ is one-to-one since $\psi^{-1}(\emptyset) = 0$. Thus an appropriate subset of models is identified with a proposition in R, namely, a proposition a is identified with the set of models ω in which a is true.

2. The characterization of maximality in Lemma 1 is a property shared by atoms of R. For $a \in R$, $a \neq 0$, and an atom α, we have either $\alpha \leq a$ or $\alpha \leq a'$ (but not both). In fact, the principal filter $R \vee \alpha = \{r \vee \alpha : r \in R\}$ generated by an atom α is maximal. Moreover, α is the unique atom in $R \vee \alpha$. In general, the class of all maximal filters Ω of R is larger than that of these principal maximal filters. However, if the ring R is such that every maximal filter is principal, then they coincide. That is, $R \vee a$ is maximal if and only if a is an atom. Indeed, if $b < a$, then $R \vee b$ properly contains $R \vee a$, b being in the former and not in the latter. For example, if R is finite, then models of R can be identified with atoms of R.

We continue now with the basics of C_2. For deduction, we consider the concept of *logical entailment relation*, denoted by $\models$. Roughly speaking b logically entails a, in symbols $b \models a$, if whenever b is true, a is true. In our setting here, this means that $b \models a$ if and only if for $\omega \in \Omega$, if $b \in \omega$, then $a \in \omega$. The following fact is well-known.

Lemma 2. *$b \models a$ if and only if $b \leq a$.*

Proof. Suppose $b \leq a$. For $\omega \in \Omega$ such that $b \in \omega$, we have $a = b \vee a \in \omega$, since ω is a filter.

Conversely, suppose $b \models a$. For each $\omega \in \Omega$ such that $b \in \omega$, we have, by hypothesis, $a \in \omega$, and hence $ab \in \omega$ since ω is a filter. We are going to show that

$b = ab$. Suppose $ab < b$, that is, $b(ab)' \neq 0$. But then, there is $\gamma \in \Omega$ such that $b(ab)' \in \gamma$. Now $b(ab)' \leq b$ implying $b \in \gamma$, $b(ab)' \leq (ab)'$ implying $(ab)' \in \gamma$, that is, $ab \notin \gamma$ since ω is maximal, which is a contradiction.

Remarks

1. In the proof above, we have used the following well-known fact. If $x \in R$ and $x \neq 0$, then there is a maximal filter ω such that $x \in \omega$. $R \vee x$ is a filter containing x, and this filter can be enlarged to a maximal one. That statement is not at all obvious, involving set theoretical niceties such as Zorn's lemma. It should be noted also that for each element $x \neq 1$, there exists a maximal filter not containing x. Indeed, any maximal filter containing x' has that property. In particular, the only element contained in every maximal filter is 1.

2. A simple proof that every non-zero element of R is contained in a maximal filter, in the case of atomic R, goes as follows. As noted, $R \vee x$ is a filter containing x, and since R is atomic, there is an atom y with $y \leq x$. Then $R \vee y$ is a maximal filter containing x.

3. It is obvious that $b \leq a$ if and only if $b \rightarrow a = b' \vee a = 1$, that is, $b \rightarrow a$ is a tautology. (An element x is a tautology if for every $\omega \in \Omega$, $x \in \omega$. Thus the only tautology is 1). Lemma 2 expressed the logical entailment relation $\models$ in classical two-valued logic in terms of the (partial) order relation $\leq$. This explains the *monotonicity* of $\models$ (due to the transitivity property of $\leq$). For more details, see Chapter 8.

Now to Probability Logic (PL). PL, as a multi-valued logic, has been treated, for example, in Rescher (1969), Hailperin (1984), and Nilsson (1986). See also Goodman and Nguyen (1985). The formal language of PL is the same as that of C_2. Thus the base space of PL is also a Boolean ring R. As far as AI is concerned, there is a need to generalize C_2 to PL in order to reason with uncertain information, such as in expert systems.

For each sentence $a \in R$, there are two sets of "possible worlds" (that is, models): $\{\omega : a \in \omega\}$, and $\{\omega : a \notin \omega\}$. Not knowing the actual model, one considers the probability of a being true as a "truth value" for a. This is obviously a generalization of C_2. In view of the axioms of probability measures on R, PL, with truth-space the unit interval $[0, 1]$, is a non-truth functional system. A *model* for PL is simply a probability measure P on R.

In view of Stone's Representation Theorem (in terms of maximal filters of R), models for PL can be also viewed as probability measures on a class of subsets of models in C_2. Also, with its axioms, each probability measure P on R acts like a "homomorphism-like" map. As in classical deduction, the concept of *probabilistic entailment relation* is crucial for probabilistic reasoning in intelligent systems (for example, Pearl, 1988; Neapolitan, 1990). See also Hailperin (1984), Nilsson (1986).

We say that a is *probabilistically entailed by* b, in symbols $b \overset{P}{\vDash} a$, if for all probability measures on R, $P(b) \leq P(a)$. In view of Lemma 2 of Section 2.2, this is equivalent to $b \leq a$ or $b \vDash a$. Also, $a \in R$ is a *probability tautology* if $P(a) = 1$, for all probability measures P on R. Again, by Lemma 2 of Section 2.2, this mean that $a = 1$.

At a practical level, probabilistic entailment is defined as the computation of the probability of a sentence in terms of the probability values of other sentences. As Hawthorne (1988) stated clearly, this entailment is in fact a "partial" entailment, that is, entailment with "degrees." This is precisely the problem of combination of (probabilistic) evidence. The decision as whether or not to "infer" a from the b_i's depends upon the magnitude of $P(a)$. A computational procedure for this problem is given in Nilsson (1986). See also McLeish (1988), and Section 5.2. For discussions concerning PL and non-monotonic logics, see for example, Grosof (1988), Hawthorne (1988), and Chapter 8, Section 8.2 .

Probability logic is sound and complete. We close this section with the concepts of truth semantics and of probabilistic entailment in the conditional case (Adams, 1975). This will be served as a comparison with our development of conditional probability logic in the next section.

First, we take this opportunity to clarify several basic aspects in Adams' book, in view of the mathematical development of the conditional space $R|R$ and its associated three-valued logic (Chapters 2, 3). By Lewis' Triviality Result, it is seen that if we assign conditional probabilities to "indicative conditionals," then these conditionals, at the syntax level, are not, in general, elements of the Boolean ring R. This fact is expressed in Pearl's book as "conditionals are non-propositional" or "... classical logic does not possess an operator equivalent to the conditioning bar $(\cdot\,|\,\cdot)$ in probability," (Pearl, 1988, p. 475, 482). In Adams' book, it is expressed as "conditional propositions are not assumed to correspond to subsets of a sample space," and as "these objects do not have truth values" (Adams, 1975, Preface and p. 9). It becomes clear that, under the fundamental assumption of Adams' work (p. 3), namely "the probability of an indicative conditional is a conditional probability," a conditional "if b is the case then a is", is a subset of R rather than an element of R. As far as truth values are concerned, it is apparent that Adams was referring to classical two-valued logic. Each conditional $(a|b)$ does have truth values, namely true (1), false (0) or undefined (u). As such, we agree with Adams that "probabilities of conditionals are not equal to their probabilities of being true." All the above can be proved in our representation of conditional events $(a|b)$ as cosets of $R|R$.

Let P be a probability on R. Of course $P(a|b)$ is a function of $P(ab)$ and $P(b)$ (when $P(b) > 0$), and it is true that the truth-values of $(a|b)$, denoted as $\hat{t}(a|b)$, is a function of $t(ab)$ and $t(b)$. Indeed,

$$\hat{t}(a|b) = \begin{cases} 1 \text{ if } t(ab) = 1 \\ 0 \text{ if } t(a'b) = 1 \\ u \text{ if } t(b') = 1 \end{cases}$$

where $t : R \to \{0, 1\}$ is a Boolean homomorphism. The knowledge of $t(ab)$ and $t(b)$ completely specifies $\hat{t}(a|b)$, since $\{ab, a'b, b'\}$ is a partition of 1.

The point is this. Since $(a|b)$ is not "Boolean," its truth-values should not be restricted to $\{0, 1\}$. we see that, with the truth-space being $\{0, 1, u\}$, conditionals are truth-functional and their probabilities are conditional probabilities. On the other hand, contrary to Adams' attitude concerning Lewis Triviality Result (Adams, p. 35), namely "The author's very tentative opinion on the "right way out" of the triviality argument is that we should regard the inapplicability of probability to compounds of conditionals as a fundamental limitation of probability, on a par with the inapplicability of truth to simple conditionals. What is needed at the present stage is less mathematical theorizing than close examination of the phenomenon of inference involving these problematic constructions, ...". We have resolved these problems from a mathematical analysis. Indeed, first, there is no problem with compounds of conditionals, since there is no need to assign probabilities directly to such objects. Simple conditionals have truth-values in $\{0, 1, u\}$, and, as cosets of the ring R, have well-defined probabilities as conditional probabilities (see Chapter 5). Viewing $R|R$ as the space of conditionals with three-valued logic, we can derive basic connectives on it (see Section 3.4). Given a system of truth tables in a three-valued logic, there corresponds a system of connectives $\wedge$, $\vee$, $'$, say, on $R|R$. These connectives are *operators* on $R|R$, that is, any compound of conditionals is a simple conditional, so that probability is assigned in the same way as for simple conditionals.

In our notation, R is a *factual language*, and $R|R$ is its *conditional extension*, and $(a|b)$ is $b \Rightarrow a$, in Adams' notation for conditionals. Let $t : R \to \{0, 1\}$ be a truth function. Adams considered the "truth-conditional semantics," that is, truth evaluations on $R|R$ as follows.

α) $(a|b)$ is "verified" under t if $t(a) = t(b) = 1$,

β) $(a|b)$ is "falsified" under t if $t(b) = 1$ and $t(a) = 0$.

But, in our development of three-valued logic for $R|R$, α) and β) say nothing more than the truth-values of $(a|b)$ are 1 and 0, respectively, in $\{0, 1, u\}$. Of course, when a conditional $(a|b)$ is neither verified nor falsified, its truth-value is u. It is this "non-verification values" u which completes the discussion concerning semantics of conditionals.

Finally, as mentioned in previous chapters, although conditionals are not treated as mathematical entities in Adams' book, Adams did propose basic connectives among them, namely "contrary," "quasi-conjunction" and "quasi-disjunction" (Adams, 1975, p. 46-47). These connectives were proposed earlier by Schay (Schay, 1968), and were rediscovered, in an independent work, later by Calabrese (Calabrese, 1987). These connectives correspond precisely to Sobocinski's three-valued logic. (See Section 3.5.)

Now to Adams' ε-semantics. From the formal language $\mathscr{L}$ (or R), of classical two-valued logic, consider its extension $\hat{\mathscr{L}}$ to "conditional formulas," denoted $\hat{\mathscr{L}} = \{a \Rightarrow b,\ a, b \in \mathscr{L}\}$. $a \Rightarrow b$ stands for an *indicative conditional* of the form "if a is the case then b is" in natural language, for example, in ordinary English. In the study of probabilistic semantics for default reasoning (Pearl, 1988, Chapter 10), $\hat{\mathscr{L}}$ is the set of default statements which are "non-propositional" in the sense that they involve the "arrow" $\Rightarrow$ connecting two propositional formulas. So $a \Rightarrow b$ is non "Boolean," that is, $a \Rightarrow b$ is not an element of $\mathscr{L}$ or of the Boolean ring R. See also Dubois and Prade (1989) for the modeling of default rules by conditionals. Also, here $\Rightarrow$ is not the material implication connective $\rightarrow$. In fact, except for a mathematical representation of the object $a \Rightarrow b$, Adams' intention was to provide a semantic evaluation map compatible with conditional probability for $\hat{\mathscr{L}}$. Basic connectives on $\hat{\mathscr{L}}$ are defined as follows (see Chapters 1, 4).

$$(a \Rightarrow b)' = (a \Rightarrow b'),$$

$$(a \Rightarrow b) \wedge (c \Rightarrow d) = (a \wedge c \Rightarrow (a \rightarrow b)(c \rightarrow d)),$$

$$(a \Rightarrow b) \vee (c \Rightarrow d) = (a \vee b \Rightarrow ab \vee cd).$$

As before, a probability model is a probability measure P on $\mathscr{L}$. The associated "truth conditional semantics" for $\hat{\mathscr{L}}$ is defined by

$$\hat{P} : \hat{\mathscr{L}} \rightarrow [0, 1], \quad \hat{P}(a \Rightarrow b) = P(b|a).$$

The set of conditional formulas $\{(a_i \Rightarrow b_i),\ i = 1, ..., n\}$ is said to entail the conditional formula $c \Rightarrow d$ if and only if for all $\varepsilon > 0$, there is $\delta > 0$ such that for all P on $\mathscr{L}$ for which $P(a_i) > 0,\ i = 1, ..., n$ and $P(c) > 0$), if $P(b_i|a_i) \geq 1 - \delta$, for $i = 1, ..., n$, then $P(d|c) >, 1 - \varepsilon$. This concept of entailment in Adams' conditional probability logic is suitable for default reasoning in AI in which $\mathscr{L}$ is a collection of propositions and $\hat{\mathscr{L}}$ is a set of default statements. Indeed, for a default statement $a \Rightarrow b$ = "almost all a's are b's," one translates "almost all" into "$P(b|a)$ is arbitrary close to 1, short of actually being 1" (Pearl, 1988, p. 480); moreover, the set of defaults

$$\{(a_i \Rightarrow b_i), i = 1, ..., n\}$$

logically entails $(c \Rightarrow d)$ if when $P(b_i|a_i)$ is "high," $i = 1, ..., n$, then $P(d|c)$ is also "high." For more details on this "ε-semantics," we refer the reader to Pearl's book (1988).

6.2 Syntax and basic properties.

Let the Boolean ring R be the base space for classical two-valued logic (also for probability logic). The base space for the conditional probability logic (CPL) we are going to develop is the mathematical conditional extension $R|R$ with its algebraic structure established in Chapters 2, 3 and 4. Now elements of $R|R$, that is, "conditional formulas," are mathematical entities, and we can describe special elements of $R|R$ as in the case of R. Specifically, we are going to describe syntactically "contradictions" and tautologies on $R|R$, in a manner compatible with truth conditional semantics in Section 6.3. By the same token, various characterizations of implicative relations in CPL are given, generalizing those of material implication in classical logic.

First, in probability logic, an element $a \in R$ is called a "contradiction" or a "P-tautology" according to $P(a) = 0$ or $P(a) = 1$ for *all* probability measures P on R. By Lemma 1 of Section 2.2, these are equivalent to $a = 0$ or $a = 1$. The counterparts of 0 and 1 on $R|R$ are now described. Observe that $R|R = \bigcup_{b \in R} R/Rb'$, where each $R|Rb'$ is a Boolean ring with its contradiction and tautology $(0|b)$ and $(1|b)$, respectively, provided $b \neq 0$. Thus,

Definition 1. *The classes of zero-type conditionals and unity-type conditionals are, respectively*

$$\mathscr{Z} = \{(0|b), b \in R\backslash\{0\}\}$$

and

$$\mathscr{N} = \{(1|b), b \in R\backslash\{0\}\} .$$

In Section 6.3, we will show that these concepts are compatible with truth conditional semantics on $R|R$. The following theorem summarizes basic properties of $\mathscr{Z}$ and $\mathscr{N}$.

Theorem 1.

(i) *Both* $\mathcal{Z}$ *and* $\mathcal{U}$ *are closed under* $\cdot$ *and* $\vee$ *on* $R|R$,

(ii) $\mathcal{Z}$ *(resp.* $\mathcal{U}$*) has the ideal-like (resp. filter-like) property:* $(R|R)\cdot \mathcal{Z} = \mathcal{Z}$ *(resp.* $(R|R) \vee \mathcal{U} = \mathcal{U}$*),*

(iii) $\mathcal{Z} \cup \{(0|0)\}$ *is closed under* $\cdot, \vee, +$ *on* $R|R$,

(iv) $\mathcal{Z}$ *and* $\mathcal{U}$ *are "complementary" in the sense that*

$$\mathcal{Z} = \{(b|b)' : (b|b) \in \mathcal{U}\}\text{ , and}$$

$$\mathcal{N} = \{(0|b)' : (0|b) \in \mathcal{Z}\}\,.$$

Proof. (i), (iii), and (iv) are obvious from the definitions of the operations on $R|R$. Since

$$(a|b)\cdot(0|c) = (0|a'b \vee c) \in \mathcal{Z},$$

$$(1|1)(0|b) = (0|b),$$

$$(a|b) \vee (c|c) = (ab|b) \vee (c|c) = (ab \vee c|ab \vee c) \in \mathcal{U},$$

and

$$(0|0) \vee (b|b) = (b|b),$$

part (ii) holds. □

It is known in classical logic that the material implication $b \to a$ is a tautology (that is, $b \to a = 1$) if and only if $b \leq a$ (that is, b "strictly" implies a). This fact is a characterization of the binary Boolean operator $\to$. In other words, $\to$ is the only binary Boolean operator on R having this property. Indeed, it is obvious that if $f: R^2 \to R$, $f(a, b) = b \to a = b' \vee a$, then $f(a, b) = 1$ whenever $b \leq a$. Conversely, if $f: R^2 \to R$ is such that

$$f(a, b) = 1 \text{ if and only if } b \leq a,$$

then

$$f(1, 1) = f(1, 0) = f(0, 0) = 1,$$

and $f(0, 1) = 0$, so that the normal disjunctive form of $f(a, b)$ reduces to

$$f(a, b) = ab \vee ab' \vee a'b' = a \vee a'b' = a \vee b'\,.$$

The situation in conditional logic is somewhat different in the sense that there are various conditional Boolean polynomials in two variables satisfying the counter-part of the equivalence betweeen strict implication in two-valued logic and being a tautology. Specifically, strict implication in two-valued logic is replaced by the order relation on $R|R$ (see Chapter 3), and tautologies in conditional logic are elements of $\mathcal{U}$ that is, are of the form $(b|b)$ with $b \neq 0$.

We are going to characterize conditional Boolean polynomials f in two variables satisfying the following equivalent condition. For any $a, b, c, d \in R$ with $b, d \neq 0$,

$$f((a|b), (c|d)) \in \mathcal{U}$$

if and only if

$$(a|b) \leq (c|d).$$

We may assume without loss of generality that f is of the form

$$f = (\alpha|\beta) = (\alpha|\alpha \vee \gamma),$$

where $\alpha, \gamma : R^4 \to R$ are Boolean functions, and $\alpha\gamma \equiv 0$. Theorems 2 and 3 below shed light not only on conditional logical operations taking values in $\mathcal{U}$, but also are needed in proving that CPL is sound and complete (Section 6.4).

Theorem 2. *Let $f : (R|R)^2 \to R|R$ be a conditional Boolean polynomial in two variables. The following are equivalent.*

(i) *For $a, b, c, d \in R$, with $b, d \neq 0$,*

$$f((a|b), (c|d)) \in \mathcal{U} \text{ if and only if } (c|d) \leq (a|b).$$

(ii) *f is of the form $f = (\alpha|\beta) = (\alpha|\alpha \vee \eta)$, where*

$$\eta(a, b, c, d) = (ab)'(cd) \vee (a'b)(c'd)',$$

and α is a Boolean function such that $\alpha \leq \eta'$, and $\alpha \neq 0$ when $\eta = 0$.

Proof. To prove that (i) implies (ii), we use the criterion that $(c|d) \leq (a|b)$ if and only if $cd \leq ab$ and $a'b \leq c'd$. This is the same as

$$(cd)(ab)' = 0 = (a'b)(c'd)',$$

or

$$(ab)'(cd) \vee (a'b)(c'd)' = 0.$$

Thus $\eta = \eta(a, b, c, d) = (ab)'(cd) \vee (a'b)(c'd)' = 0$ if and only if $(c|d) \leq (a|b)$. For

(i) to hold, $f = (\alpha | \alpha \vee \gamma)$ has to be such that $\gamma \equiv \eta$, since otherwise it is possible that simultaneously $\gamma(a, b, c, d) = 0$, $\alpha(a, b, c, d) \neq 0$, and $\eta(a, b, c, d) \neq 0$, contradicting (i).

That (ii) implies (i) is easy. □

The precise forms as well as the total number of fs in (ii) can be determined as follows. Let

$$w_i(a|b) = \begin{cases} ab & \text{if } i = 1 \\ a'b & \text{if } i = 0 \\ b' & \text{if } i = u\,. \end{cases}$$

We have

$$\eta(a, b, c, d) = (ab)'(cd) \vee (a'b)(c'd)'$$

$$= a'bcd \vee b'cd \vee a'bd'$$

$$= w_0(a|b)w_1(c|d) \vee w_u(a|b)w_1(c|d) \vee w_0(a|b)w_u(c|d)$$

$$= \vee_{(i,j)\in J}\, w_i(a|b)w_j(c|d),$$

where $J = \{(0,1), (u,1), (0,u)\}$. Thus $\alpha(a, b, c, d)$ must be of the form

$$\vee_{(i,j)\in K}\, w_i(a|b)w_j(c|d),$$

where

$$K = \{(0,0), (u,0), (1,0), (1,1), (1,u), (u,u)\}.$$

As examples, for

$$K = \{(0,0), (u,0), (1,0), (1,1), (1,u)\},$$

$$\alpha = a'bc'd \vee abc'd \vee abcd \vee abd' \vee b'c'd = ab \vee c'd.$$

When $\eta = 0$,

$$\eta' = ab \vee c'd \vee b'd' = 1.$$

But for $b, d \neq 0$, $b'd' < 1$, so that $ab \vee c'd \neq 0$. Here, $ab \vee c'd \vee \eta = b \vee d$. Thus f is of the form

$$f((a|b), (c|d)) = (ab \vee c'd | b \vee d).$$

For

$$K = (0,0), (u,0), (1,0), (1,1), (1,u), (u,u)\},$$

$\alpha = ab \vee c'd \vee b'd'$, which is not 0 when $\eta = 0$. In fact $\alpha = 1$ when $\eta = 0$. Thus

$$f((a|b), (c|d)) = (ab \vee c'd \vee b'd' \,|\, 1)$$

since for all $a, b, c, d \in R$,

$$ab \vee c'd \vee b'd' \vee \eta = 1.$$

These two forms have interesting interpretations. The last,

$$f_1(a|b), (c|d)) = (ab \vee c'd \vee b'd' \,|\, 1)),$$

is the consequent of Lukasiewicz's implication (see Section 3.4), where the consequent of a conditional $(a|b)$ is defined to be $C(a|b) = ab$.

Using Theorem 3, Section 3.4, it can be checked that the first form

$$f_2((a|b), (c|d)) = (ab \vee c'd|b \vee d)$$

corresponds to Sobocinski's truth table for implication. This truth table is given in Rescher (1969, p. 70) with the sign + (respectively -) in front of the truth valuess to indicate "designated" (respectively, "anti-designated") values for consideration of tautologies (respectively, contradictions) in multi-valued logic. We will discuss this further in Section 6.3. Adams, Calabrese, and one of Schay's conditional disjunctions $\vee_0$ are all defined to be

$$(a|b) \vee_0 (c|d) = (ab \vee cd|b \vee d).$$

Thus

$$f_2((a|b), (c|d)) = (a|b) \vee_0 (c|d)'.$$

We will return to this observation in Section 6.4.

Not all subsets K of

$$\{(0,0), (u,0), (1,0), (1,1), (1,u), (u,u)\}$$

lead to α's satisfying condition (ii) of Theorem 2. For example, if

$$K = \{(1,0), (1,1), (1,u), (u,u)\},$$

then

$$\alpha = abd' \vee abcd \vee abc'd \vee b'd' = ab \vee b'd'.$$

Taking $a = c = 0$ and $b = d = 1$, we have,

$$ab' \vee c'd \vee b'd' = 1,$$

so that $\eta(0, 1, 0, 1) = 0$. But $\alpha(0, 1, 0, 1) = 0$. Thus $\alpha(a, b, c, d) = ab \vee b'd'$ does not satisfy our condition (ii).

We now look closer at f_1 and f_2. First, since $f_1 = (\alpha | \alpha \vee \eta)$ with $\alpha = \eta'$, we see that for all $a, b, c, d \in R$, f_1 satisfies

$$f((a|b), (c|d)) \in \mathcal{U} \text{ if and only if } (c|d) \leq (a|b). \qquad (*)$$

Now f_2 does not satisfy (*). Indeed, when $\eta = 0$, we have $abc'd = b \vee d$. This equality holds also when $b = d = 0$, but then $f_2((a|0), (c|0)) \notin \mathcal{U}$. However, f_2 satisfies

$$f((a|b), (c|d)) \in \mathcal{U} \text{ if and only if } (c|d) \leq (a|b) \text{ and } b \text{ or } d \neq 0 \qquad (**).$$

Indeed, when $(c|d) \leq (a|b)$, we have $ab \vee c'd = b \vee d$. If b or d is $\neq 0$, then $ab \vee c'd = b \vee d \neq 0$, and hence $f_2((a|b), (c|d)) \in \mathcal{U}$. Conversely, if $f_2((a|b), (c|d)) \in \mathcal{U}$, then $ab \vee c'd = b \vee d \neq 0$, implying that $\eta = 0$ and b or $d \neq 0$. On the other hand f_1 does not satisfy (**), since $f_1((a|0), (c|0)) \in \mathcal{U}$.

It turns out that (*) and (**) characterize f_1 and f_2, respectively. Consider first the condition (*). As before, $f = (\alpha | \alpha \vee \eta)$, where

$$\alpha(a, b, c, d) = \vee_{(i,j)\in K} w_i(a|b)w_j(c|d),$$

with

$$K \subseteq \{(0,0), (u,0), (1,0), (1,1), (1,u), (u,u)\}.$$

We are going to show that if K is a strict subset, then there is an $(a, b, c, d) \in R^4$ such that $\alpha(a, b, c, d) = 0$. We only have to look at subsets of $ab \vee c'd \vee b'd' = \eta'$. We already know from above that if a is $ab \vee c'd$ or $ab \vee b'd'$ or $c'd \vee b'$, then f will not satisfy (*). Thus it suffices to consider subsets of the form $ab \vee c'd \vee xyb'd'$ or $ab \vee xyc'd \vee b'd'$ or $xyab \vee c'd \vee b'd'$, where x and y can be one of a, b, c, d, or their complements. For example, in the case $ab \vee c'd \vee xyb'd'$ where $xyb'd' \neq b'd'$, then when $\eta = 0$, we have $ab \vee c'd \vee xyb'd' = ((xy)'b'd')'$, and it is easy to pick x, y, b, d so that $(xy)'b'd' = 1$. The other details are left to the reader.

Consider now the condition (**). For (*) to hold, $f = (\alpha | \alpha \vee \eta)$, where

$$\alpha(a, b, c, d) = \vee_{(i,j)\in K} w_i(a|b)w_j(c|d),$$

with

$$K \subseteq \{(0,0), (u,0), (1,0), (1,1), (1,u), (u,u)\}.$$

We are going to specify K so that (**) holds. We need to pick α so that $\alpha > 0$ is equivalent to b or d being > 0. Now b or $d > 0$ if and only if $b \vee d > 0$. Using the decomposition of $b \vee d$ in terms of the $w_i(a|b)w_j(c|d)$'s, we see that $b \vee d$ if and only if $w_i(a|b)w_j(c|d) > 0$ for some $(i, j) \neq (u, u)$. But when $(c|d) \leq (a|b)$, that is, when $\eta = 0$, we have $w_i(a|b)w_j(c|d) = 0$ for all $(i, j) \in \{(0, 1), (u, 1), (0, u)\}$. Thus $\alpha(a, b, c, d) > 0$ when $\eta = 0$ and b or $d \neq 0$ only for

$$K \supseteq \{(0,0), (u,0), (1,0), (1,1), (1,u)\}.$$

But the upper bound of K is $\{(0,0), (u,0), (1,0), (1,1), (1,u), (u,u)\}$, and it leads to f_1, which does not satisfy (**). Hence K must be $\{(0,0), (u,0), (1,0), (1,1), (1,u)\}$, which yields f_2.

In classical two-valued logic, the equivalence relation $\leftrightarrow$ defined by

$$a \leftrightarrow b = (a \rightarrow b) \wedge (b \rightarrow a)$$

is characterized as the only binary Boolean operation f such that

$$f(a, b) = 1 \text{ if and only if } a = b.$$

Using the definition $a \rightarrow b = a' \vee b$, this is routine to check. The counterpart in conditional logic is expressed in the following theorem.

Theorem 3. *Let $f : (R|R)^2 \rightarrow R|R$ be a conditional Boolean polynomial in two variables. The following are equivalent.*

(i) For any $a, b, c, d \in R$ with $b, d \neq 0$,

$$f((a|b), (c|d)) \in \mathcal{U} \text{ if and only if } (a|b) = (c|d),$$

(ii) f is of the form $(\alpha|\alpha \vee \xi)$, where

$$\xi(a,b,c,d) = (ab + cd) \vee (b + d),$$

α is Boolean, $\alpha \leq \xi'$, and $\alpha \neq 0$, when $\xi = 0$.

Proof. First, $(a|b) = (c|d)$ if and only if $ab = cd$ and $b = d$. This is the same as $ab + cd = b + d = 0$, or $(ab + cd) \vee (b + d) = 0$. Let

$$\xi = \xi(a,b,c,d) = (ab + cd) \vee (b + d).$$

Then

$$\xi = ab'cd \vee abd' \vee a'bcd \vee b'cd \vee a'bd' \vee b'c'd$$

$$= \vee_{(i,j)\in J}\, w_i(a|b)w_j(c|d),$$

where

$$J = \{(u,0), (1,0), (u,1), (1,u), (0,1), (0,u)\}.$$

Now, that (ii) implies (i) is obvious, and (i) implies (ii) since $f = (\alpha|\alpha \vee \gamma)$ will satisfy (i) if $\gamma = \xi$. □

The specific form of α is

$$\alpha = \vee_{(i,j)\in I}\, w_i(a|b)w_j(c|d),$$

where

$$I \subseteq \{0, 1, u\}^2 - J = \{(0,0), (1,1), (u,u)\}.$$

Again, not all subsets I lead to an α satisfying (ii). Two interesting candidates are

$$I = \{(0,0), (1,1), (u,u)\},$$

and

$$I = \{(1,1), (0,0)\}.$$

For the first,

$$\alpha = a'bc'd \vee abcd \vee b'd'.$$

Here, $\beta = \alpha \vee \xi = \xi' \vee \xi = 1$, so that

$$f_3((a|b), (c|d)) = abcd \vee a'bc'd \vee b'd'.$$

When $\xi = 0$, $\alpha = 1$, and hence $\neq 0$.

For $I = \{(1,1), (0,0)\}$, $\alpha = abcd \vee a'bc'd$. When $\xi = 0$,

$$\xi' = 1 = abcd \vee a'bc'd \vee b'd'.$$

But for $b \neq 0$, $d \neq 0$, we have $b'd' < 1$, so that $abcd \vee a'bc'd \neq 0$ for all $a, c \in R$. Here $\beta = \alpha \vee \gamma = b \vee d$, and

$$f_4((a|b), (c|d)) = (abcd \vee a'bc'd|b \vee d).$$

The conditional polynomial f_3 corresponds to the consequent of Lukasiewicz's three-valued logic equivalence, while f_4 is the syntax of Sobocinski's three-valued logic equivalence. See Chapter 3 for more details.

It turns out that f_3 and f_4 are the only candidates. Indeed, for other $I \subsetneq \{(0,0), (1,1), (u,u)\}$, one can find a, b, c, d, with $b \neq 0 \neq d$, such that $\alpha(a, b, c, d) = 0$ when $\xi(a, b, c, d) = 0$. For example, if $I = \{(1,1), (u,u)\}$, then $\alpha = abcd \vee b'd'$. Taking $a = c = 0$ and $b = d = 1$, we get $\xi(0, 1, 0, 1) = 0$, and $\alpha(0, 1, 0, 1) = 0$.

As a final note, f_2 and f_4 satisfy

$$f((a|b), (a|b)) = (b|b).$$

6.3 Truth conditional semantics

This section consists of extending the basics of classical two-valued logic (C_2) and Probability Logic (PL) to Conditional Logic (CL) and Conditional Probability Logic (CPL). By Conditional Logic, we mean Lukasiewicz's three-valued logic on the conditional space $R|R$, where R is a Boolean ring, or equivalently, $R|R$ equipped with logical operations developed in Chapters 3 and 4. By Conditional Probability Logic, we mean a multi-valued logic with base space $R|R$ on which truth-values are conditional probabilities. The base space of CL is a (special) Stone algebra $R|R$ (see Chapter 4). Similarly, the truth-space of CL is the Stone algebra $\{0, u, 1\}$, with $0 < u < 1$, with the following operators. (See Section 3.4 for the appearance of $\{0, u, 1\}$ as the truth space for $R|R$.) We use the same notation $'$, $\wedge$, and $\vee$ on $R|R$. In view of Lukasiewicz's truth tables, for $i, j \in \{0, u, 1\}$, we have

$$i \wedge j = min\{i, j\}, \quad i \vee j = max\{i, j\},$$

$$0' = 1, \quad 1' = 0, \quad u' = u.$$

The pseudo-complementation on $\{0, u, 1\}$ is $0^* = 1, u^* = 0, 1^* = 0$, which does satisfy Stone's identity

$$i^* \vee i^{**} = 1$$

for all $i \in \{0, u, 1\}$.

First, we formulate the concept of a model in CL.

Definition 1. *A model in CL is a homomorphism from* $R|R$ *to* $\{0, u, 1\}$, *that is, a map preserving the operators* $'$, $\wedge$, *and* $\vee$.

It turns out that models in CL can be built from those in C_2. Specifically,

Theorem 1.

(i) If $\omega \in \Omega$ *is a maximal filter of* R*, then the map* $h_\omega : R|R \to \{0, u, 1\}$ *defined by*

$$h_\omega(a|b) = \begin{cases} 1 \text{ if } ab \in \omega \\ 0 \text{ if } a'b \in \omega \\ u \text{ if } b' \in \omega \end{cases}$$

is a homomorphism.

(ii) If $h : R|R \to \{0, u, 1\}$ *is a homomorphism, then there is an* $\omega \in \Omega$ *such that* $h \equiv h_\omega$.

Proof. Note that since $\{ab, a'b, b'\}$ forms a partition of 1, and ω is a maximal filter of R, h_ω is well-defined. For the proof of (i) we have

$$h_\omega((a|b)') = h_\omega(a'|b) = \begin{cases} 1 \text{ if } a'b \in \omega \\ 0 \text{ if } ab \in \omega \\ u \text{ if } b' \in \omega \end{cases}.$$

In view of Lukasiewicz's negation on $\{0, u, 1\}$ (see Section 3.5), we get

$$h_\omega((a|b))' = [h_\omega(a|b)]'.$$

By DeMorgan's laws on $R|R$ (Theorem 3, Section 4.1) and the fact that $'$ is involutive, it remains only to show that for $(a|b), (c|d) \in R|R$,

$$(*) \qquad h_\omega[(a|b), (c|d)] = h_\omega(a|b)h_\omega(c|d).$$

Now

$$(a|b)(c|d) = (ac|a'b \vee c'd \vee bd).$$

Since

$$ac(a'b \vee c'd \vee bd) = abcd,$$

and $abcd \in \omega$ if and only if $ab \in \omega$ and $cd \in \omega$, (*) is true for the value 1. For the value 0, we have

$$(ac)'(a'b \vee c'd \vee bd) = a'b \vee c'd \in \omega$$

if and only if either $a'b$ or $c'd$ (or both) $\in \omega$. Thus, in view of Lukasiewicz's conjunction on $\{0, u, 1\}$ (see Section 3.5), (*) is true.

Finally, the following are equivalent:

(1) $h_\omega[(a|b)(c|d)] = u$,
(2) $a'b \vee c'd \vee bd \notin \omega$,
(3) $a'b, c'd, bd \notin \omega$,
(4) $a \vee b', c \vee d', b' \vee d' \in \omega$,
(5) (a or $b' \in \omega$) and (c or $d' \in \omega$).

The case "only a and c are in ω" is excluded, since then $abcd \in \omega$, contradicting the condition $h_\omega[(a|b)(c|d)] = u$. All remaining cases correspond to $h_\omega(a|b)h_\omega(c|d) = u$. For example, if only a and d' are in ω, then $ab \in \omega$, $d' \in \omega$, and hence $h_\omega(a|b)h_\omega(c|d) = 1 \cdot u = u$.

To prove (ii), let $h : R|R \to \{0, u, 1\}$ be a homomorphism. Let g be the restriction of h to R viewing R as $R|1$. This restriction g can only take values in $\{0, 1\}$. Indeed if there is an $a \in R$ such that $g(a) = u$, then since g is obviously a homomorphism from R to $\{0, u, 1\}$, we have

$$0 = g(0) = g(aa') = g(a)g(a') = [g(a)][g(a)]' = uu' = uu = u,$$

which is impossible. Thus g is a Boolean homomorphism between R and $\{0, 1\}$, and hence is the indicator function 1_ω of some maximal filter ω of R.

It remains to show that $h \equiv h_\omega$. Observe that

$$(a|b) = (ab|1) \vee [(b'|1)(0|0)]$$

and

$$(0|0)' = (1|0) = (0|0),$$

which implies that $h(0|0) = h(0|0)' = [h(0|0)]' = u$ since u is the unique element in $\{0, u, 1\}$ such that $u' = u$. Thus

$$h(a|b) = 1_\omega(ab) \vee 1_\omega(b')\cdot u\,.$$

From this, since $h_\omega(b')\cdot u \leq u$, $h(a|b) = 1$ if and only if $1_\omega(ab) = 1$, if and only if $h_\omega(a|b) = 1$. Next, $h(a|b) = 0$ if and only if $1_\omega(ab) = 1_\omega(b') = 0$, if and only if $1_\omega(a'b) = 1$, if and only if $h_\omega(a|b) = 0$. Finally, $h(a|b) = u$ if and only if $1_\omega(ab) = 0$ and $1_\omega(b') = 1$, if and only if $h_\omega(a|b) = u$. □

In view of the Theorem above, models in CL are precisely $\{h_\omega, \omega \in \Omega\}$.

We investigate now two possible counterparts of maximal filters in the case of Stone algebras. First, consider $h_\omega^{-1}(1)$. Since $(0|0) = (0|0)'$ does not belong to any $h_\omega^{-1}(1)$, the characterization of maximality of (Boolean) filters does not hold for $h_\omega^{-1}(1)$.

However, other properties of ω remain valid for $h_\omega^{-1}(1)$. In particular, each set $h_\omega^{-1}(1)$ is a filter in the lattice $(R|R, \wedge, \vee)$. That is, if $(a|b)$, $(c|d) \in h_\omega^{-1}(1)$, then

$$(a|b) \wedge (c|d) \in h_\omega^{-1}(1),$$

and if $(a|b) \in h_\omega^{-1}(1)$ and $(c|d) \in R|R$, then

$$(a|b) \vee (c|d) \in h_\omega^{-1}(1).$$

In fact, since $h_\omega : R|R \to \{0, u, 1\}$ is a homomorphism, each $h_\omega^{-1}(1)$ satisfies the following stronger conditions:

(1) $(a|b), (c|d) \in h_\omega^{-1}(1)$ if and only if $(a|b) \wedge (c|d) \in h_\omega^{-1}(1)$,

(2) $(a|b) \vee (c|d) \in h_\omega^{-1}(1)$ if and only if $(a|b) \in h_\omega^{-1}(1)$ or $(c|d) \in h_\omega^{-1}(1)$, and

(3) $(1|1) \in h_\omega^{-1}(1)$, $(0|1) \notin h_\omega^{-1}(1)$, $(0|0) \notin h_\omega^{-1}(1)$.

Moreover, the class $\mathscr{A}_1$ of filters of $R|R$ satisfying (1), (2) and (3) are the $h_\omega^{-1}(1)$, $\omega \in \Omega$. To see this, let $A \subseteq R|R$ satisfy (1), (2) and (3), and set $\omega = A \cap R$, where R is identified with $R|1$. ω is obviously a filter in R. Moreover, for $a \in R$, either $(a|1)$ or $(a'|1)$ is in A, since otherwise,

$$(a|1) \vee (a'|1) = (1|1)$$

will not be in A, by (2), a contradiction. Thus ω is maximal. It remains to verify that $A = h_\omega^{-1}(1)$. If $(a|b) \in h_\omega^{-1}(1)$, that is, $ab \in \omega = A \cap R$, then $(ab|1) \in A$. But $(a|b) \wedge (ab|1) = (ab|1)$, so that $(a|b) \in A$ by (1). Conversely, if $(a|b) \in A$, then write $(a|b) = ab \vee (b' \cdot (0|0)) \in A$. By (2), we have $ab \in A$ or $b' \cdot (0|0) \in A$. But $b' \cdot (0|0) \in A$ holds only if $b' \in A$ and $(0|0) \in A$, by (1). However, by (3), $(0|0) \notin A$, thus only $ab \in A$ holds, that is, $ab \in \omega$, so that $h_\omega(a|b) = 1$. □

Consider now $h_\omega^{-1}(\{u, 1\})$, $\omega \in \Omega$. Since $h_\omega : R|R \to \{0, u, 1\}$ is a homomorphism, the following facts are easy to derive:

(i) $h_\omega^{-1}(\{u, 1\}) \cap R = \omega$, a maximal filter of R.

(ii) for $(a|b) \in (R|R)$, $(a|b) \in h_\omega^{-1}(\{u, 1\})$ or $(a|b)' \in h_\omega^{-1}(\{u, 1\})$, or both.

(iii) If $(a|b) \in h_\omega^{-1}(\{u, 1\})$, then for $(c|d) \in (R|R)$, $(a|b) \vee (c|d) \in h_\omega^{-1}(\{u, 1\})$.

(iv) $(a|b) \wedge (c|d) \in h_\omega^{-1}(\{u, 1\})$ if and only if both $(a|b)$ and $(c|d)$ are in $h_\omega^{-1}(\{u, 1\})$.

(v) $(a|b) \vee (c|d) \in h_\omega^{-1}(\{u, 1\})$ if and only if $(a|b) \in h_\omega^{-1}(\{u, 1\})$ or $(c|d) \in h_\omega^{-1}(\{u, 1\})$.

(vi) for $b \in R$, $(1|b) \in h_\omega^{-1}(\{u, 1\})$.

(vii) $(a|b) \in h_\omega^{-1}(\{u, 1\})$ if and only if $b \to a = b' \vee a \in \omega$.

As in the case of $h_\omega^{-1}(1)$, the class $\mathcal{A}_2$ of filters of $R|R$ satisfying (i)-(vii) above is precisely $\{h_\omega^{-1}(\{u, 1\}), \omega \in \Omega\}$.

Remark. Since $(0|1) \notin h_\omega^{-1}(\{u, 1\})$, the filter $h_\omega^{-1}(\{u, 1\})$ is proper. If $A \subseteq R|R$ is a filter satisfying (i)-(vii), and $h_\omega^{-1}(\{u, 1\}) \subseteq A$, then $h_\omega^{-1}(\{u, a\}) = A$, that is, $h_\omega^{-1}(\{u, 1\})$ is "maximal." Indeed, we have $\omega \subseteq A \cap R = a$ maximal filter of R by (i). But then $\omega = A \cap R$. From (iii) and the above observation, if $(a|b) \in A$ then $(b \to a) \in \omega$, implying that $(a|b) \in h_\omega^{-1}(\{u, 1\})$.

We specify now basic semantic concepts of CL. Recall again that Ω is the class of models (maximal filters) of R. In order to define the concept of *tautologies* in terms of models of $R|R$, we need to specify the class of "designated truth values" (Rescher, 1969, p. 66-71). Indeed, as in any multi-valued logic, among the truth-values $0, u, 1$, we have to classify (or designate) certain of these values as "truth-like" values (for the concept of *contradictions*, the dual concept is "false-like" values or "antidesignated" values). Thus, if 1 is the only designated value, then $a|b \in R|R$ is a tautology if it is "true" in all models of $R|R$, that is, for $\omega \in \Omega$, $(a|b) \in h_\omega^{-1}(1)$. It is clear that $(1|1)$ is the only tautology in this sense. Indeed, $(a|b) \in h_\omega^{-1}(1)$ if and only if $ab \in \omega$. Thus if $0 < ab < 1$, then $(ab)' \neq 0$, so that there is some $\omega \in \Omega$ such that $(ab)' \in \omega$ and hence $ab \notin \omega$. Of course, if $ab = 1$ then $b \geq ab$ implying that $b = 1$, and $(a|b) = (ab|b) = (1|1)$.

If $\{u, 1\}$ is the set of designated truth-values, then $(a|b)$ is a tautology if for $\omega \in \Omega$, $(a|b) \in h_\omega^{-1}(\{u, 1\})$.

Now $0 < ab < b$, then $a'b \neq 0$, so that there is some $\gamma \in \Omega$ such that $a'b \in \gamma$, so that $(a|b) \notin h_\gamma^{-1}(\{u, 1\})$. Thus $ab = b \neq 0$, that is, $(a|b) = (ab|b) = (b|b) = (1|b)$, and hence, the class of $\{u, 1\}$-tautologies is $\{(1|b), b \in R\backslash\{0\}\}$ which is the class of unity-type conditionals $\mathcal{U}$ investigated in Section 6.1. Note that for $\omega \in \Omega$, $h_\omega(0|0) = u$, so that formally $(0|0)$ is also a tautology. To exclude $(0|0)$, one should require that $(a|b)$ is a $\{u, 1\}$-tautology if $(a|b) \in h_\omega^{-1}(\{u, 1\})$ for $\omega \in \Omega$, and there is at least one $\gamma \in \Omega$ such that $h_\gamma(a|b) = 1$.

The concept of *entailment relation* in CL is formalized as follows. We say that $(a|b)$ logically entails $(c|d)$, in symbols,

$$(a|b) \overset{CL}{\vDash} (c|d),$$

if for $\omega \in \Omega$, $(c|d) \in h_\omega^{-1}(1)$, whenever $(a|b) \in h_\omega^{-1}(1)$ and $(c|d) \in h_\omega^{-1}(\{u, 1\})$ whenever $(a|b) \in h_\omega^{-1}(\{u, 1\})$. Roughly speaking, $(a|b)$ entails $(c|d)$ if the truth-value of $(c|d)$ is greater (or equal) than that of $(a|b)$. More precisely,

Theorem 2. *The following are equivalent.*

(*i*) $(a|b) \overset{CL}{\vDash} (c|d)$,

(*ii*) for $\omega \in \Omega$, $h_\omega(a|b) \le h_\omega(c|d)$, and

(*iii*) $(a|b) \le (c|d)$.

Proof. That (i) and (ii) are equivalent is obvious. To get the equivalence of (i) and (iii), note that $h_\omega(a|b) = 1$ if and only if $ab \in \omega$, and $h_\omega(a|b) \in \{u, 1\}$ if and only if $b' \vee a \in \omega$. This can be rephrased. For $\omega \in \Omega$, $ab \in \omega$ implies $cd \in \omega$, and for $\omega \in \Omega$, $b' \vee a \in \omega$ implies $d' \vee c \in \omega$. By Lemma 2 of Section 6.1, these statements are equivalent to $ab \le cd$ and $b' \vee a \le d' \vee c$, or $ab \le cd$ and $c'd \le a'b$, which means (iii). (See Theorem 1, Section 3.3). □

As in the case of C_2, the logical entailment relation $\vDash$ in CL is monotone. This follows readily from the fact that h_ω is a homomorphism. See, however, Chapter 8.

Now to Conditional Probability Logic (CPL). One of the practical motivation for considering conditional probabilities lies in the construction of Bayesian (causal) networks (for example, Lauritzen and Spiegelhalter, 1988). For quantifying rules in intelligent systems with other uncertainty measures, see for example, Dubois and Prade, 1990. As in the case of PL, if P is a probability measure on R, then $P(a|b) = r$ means that a is "time" in 100 r% of the "possible worlds" in which b is "true." A *model of CPL* is an extension $\hat{P} : R|R \to [0,1]$ of a probability measure P on R, defined by

$$\hat{P}((a|b)) = P(a|b), \text{ for } P(b) \ne 0.$$

As in the case of probability models, $\hat{P}$ has the flavor of a "homomorphism-like" map. See also the previous discussion concerning Adams' ε-semantics. We write $\hat{P}$ simply as P.

If 1 is the only designated truth-value, then $(a|b) \in R|R$ is a *CPL-tautology* if $P(a|b) = 1$ for all P on R such that $P(b) \neq 0$. The class of CPL-tautologies is precisely that of unity-type conditionals $\mathcal{U}$ of Section 6.1. Indeed, if $P(a|b) = 1$ for all P, then $P(ab) = P(b)$, for all P, and hence $ab = b$ (Lemma 1 of Section 2.2), so that $(a|b) = (ab|b) = (b|b) = (1|b)$. The converse is obvious. In the same vein, $P(a|b) = 0$ for all P if and only if $(a|b) = (0|b) \in \mathcal{Z}$, the class of zero-type conditionals in Section 6.1.

If $\{u, 1\}$ is the set of designated truth-values, then $(a|b)$ is a $\{u, 1\}$-tautology if $(a|b) \in h_{\omega}^{-1}(1)$ for at least one $\omega \in \Omega$. This class of $\{u, 1\}$-tautologies also coincides with $\mathcal{U}$. Indeed, let $(1|b) \in \mathcal{U}, b \neq 0$. We have $(1|b) \in h_{\omega}^{-1}(\{u, 1\})$, for $\omega \in \Omega$ since $bb' = 0$. Next, since $b \neq 0$, there is some $\gamma \in \Omega$ such that $b \in \gamma$, that is, $(1|b) \in h_{\omega}^{-1}(1)$.

Conversely, let $(a|b)$ be a $\{u, 1\}$-tautology. We have $a'b = 0$, that is, $b \leq a$. Hence $(a|b) = (ab|b) = (b|b)$ with $b \neq 0$, since by hypothesis, there is $\gamma \in \Omega$ such that $b \in \gamma$.

The following theorem summarizes basic relations among all above concepts, the proof of which follows simply by inspection.

Theorem 3.

(i) *The following are equivalent.*

α) $(a|b) = (c|d)$,

β) *for* $\omega \in \Omega$, $h_{\omega}(a|b) = h_{\omega}(c|d)$,

γ) *for* $j = 3$ *or* 4, $f_j((a|b), (c|d))$ *is a CPL-tautology* (f_3, f_4 *of Theorem 3, Section 6.2*).

δ) *for* $j = 1$ *or* 2, $f_j((a|b), (c|d))$ *and* $f_j((c|d), (a|b))$ *are CPL-tautologies* (f_1, f_2 *of Theorem 2, Section 6.2*).

(ii) *The following are equivalent.*

α) $P(a|b) = P(c|d)$ *for* P *such that* $P(b), P(d) \neq 0$,

β) *for* $j = 3$ *or* 4, $(a|b)$ *and* $(c|d)$ *are CPL-tautologies, or* $(a|b)'$ *and* $(c|d)'$ *are CPL-tautologies, or* $f_j((a|b), (c|d))$ *is a CPL-tautology,*

γ) $(a|b) = (c|d)$ *or* $(a|b), (c|d) \in \mathcal{U}$ *or* $(a|b), (c|d) \in \mathcal{Z}$.

(iii) *The following are equivalent.*

α) $(a|b) \le (c|d)$,

β) *for* $\omega \in \Omega$, $h_\omega(a|b) \le h_\omega(c|d)$,

γ) *for* $j = 1$ *or* 2, $f_j((a|b), (c|d))$ *is a CPL-tautology.*

(iv) *The following are equivalent.*

α) $P(a|b) \le P(c|d)$ *for* P *such that* $P(b), P(d) \ne 0$,

β) $(a|b) \le (c|d)$ *or* $(a|b) \in \mathcal{Z}$ *or* $(c|d) \in \mathcal{U}$,

γ) *for* $j = 1$ *or* 2, $f_j((a|b), (c|d))$ *or* $(a|b)'$ *is a CPL-tautology, or* $(c|d)$ *is a CPL-tautology.*

In C_2, $b \models a$ if and only if $b \to a = b' \vee a = 1$. The counterpart of this equivalence in CL is that $(c|d) \models (a|b)$ if and only if $f_1((a|b), (c|d))$ or $f_2((a|b), (c|d))$ is a (CPL)-tautology (that is, in $\mathcal{U}$). Note that f_1 and f_2 play the role of material implication on R. The equivalence above follows from Theorem 2 of Section 6.2.

Finally, the logical entailment relation in CPL is defined by

$$(c|d) \overset{CPL}{\models} (a|b) \quad \text{if} \quad P(c|d) \le P(a|b)$$

for P such that $P(b)$ and $P(d) \ne 0$.

In view of Theorem 3, (iv), $(c|d) \overset{CPL}{\models} (a|b)$ if and only if

(1) $(c|d) \le (a|b)$, or

(2) $cd = 0$, or

(3) $b \le a$.

We summarize the four logical systems discussed in this chapter.

Classical two-valued Logic (C_2).

Alphabet/Base space: R

Logical operators and relations: $(\cdot)'$, $\wedge$, $\vee$, $\le$, $\to$, ...

Equational axioms: Axioms of Boolean ring R

Truth space: $\{0, 1\}$, designated value: 1

Models: $\Omega = \{\text{maximal filters of } R\}$

Tautologies: 1

Conditional Logic (CL)

Alphabet/Base: $R|R$

Logical operators and relations: See Chapters 3 and 4: (Lukasiewicz's three-valued logic)

Equational axioms: Axioms of abstract conditional space (Chapter 4)

Truth space: $\{0, u, 1\}$, designated values $\{u, 1\}$

Models: {homomorphisms h_ω, $\omega \in \Omega$}

Tautologies: $\mathcal{U} = \{(1|b), b \in R\}$

Probability Logic (PL)

Alphabet/Base space: R

Logical operators and relations: same as C_2

Equational axioms: same as C_2

Truth space: $[0, 1]$ designated value: 1

Models: $\{P : R \to [0, 1]$, probability measures$\}$

Tautologies: 1

Conditional Probability Logic (CPL)

Alphabet/Base space: $R|R$

Logical operators and relations: same as CL

Equational axioms: same as CL

Truth space: $[0, 1]$, designated value: 1

Models: {extended P from R to $R|R$}

Tautologies: $\mathcal{U} = \{ (1|b) : b \in R \setminus \{0\}\}$

6.4 Additional properties of CPL

Although the concrete base space $R|R$ is sufficient for applications, we present, however, in this section basic properties of CPL in a more general setting. Recall from Chapter 4 that the abstraction of $R|R$ is an abstract conditional space L in which its skeletal set L^* plays the role of R, and L is isomorphic to the concrete realization $(L^*|L^*)$.

As usual, the logical structure of CPL (L) is given by sets of rules $(Rul(L))$, deducts $(Ded(L))$, models $(Mod(L))$, semantic evaluations $(P(L)$ = all probability measures on $L^*)$, tautologies ($\mathcal{U}(L)$ where $\mathcal{U}(L) = \{(b|b), b \in L^*\setminus\{0\}\})$, and axioms $(Ax(L)$ = axioms of L as an algebraic structure, together with a set of logical connectives

$f^{(o)}$ on $L^*|L^*$). Note that, as a base space, $(L^*|L^*)$ is a three-valued logical system. When $f^{(o)}$ is our set of logical connectives developed in Chapters 3 and 4, the corresponding three-valued logic is Lukasiewicz's. Different choices of $f^{(o)}$ lead to different three-valued logical systems.

In order to investigate deducts and tautologies, it is necessary to be able to identify certain deducts as tautologies and conversely. Specifically, first deducts here are of the form $h(\alpha) = g(\alpha)$ or $h(\alpha) \leq g(\alpha)$ where α is any collection of conditional event variables and h, g are combinations of logical operators of L. In view of the remarks following Theorems 2 and 3 of Section 6.3, we can make the following identifications:

$$(h(\alpha) = g(\alpha)) \mapsto f_i(h(\alpha), g(\alpha)), \quad i = 3 \text{ or } 4,$$

$$(h(\alpha) \leq g(\alpha)) \mapsto f_i(h(\alpha), g(\alpha)), \quad i = 1 \text{ or } 2,$$

In the case of $R|R$, we have,

$$f_2((a|b), (c|d)) = e \cdot ((c|d) \Rightarrow (a|b)),$$

where $e = ab \vee c'd \vee bd \vee b'd'$ and $\Rightarrow$ is the extended material implication on $R|R$, that is, $(c|d) \Rightarrow (a|b) = (c|d)' \vee (a|b)$. Note that f_2 is Sobocinski's material implication.

Using the notation of Chapter 4, it can be checked that the same situation holds in the general case. Specifically, on L, we have

$$f_2(\beta, \alpha) = e \cdot (\beta \Rightarrow \alpha),$$

where

$$e = \beta^* \vee \alpha'^* \vee ((\alpha \cdot \alpha')^* \Leftrightarrow (\beta \cdot \beta')^*).$$

Thus f_2 is definable in terms of the primitive operators of L. We are now ready to prove the following.

Theorem 1. *CPL is sound and complete.*

Proof. Using the above identifications, any deduct of L (in the form of equality or inequality) is a single conditional event. By Theorem 1 of Section 6.3, its identification is a tautology if and only if its corresponding deduct represents a true relation (equality or inequality) which is obvious here.

For completeness, first note that, for $\alpha \in L$, $\alpha = f_2(\alpha, \alpha)$. In particular, if $\alpha \in \mathcal{U}(L)$, then $f_2(\alpha, \alpha)$ is a tautology. Using the identification $\alpha \leq \alpha \mapsto \alpha$ (as a deduct), α is itself a deduct. □

Remarks

1. Suppose that instead of identifying equational axioms and resulting deduct forms as above, one replaces formally all axioms by the corresponding single conditional event forms depending on the f_i's. Thus to completely axiomatize all relevant expressions, avoiding the introduction of external equality, single conditional event forms must be introduced as axioms characterizing f_2 and f_4, in part. Therefore, one can ask whether it is true that the added axioms in combination with new *Rul*(*L*) would yield *Ded*(*L*) as interpreted in the above identification from the equational axiom approach. Here, the added axioms are

$$(\text{for all } \alpha, \beta \in L)\ (f_4(f_2(\beta, \alpha), e\cdot(\beta \Rightarrow \alpha)),$$

$$(\text{for all } \alpha, \beta \in L)\ (f_4(f_4(\alpha, \beta), f_2(\alpha, \beta)\cdot f_2(\beta, \alpha)))$$

and new *Rul*(*L*) is given by using f_2 and f_4 analogously as the derived inequality (partial ordering) $\leq$ and equality $=$:

For all $\alpha, \beta, \gamma \in L$,

If $f_j(\alpha, \beta), f_j(\beta, \gamma)$ are deducts, then so is $f_j(\alpha, \gamma)$, $j = 2$ or 4.
$f_j(\alpha, \alpha)$ is always a deduct, $j = 2$ or 4 (this can also be an axiom).
If $f_4(\alpha, \beta)$ is a deduct, then so is $f_4(\beta, \alpha)$.

In a related vein, Rescher (1969, p. 66-67) discusses changing Lukasiewicz $\mathcal{L}_3$ from a one designated truth value logic $\{u, 1\}$, making a significant enlargement of the class of possible tautologies for the logic. Rescher states that the axiomatization of this new logic is an open issue.

In view of our results in Chapter 3 and 6.2, together with the identifications in L, and with $\mathcal{L}_3$ augmented with Slupecki's τ-operator (Rescher, 1969, p. 163), and L augmented with $\Rightarrow^{\mathcal{L}_3}$, etc., necessarily with two designated truth values, Theorem 1 in this section seems to point to the possible axiomatization of the logic Rescher considers via the structure of L. But, more work is needed on this.

2. Abstract conditional spaces appear to be related to "implicative" algebras in general (for example, Rasiowa, 1974). From an examination of the axioms describing them, the more specialized pseudo-Boolean or quasi-pseudo-Boolean algebras may also be related. The relations need to be also explored to determine any mutual benefit of results.

3. By comparison of truth values of various conditional operators (Chapter 3) with Sobocinski's truth tables given in Rescher (1969, p. 70), it follows that Sobocinski's logic with designated values $\{u, 1\}$ coincides with the choice of negation, conjunction and disjunction, as Schay-Adams-Calabrese have independently done. (See also Dubois and Prade, 1989.) Unlike Lukasiewicz's implication operator, Sobocinski's implication is a material implication formed out of truth tables $\psi_{(\cdot)'}$ (negation) and $\psi_{\vee}$ (Schay), that is, $j \rightarrow i$ has truth table given by $\psi_{\vee}(\psi_{(\cdot)'}(j), i)$, $i, j \in \{0, u, 1\}$. See also Sobocinski's original work (Sobocinski, 1952) where it is shown, as an alternative to Wajsberg's well-known full axiomatization of $\mathcal{L}_3$ -- or the associated expanded axiomatization for Slupecki's extension of $\mathcal{L}_3$ via his τ-operator, corresponding to the special element u_o (or $(0|0)$ in the concrete case of R; see Rescher (1969, p. 155), that Sobocinski's system can be fully axiomatized. Furthermore, as a justification for Sobocinski's approach, by an analogous extension as Slupecki's, the resulting logic is seen to be also truth functionally operator-complete, being the only other known such system. (See also Rose (1953) and Schalz (1959)). Specifically, note that Slupecki's extension of $\mathcal{L}_3$ being truth functionally operator-complete translates, via Theorem 2 of Section 3.4, into the fact that $(R|R, \cdot, \vee, \overset{\mathcal{L}_3}{\Rightarrow}, (0|0))$ is a truth functionally operator-complete system relative to all possible extended Boolean conditional operators. Indeed, going back to $\mathcal{L}_3$, since max ("or"), min ("and") and $1-\cdot$ ("not") can all be shown to be compounds only of $\overset{\mathcal{L}_3}{\Rightarrow}$, so that $(\overset{\mathcal{L}_3}{\Rightarrow}, u)$ is sufficient to span operationally all three-valued truth-functional operators, hence by Theorem 2 of Section 3.4 again, the corresponding conditional operators must, likewise, span all possible extended Boolean conditional operators. Similarly, the enlarged Schay-Adams-Calabrese system, corresponding to the enlarged Sobocinski logic, is truth functionally operator-complete.

Thus, via Theorem 2 of Section 3.4, one can now justify the Schay-Adam-Calabrese approach to conditional event algebra as being equivalent to Sobocinski's logic, however, as noted earlier in Section 3.5, quite distinct from Lukasiewicz's logic, the monotone bound violations for conjunction and disjunction notwithstanding! See, however, Chapter 8, Section 8.2.

4. Using the technique of Theorem 2, Section 3.4, we obtain the following three-valued truth tables for corresponding conditional operators:

(i) It will be shown in Chapter 8, Theorem 2, Section 8.1 that $R|R$ is a relatively pseudo-complemented lattice with *relative* pseudo-complementation $\rightarrow$ given by:

$$(c|d) \rightarrow (a|b) = (ab \vee c'd \vee b'd' \,|\, b \vee c'd \vee b'd') .$$

Its truth table is

$$\psi(i,j) = \begin{cases} 1 \text{ for } (i,j) \in \{(0,0),(0,u),(0,1),(u,u),(u,1),(1,1)\} \\ u \text{ for } (i,j) = (1,u) \\ 0 \text{ for } (i,j) \in \{(u,0),(1,0)\}\ . \end{cases}$$

(ii) In particular, the pseudo-complementation operator of $(R|R)$ is:

$$(a|b)^* = (a'b|1)$$

with truth table

$$\psi(0) = 1,\ \ \psi(u) = 0 = \psi(1),$$

which is that of negation in Heyting's three-valued logic (as mentioned in Section 3.5).

(iii) The material implication on $(R|R)$, $(c|d) \Rightarrow (a|b) = (c|d)' \vee (a|b)$, has truth table given by

$$\psi(i,j) = \begin{cases} 1 \text{ for } (i,j) \in \{(0,0),(0,u),(0,1),(u,1),(1,1)\} \\ u \text{ for } (i,j) \in \{(u,0),(u,u),(1,u)\} \\ 0 \text{ for } (i,j) = (1,0)\ . \end{cases}$$

(iv) Slupecki's τ-operator (Rescher, 1969, p. 163) has the following truth table corresponding to the constant function u, that is, $\psi(i) = u$, for $i \in \{0, u, 1\}$.

CHAPTER 7

FUZZY CONDITIONALS

This chapter is devoted to the extension of the measure-free conditioning concept to the fuzzy case. Motivated by a random set connection and by the concept of generalized indicator function of conditional events, a form of membership functions for fuzzy conditionals is proposed. It turns out that fuzzy conditionals are interval-valued fuzzy sets. Syntax considerations, as well as probability qualification of fuzzy conditionals, are investigated. Prior to a formal development of fuzzy conditionals, basic aspects of fuzziness and fuzzy logics are reviewed.

7.1 Generalities on fuzziness

The reader is referred to Klir and Folger (1988) for an introduction to the theory of fuzzy sets, and to Zadeh (1988) for an excellent exposition of fuzzy logic and its applications.

Human communication is based on natural language. Natural language contains fuzzy concepts such as "high," "almost," "likely," "intelligence," etc. From a human viewpoint, fuzziness is well-understood, and can be taken as a primitive notion. The uncertainty in fuzziness is much more complex than that in randomness. Indeed, imprecision, subjectivity, and context dependency surround each fuzzy label in natural language. The imprecision and the context dependency of the above examples of fuzzy labels are clear. By subjectivity, we mean that individuals might "understand" a fuzzy label in different ways. In other words, fuzzy concepts are not universal (or objective), as opposed to, say, mathematical concepts. This is perhaps the main source of difficulty in trying to formulate a semantic (meaning) information theory. See also MacLennan (1988).

Fortunately, there exists such a "thing" as common sense knowledge which allows us to approximate fuzzy concepts in a reasonable fashion. Consider, for example, the information "the temperature is high." A little reflection will reveal that, underlying this statement, there are: a universe of discourse X, namely the range of the (variable) temperature; the variable "temperature" ξ itself; and the fuzzy predicate a = "high." Thus, the above information is translated into "ξ is a." For this translation to be part of a knowledge representation language, we need to model a more concretely. With respect to X, a is "inside" X. The standard approach to translate this vague idea into mathematics is to regard a as a sort of subset of X. Specifically, the imprecision in the word "high" forces us to consider a as a generalized subset of X, in the sense that membership in a ranges over the unit interval $[0, 1]$ rather than just $\{0, 1\}$ as in the case of ordinary subsets of X. Generalizing the concept of indicator functions of ordinary sets, a semantic modeling of the fuzzy concept a is given by a membership function $\mu_a : X \to [0, 1]$, where, for each $x \in X$, $\mu_a(x)$ is to be interpreted as the degree to which x is compatible with the meaning of a. Also, $\mu_a(x)$ can be interpreted as the truth value of the proposition "x is a member of a." Defining this way, a is referred to as a fuzzy subset of X. (See Section 7.3 for a syntactic approach to fuzzy sets.)

Now, the subjectivity becomes apparent. For the same a, different individuals can assign different maps μ_a. The situation is diametrically opposite to that in random analysis where each random phenomenon is governed by one and only one distribution law. When a random law is unknown, one can try to gather relevant statistical data to estimate it or to test about it. This is possible since the law in question is unique.

From the above, we see that, to each fuzzy concept a (relative to X), there are different interpretations of its meaning representation at the mathematical level. This flexibility is sometimes beneficial. For example, users of a consulting system can input their own perception about fuzzy concepts.

At the level of application, a common sense membership function μ_a might be desirable. This μ_a can be obtained in various ways. For example, by bias of profession, a statistician might immediately think about getting μ_a by collecting data, say, in the form of questionnaires and by constructing μ_a based upon a frequency approach. Perhaps, this objective approach to constructing membership functions of fuzzy sets has triggered statements such as "probability theory can handle fuzziness." We emphasize the fact that, while at the practical level, a probabilistic approach to modeling fuzzy concepts is reasonable (but not the only one), the primitive concept of fuzziness is clearly different from that of randomness. In fact, a coexistence of these two notions is useful in machine intelligence. Moreover, fuzziness has the luxury of producing membership functions from human perception, when statistical data are not available. However, at the membership function level, there is a specific relationship between fuzziness and probability theory via the concept of random sets. This relationship shows that fuzziness is a weak specification of random sets through the one-point coverage function. (See Section 7.4 for an application of this relationship.)

It is appropriate here to say a few words about uncertainty. Statements like "all statisticians agree on the use of probability to model uncertainty" (French, 1990) should be clarified a little further. By uncertainty in statistics, we mean a very specific type of uncertainty, namely randomness. It is now well-accepted that, outside of statistics, especially in AI models, there is a clear distinction between uncertainty and probability (see for example, Bellman, 1978; Levi, 1973; Neapolitan, 1990). More specifically, probability theory models one type of uncertainty, while in general decision theory, other types of uncertainty may surface. Of course, by analogy with randomness, one can try to use statistical methodologies and techniques to model or to approximate other types of uncertainty (see for example, Mosteller and Youtz, 1990). But the intrinsic property of each type of uncertainty remains unchanged (see the comments of N. Cliff following the article of Mosteller and Youtz, p. 16-18). In our view, other non-probabilistic approaches to uncertainty modeling are not alternatives to statistical tools. Rather they address different problems in which the uncertainty involved is not statistical in nature (see for example, Neapolitan, 1990). This is similar to the situation in quantum probability (for example, Gudder, 1988). The concept of fuzziness, as an example, is best explained in the context of semantic processing of natural languages. (See again Neapolitan, 1990; also Levi, 1973). There are various reasons for ad-hoc uncertainty modeling to be attractive to designers of intelligent machines. This is a healthy sign in view of AI problems. For the problem of admissibility of uncertainty measures in expert systems, see Goodman, Nguyen and Rogers (1990).

So far, we have discussed the problem of meaning representation of fuzzy concepts. Whatever approaches are taken, we have a collection of membership function μ_a, $a \in A$, say, in a knowledge base of some system. The problem of interest is how to combine them in order to extract information for decision processes. This is basically the problem of using logic as a formal tool in artificial intelligence (see for example, Ramsay, 1988). More specifically, a formal logic will provide us with a way of constructing a meaning representation language in which facts, rules and deduction (for inference) can be stated. In this spirit, we are going to look at logical aspects of fuzzy sets.

7.2 Fuzzy logics

Roughly speaking, fuzzy logic is a knowledge representation language in which facts and rules involving fuzzy information can be represented mathematically, and in which inference with fuzzy data can be described logically. Fuzzy logic is essentially a logic that models the fuzziness in natural language.

To avoid confusion, it is necessary to classify different types of fuzzy logics. First-order fuzzy logics refer to logics of fuzzy sets in which the truth space is the unit interval $[0, 1]$. A fuzzy logic is called second-order if its truth space is the space of fuzzy subsets of $[0, 1]$. In any case, fuzzy logics are multivalued logics.

A standard first-order fuzzy logic is proposed by Zadeh by specifying semantic operations among fuzzy subsets of X as follows. The class of all fuzzy subsets of X is the set of maps $\mathcal{F}(X) = \{f : X \to [0, 1]\}$ from X into $[0, 1]$. For $f \in \mathcal{F}(X)$, the negation f' of f is defined to by $f'(x) = 1 - f(x)$. For $f, g \in \mathcal{F}(X)$, the "truth tables" for the connectives "and," "or" are, respectively

$$(f \wedge g)(x) = f(x) \wedge g(x) = min(f(x), g(x)),$$

and

$$(f \vee g)(x) = f(x) \vee g(x) = max(f(x), g(x)).$$

With respect to the truth space $[0, 1]$, these are Lukasiewicz's logical operations. Of course, this standard first-order fuzzy logic generalizes classical two-valued logic. See Klir and Folger, 1988, Section 1.6, for details. This approach is semantic in the sense that the objects under study are membership functions, generalizing indicator functions of ordinary subsets of X, rather than their counterparts of ordinary subsets of X. This point will be made precise in the next section.

When the above logical operations are applied to fuzzy subsets of $[0, 1]$ viewed as truth values in a second-order fuzzy logic, the resulting logic is referred in the literature simply as fuzzy logic. See Zadeh (1988) for additional details. When the truth space is $[0, 1]$, one can model the basic connectives "not," "and," "or" in various ways, extending however classical two-valued truth tables of these connectives. For negation (or fuzzy complement), one can use any *negation operator* , that is, any function

$$N : [0, 1] \to [0, 1]$$

such that

(i) $N(0) = 1, N(1) = 0,$
(ii) N is continuous,
(iii) N is involutive, that is, $N(N(x)) = x$ for $x \in [0, 1]$, and
(iv) N is non-increasing.

See for example, Bonissone and Decker, 1988.

For conjunction, it turns out that the class of *t-norms* (see Schweizer and Sklar, 1983) is appropriate to represent conjunction operators, where a *t*-norm is a binary operation T on $[0, 1]$ such that

(i) T is associative,
(ii) T is commutative
(iii) T is nondecreasing in each place, that is, if $x \leq y$ and $u \leq v$ then $T(x, u) \leq T(y, v)$, and

(iv) for $x \in [0, 1]$, $T(x, 1) = x$.

Note that (iii) and (iv) imply that $T(0, x) = 0$ for $x \in [0, 1]$, in particular, $T(0, 0) = 0$. Indeed, for $x \in [0, 1]$, $T(0, x) \leq T(0, 1) = 0$. A *t*-norm T is "Boolean-like" in the sense that its restriction to the vertices of $[0, 1]^2$ is a Boolean function, that is, $T(x, y) = 0$ or 1 whenever x and y are 0 or 1. These functions are used, for example, in neural networks to model activation functions of the units in the network. See for example, Williams (1986). Here, values in $[0, 1]$ are viewed as degrees of activation. Note that the associativity of *t*-norms is essential in extending these binary operations to *n*-ary operations on $[0, 1]$, $n \geq 2$.

Some common examples of *t*-norms are these:

$$T_1(x, y) = min\{x, y\},$$

$$T_2(x, y) = xy,$$

$$T_3(x, y) = max\{x + y - 1, 0\}.$$

For disjunction, the class of *t-conorms* is appropriate. A *t*-conorm S is a binary operation on $[0, 1]$ such that

(i) S is associative,
(ii) S is commutative,
(iii) S is nondecreasing in each place, and
(iv) $S(0, x) = x$j and $S(1, x) = 1$, for all $x \in [0, 1]$.

t-norms and *t*-conorms are dual in the following sense. If T is a *t*-norm, then

$$S(x, y) = 1 - T(1 - x, 1 - y)$$

is a *t*-conorm, and if S is a *t*-conorm, then

$$T(x, y) = 1 - S(1 - x, 1 - y) \text{ is a } t\text{-norm.}$$

If the negation operator N is defined by $N(x) = 1 - x$, then dual *t*-norms and *t*-conorms are related to N. Each triple (N, T, S) defines a first-order fuzzy logic. Thus, one can speak of fuzzy logics (in the plural).

The *t*-conorms dual to T_1, T_2, T_3 are

$$S_1(x, y) = max\{x, y\},$$

$$S_2(x, y) = x + y - xy,$$

$$S_3(x, y) = min\{x + y, 1\}.$$

The triple (N, T, S) given by

$$N(x) = 1 - x,$$

$$T(x, y) = min\{x, y\}, \text{ and}$$

$$S(x, y) = max\{x, y\},$$

forms the collection of basic operations on fuzzy sets. It is interesting to note that some t-norms admit probabilistic interpretations. For example, if the t-norm T is such that for $x \le u$ and $y \le v$,

$$T(u, y) - T(x, y) \le T(u, v) - T(x, v) , \qquad *$$

then, T is a two-dimensional *copula* (see Schweizer and Sklar, 1983). That is,

$$T : [0, 1]^2 \to [0, 1]$$

satisfies (*) and

a) $T(0, x) = T(x, 0) = 0$, for $x \in [0, 1]$, and
b) $T(1, x) - T(x, 1) = x$, for $x \in [0, 1]$.

t-norms satisfy all axioms of two-dimensional copulas (or copulas for short) except possibly (*). In general, copulas are not associative. The probabilistic interpretation of copulas is this. The distribution function of a random variable ξ that is uniformly distributed on $[0, 1]$ is

$$F(x) = \begin{cases} 0 \text{ for } x < 0 \\ x \text{ for } 0 \le x \le 1 \\ 1 \text{ for } x > 1 \end{cases} .$$

If (ξ, η) is a random vector with joint distribution function

$$G(x, y) = P(\xi \le x, \eta \le y),$$

then the marginal distributions are

$$F_\xi(x) = P(\xi \le x, \eta \le +\infty) = G(x, +\infty) ,$$

and

$$F_\eta(y) = P(\xi \leq +\infty, \eta \leq y) = G(+\infty, y) .$$

If F_ξ and F_η are both equal to F, then the restriction of G to $[0, 1]^2$ is a copula. Conversely, if T is a copula, then $H : \mathbb{R}^2 \to [0, 1]$ defined by

$$H(x, y) = T(F(x), F(y))$$

is a two-dimensional distribution function each of whose marginal distributions is F, where $\mathbb{R}$ denotes the set of real numbers.

Thus, roughly speaking, a copula is nothing more than a two-dimensional distribution function on $[0, 1]^2$ with uniform marginal distributions on $[0, 1]$. A basic result in Schweizer and Sklar (1983) is this. If H is the joint distribution function of (ξ, η), then there is a copula T such that

$$H(x, y) = T(H(x, +\infty), H(+\infty, y)), \text{ for } x, y \in \mathbb{R} .$$

The t-norms T_1, T_2, T_3 above are all copulas. For more details, see Schweizer and Sklar, 1983.

7.3 Syntax representation of fuzzy sets

Let X be a set. The power set of X is denoted by $\mathcal{P}(X)$. One can identify $\mathcal{P}(X)$ with the space $\{0, 1\}^X = \{f : X \to \{0, 1\}\}$ via the bijection

$$\varphi : \{0, 1\}^X \to \mathcal{P}(X)$$

defined by

$$\varphi(f) = f^{-1}(1) = \{x : f(x) = 1\} .$$

Two remarks are in order here. First, if $a \in \mathcal{P}(X)$, then $\varphi^{-1}(a) = 1_a$, the indicator function of a on X. If $\mathcal{P}(X)$ represents a collection of propositions, then it is the base space of classical two-valued logic or the "syntax part" of the logic. For each $x \in X$, the map $h_x : \mathcal{P}(X) \to \{0, 1\}$ defined by

$$h_x(a) = \begin{cases} 1 & \text{if } x \in a \\ 0 & \text{if } x \in a' \end{cases}$$

is a Boolean homomorphism, that is, a model of C_2. (See Chapter 6). Thus, the space of indicator functions $\{0, 1\}^X$ plays the role of "semantic part" of the logic, in the sense that, given a model h_x or simply x, a is true or false in x according to whether $1_a(x) = 1$ or 0.

Second, the above bijection φ can be written in a more explicit form:

$$f \in \{0, 1\}^X \to (f^{-1}(1), X),$$

with $f^{-1}(1) \subseteq X$. If we define, for each $t \in [0, 1]$,

$$A_t(f) = \{x : f(x) \geq t\} ,$$

then for $t > 0$,

$$A_0(f) = X, \quad A_t(f) = f^{1}(1).$$

Conversely, given (a, X), with $a \subseteq X$, then

$$1_a(x) = sup\{t \in [0, 1] : x \in A_t\},$$

where $A_0 = X, A_t = a, \forall t > 0$, is such that $A_t = A_t(1_a)$.

These facts are carried over to the fuzzy case in a straightforward manner as follows. In the standard approach, membership functions are used to model fuzzy concepts. Thus, the "semantic part" of fuzzy logic is $\mathcal{F}(X) = [0, 1]^X$. The "syntax part" is obtained as in the case of two-valued logic. Specifically, if $f \in \mathcal{F}(X)$, then the *level sets* (or α-cuts) of f are, for $\alpha \in [0, 1]$, $A_\alpha(f) = \{x : f(x) \geq \alpha\}$. (See for example, Dubois and Prade, 1980.) The family of ordinary subsets A_α, $\alpha \in [0, 1]$, of X satisfies the following properties:

(i) $\alpha \leq \beta$ implies $A_\beta \subseteq A_\alpha$,
(ii) $A_0 = X$, and
(iii) for $I \subseteq [0, 1]$, $\bigcap_{\alpha \in I} A_\alpha = A_{Sup\, I}$.

The condition (iii) is a form of left-continuity of the map $A : [0, 1] \to \mathcal{P}(X)$ defined by $\alpha \to A_\alpha$ in the sense that, for $\alpha \in [0, 1]$,

$$\lim_{\alpha \uparrow \alpha} A_\alpha = \bigcap_{\alpha < \alpha} A_\alpha = A_{\alpha^-} = A_\alpha,$$

where A_{α^+} denotes $\lim_{a \downarrow \alpha} A_\alpha = \bigcup_{\alpha > \alpha} A_\alpha$).

It turns out that these three conditions characterize the syntax part of fuzzy logic. Indeed, let us call a family $\{A_\alpha, \alpha \in [0, 1]\}$ of subsets of X a *flou set* (for example, Gentilhomme, 1968; Negoita and Ralescu, 1975) if the A_α's satisfy (i), (ii) and (iii) above. Denote the class of all flou sets of X by $\mathcal{F}\mathcal{L}(X)$, and consider the map $\varphi : \mathcal{F}(X) \to \mathcal{F}\mathcal{L}(X)$ defined by $\varphi(f) = \{A_\alpha(f), \alpha \in [0, 1]\}$. Note that, a flou set is in fact a map $A : [0, 1] \to \mathcal{P}(X)$ given by $\alpha \to A_\alpha$, and we write $A = \{A_\alpha, \alpha \in [0, 1]\}$ for simplicity. Thus, two flou sets $A = \{A_\alpha, \alpha \in [0, 1]\}$ and $B = \{B_\alpha, \alpha \in [0, 1]\}$ are equal if and only if $A_\alpha = B_\alpha$, for $a \in [0, 1]$. It is easy to check that φ is a bijection. Indeed, if $\varphi(f) = \varphi(g)$, then $A_\alpha(f) = A_\alpha(g)$, for $\alpha \in [0, 1]$. But, obviously for $x \in X$,

$$f(x) = sup\{\alpha : x \in A_\alpha(f)\} ,$$

and hence $f \equiv g$, that is, φ is one-to-one. To show that φ is onto, we take an arbitrary flou set $A = \{A_\alpha, \alpha \in [0, 1]\}$, and consider its "characteristic function"

$$\psi_A : X \to [0, 1],$$

where

$$\psi_A(x) = sup\{\alpha : x \in A_\alpha\} .$$

We are going to show that A_α is a α-level set of ψ_A for all $\alpha \in [0, 1]$. If $x \in A_\alpha$, then by construction, $\psi_A(x) \geq \alpha$. Conversely, let x be such that $\psi_A(x) \geq \alpha$ and $I_x = \{\beta : x \in A_\beta\}$, we have $\psi_A(x) = sup\, I_x$. By condition (iii), $\bigcap_{\beta \in I_x} A_\beta = A_{\psi_A(x)}$. By (ii), $A_{\psi_A(x)} \subseteq A_\alpha$. Thus $\{x : \psi_A(x) \geq \alpha\} \subseteq A_\alpha$, and the result follows.

The logical operations on $\mathcal{F}\mathcal{L}(X)$ can be defined in such a way that φ is an isomorphism. For this purpose, conjunction and disjunction are defined as follows. For $A = \{A_\alpha, \alpha \in [0, 1]\}$ and $B = \{B_\alpha, \alpha \in [0, 1]\}$,

$$A \wedge B = \{A_\alpha \cap B_\alpha, \alpha \in [0, 1]\} ,$$

and

$$A \vee B = \{A_\alpha \cup B_\alpha, \alpha \in [0, 1]\} .$$

With respect to these operations, Negoita and Ralescu (1975) have established a lattice $(\wedge, \vee)$-isomorphism between $\mathcal{F}(X)$ and $\mathcal{F}\mathcal{L}(X)$. This can be seen by observing that for $f, g : X \to [0, 1]$, we have for $\alpha \in [0, 1]$,

$$\{x : f(x) \wedge g(x) \geq \alpha\} = \{x : f(x) \geq \alpha\} \cap \{x : g(x) \geq \alpha\} ,$$

and

$$\{x : f(x) \vee g(x) \geq \alpha\} = \{x : f(x) \geq \alpha\} \cup \{x : g(x) \geq \alpha\} .$$

The "negation" operator on $\mathcal{F}\mathcal{L}(X)$ is defined as follows. From

$$A = \{A_\alpha, \alpha \in [0, 1]\},$$

look at its "characteristic function"

$$\psi_A(x) = sup\{\alpha : x \in A_\alpha\}.$$

Consider $\psi'_A(x) = 1 - \psi_A(x)$, for $x \in X$. Let $A' = \{A'_\alpha, \alpha \in [0, 1]\}$, where A'_α is the α-level set of $\psi'_A(\cdot)$, that is,

$$\begin{aligned} A'_\alpha &= \{x : \psi'_A(x) \geq \alpha\} \\ &= \{x : \psi_A(x) \leq 1 - \alpha\} \\ &= \{x : sup\{\beta : x \in A_\beta\} \leq 1 - \alpha\} \\ &= X \setminus \{x : \psi_A(x) > 1 - \alpha\} \\ &= X \setminus \bigcup_{\beta > 1-\alpha} (x : \psi_A(x) \geq \beta) \\ &= X \setminus A_{(1-\alpha)^+}. \end{aligned}$$

The negation of A is taken to be A' above.

Theorem 1. *φ is an isomorphism between $(\mathcal{F}(X), ', \wedge, \vee)$ and $(\mathcal{F}\mathcal{L}(X), ', \wedge, \vee)$.*

Proof. In view of the previous analysis, it remains only to show that φ preserves the logical operations. The preservation of $\wedge$ and $\vee$ is obvious. That of $'$ follows from the fact that if $\varphi(f) = A = \{A_\alpha, \alpha \in [0, 1]\}$, then $f = \psi_A$, and from the definition of $'$ on $\mathcal{F}\mathcal{L}(X)$. □

More concretely, flou sets can be identified with *partitions* of X as follows. By a partition of X we mean a *map* $Q : J \to \mathcal{P}(X)$, where $J \subseteq [0, 1]$, satisfying

(i) for $\alpha \in J, Q_\alpha \neq \emptyset$,
(ii) for $\alpha, \beta \in J$ with $\alpha \neq \beta, Q_\alpha \cap Q_\beta = \emptyset$, and
(iii) $\bigcup_{\alpha \in J} Q_\alpha = X$.

Note that a *usual* partition of X is the range of a partition (map) in the above sense. As in the case of flou sets, two partitions $Q^{(1)} : J_1 \to \mathcal{P}(X)$, $Q^{(2)} : J_2 \to \mathcal{P}(X)$

are equal if and only if $J_1 = J_2$ and $Q_\alpha^{(1)} = Q_\alpha^{(2)}$, for $\alpha \in J_1$.

Theorem 2. *Let* $\mathbb{P}(X)$ *be the space of all partitions of* X, *and* $\eta : \mathbb{P}(X) \to \mathcal{F}\mathcal{L}(X)$ be defined by $\eta(Q) = A$, *where* $Q : J \to \mathcal{P}(X)$ and, $A_\alpha = \bigcup_{\substack{\beta\in J\\ \beta\geq\alpha}} Q_\beta$. *Then* η *is a bijection.*

Proof. First, $\eta(Q)$ so defined is indeed a flou set. Since Q is a partition, $A_0 = X$. This is obvious by construction, for $\alpha, \beta \in [0, 1]$, if $\alpha \leq \beta$ then $A_\beta \subseteq A_\alpha$. Finally, the "left-continuity" of A is seen as follows. Let $I \subseteq [0, 1]$. If

$$x \in A_{sup\, I} = \bigcup_{\substack{\beta\in J\\ \beta\geq sup\, I}} Q_\beta,$$

then by monotonicity of A,

$$x \in \bigcap_{\alpha\in I} \Big(\bigcup_{\substack{\beta\in J\\ \beta\leq\alpha}} Q_\beta \Big).$$

Conversely, if

$$x \in \bigcap_{\alpha\in I} \Big(\bigcup_{\substack{\beta\in J\\ \beta\geq\alpha}} Q_\beta \Big),$$

then for $\alpha \in I$, there exists $\beta \in J$, $\beta \geq \alpha$ such that $x \in Q_\beta$. But Q is a partition, so there is only one value of Q that contains x, say $Q_{\beta(x)}$. Thus $\beta(x) \geq \alpha$, for $\alpha \in I$, and hence $sup\, I \leq \beta(x)$, implying that $x \in A_{sup\, I}$.

To show that η is onto, we proceed as follows. Let $A = \{A_\alpha, \alpha \in [0, 1]\}$ be a flou set of X. By Theorem 1, A is uniquely determined by its characteristic function ψ_A. Let $J_A \subseteq [0, 1]$ be the range of ψ_A, that is, $\alpha \in J_A$ if and only if

$$\{x \in X : \varphi_A(x) = \alpha\} = (\varphi_A = \alpha) = \psi_A^{-1}(\alpha) \neq \emptyset.$$

Obviously, $J_A \neq \emptyset$. For $\beta \in J_A$, define

$$Q_\beta = A_\beta \setminus A_{\beta+} = \psi_a^{-1}[\beta, 1] \setminus \psi_A^{-1}(\beta, 1] = \psi_A^{-1}(\beta),$$

where $A_{\beta+} = \bigcup_{\gamma>\beta} A_\gamma$ with $A_{1+} = \emptyset$. Obviously, $Q_\beta \neq \emptyset$ for $\beta \in J_A$. By the definition of J_A, if $\alpha, \beta \in J_A$ and $\alpha \neq \beta$, we have $Q_\alpha \cap Q_\beta = \emptyset$. Finally, if $x \in X$, then $x \in (\psi_A = \psi_A(x))$ so that $x \in Q_\alpha$ with $\alpha = \psi_A(x) \in J_A$. Thus $Q = \{Q_\beta, \beta \in J_A\}$ is a partition of X. It remains to check that $\eta(Q) = A$. But

$$A_\alpha = \psi_A^{-1}[\alpha, 1] = \bigcup_{\beta \le \alpha} \psi_A^{-1}(\beta) = \bigcup_{\substack{\beta \in J_A \\ \beta \le \alpha}} Q_\beta.$$

This last equality follows from the fact that if $\beta \geq \alpha$ and $\beta \notin J_A$, then $\psi_A^{-1}(\beta) = \emptyset$.

To show that η is one-to-one, we suppose $\eta(Q^{(1)}) = \eta(Q^{(2)}) = A$, where $Q^{(1)}$ and $Q^{(2)}$ are two partitions of X, with domains $J^{(1)}$ and $J^{(2)}$, respectively. From the above discussion, we see that $J^{(1)} = J^{(2)} =$ range of ψ_A. Also, for $\beta \in J^{(1)} = J^{(2)}$, $Q_\beta^{(1)} = Q_\beta^{(2)} = \psi_A^{-1}(\beta)$. □

Remarks. (i) In view of Theorem 2 and the algebraic structure of $\mathcal{F}\mathcal{L}(X)$, one can define logical operations on $\mathbb{P}(X)$ so that the bijection η is an isomorphism. Specifically, if $Q : J \subseteq [0, 1] \to \mathcal{P}(X)$ is a partition of X, then its "negation" is the partition $Q' : 1 - J \to \mathcal{P}(X)$ defined by $Q'_\alpha = Q_{1-\alpha}$. This is justified as follows. Let $A = \eta(Q)$. Then for $\beta \in J$, $Q_\beta = (\psi_A = \beta)$. The "complement" of ψ_A is $\psi'_A = 1 - \psi_A$. The range of ψ'_A is $1 - J$. Thus $Q'_\alpha = (\psi'_A - \alpha)$ for $\alpha \in 1 - J$, that is, $Q'_\alpha = (\psi_A = 1 - \alpha) = Q_{1-\alpha}$.

For $i = 1, 2$, let $Q^{(i)} : J^{(i)} \to \mathcal{P}(X)$ and $A^{(i)} = \eta(Q^{(i)})$. The "conjunction" of $\psi_{A^{(1)}}$ and $\psi_{A^{(2)}}$ is $\psi_{A^{(1)}} \wedge \psi_{A^{(2)}}$. Define $Q : \text{range}(\psi_{A^{(1)}} \wedge \psi_{A^{(2)}}) \to \mathcal{P}(X)$ by $Q_\alpha = (\psi_{A^{(1)}} \wedge \psi_{A^{(2)}} = \alpha)$. Q is taken to be the "conjunction" of $Q^{(1)}$ and $Q^{(2)}$. Similarly, the "disjunction" of $Q^{(1)}$ and $Q^{(2)}$ is the partition defined on the range of $\psi_{A^{(1)}} \vee \psi_{A^{(2)}}$ by $(\psi_{A^{(1)}} \vee \psi_{A^{(2)}} = \alpha)$.

(ii) A similar isomorphism can be established between $\mathcal{F}\mathcal{L}(X)$ and the class of *nested random sets* of X. Specifically, by a *nested random set* of X, we mean a random element S, defined on the probability space $(\Omega, \mathcal{A}, P)$ with values in $\mathcal{P}(X)$, of the form $S(\omega) = A_{U(\omega)}$, where $A = \{A_\alpha, \alpha \in [0, 1]\}$ is a flou set, and U is a random variable, defined on $(\Omega, \mathcal{A}, P)$, and without loss of generality, uniformly distributed on $[0, 1]$. For a fixed U, consider

$$\mathcal{R}(U) = \{A_U : A \in \mathcal{F}\mathcal{L}(X)\}.$$

It is easy to check that the map

$$\tau : \mathcal{F}\mathcal{L}(X) \to \mathcal{R}(U)$$

defined by $\tau(A) = A_U$ is an isomorphism. For more details on algebraic and probabilistic bases for fuzzy sets, see Goodman (1990).

7.4 Fuzzy conditionals

As in any logic, the concept of fuzzy entailment (or implication) in fuzzy logic is essential for inference purposes. Consider a conditional rule of the form

"If X is a then Y is b",

where a, b are fuzzy subsets of some set Ω, say, and X and Y are variables taking values in Ω. In the *theory of possibility* (Zadeh, 1978), the *possibility distribution* of X (resp. Y) is taken to be the membership function μ_a (resp. μ_b) with the interpretation that

$$Poss(X = \omega) = \mu_a(\omega), \quad \omega \in \Omega .$$

Thus, a conditional rule of the above form can be viewed as a "fuzzy conditional."

In the past, various approaches to defining *conditional possibility distributions* have been proposed. Let $f(x, y)$ denote the joint possibility distribution of (X, Y), and f_1 (respectively, f_2) denote the marginal possibility distribution of X (respectively, Y), where

$$f_1(x) = \sup_y f(x, y).$$

In Nguyen (1978), the conditional possibility distribution of Y given X is defined by

$$f(y \mid x) = f(x, y) \max\{1, f_1(x)/f_2(y)\} ,$$

and in Hisdal (1978) as

$$f(y \mid x) = \begin{cases} f(x, y) & \text{if } f_1(x) > f(x, y) , \\ [f(x, y), 1] & \text{if } f_1(x) = f(x, y) . \end{cases}$$

Bouchon (1987) proposed two types of conditional forms. Let $f : \Omega_1 \to [0, 1]$, $g : \Omega_2 \to [0, 1]$ and T be a continuous t-norm.

(i) $(f(x) \mid g(y))_T = \sup\{t : t \in [0, 1], T(g(y), t) \leq f(y)\}$ with two special cases. First, for $T(x, y) = \min\{x, y\}$,

$$(f(x) \mid g(y))_T = \begin{cases} 1 & \text{if } f(x) \geq g(y) \\ f(x) & \text{if } f(x) < g(y). \end{cases}$$

Second, for $T(x, y) = xy$,

$$(f(x)\,|\,g(y))_T = min\{f(x)/g(y),\ 1\}.$$

(ii) Let $h : [0, 1] \to \mathbb{R}^+$ be a non-increasing, continuous function with $h(0) \leq +\infty$ and $h(1) = 0$. Let $N_h(t) = h^{-1}(h|0) - h(t))$, a negation. Then

$$(f(x)\,|\,g(y))_{N_h} = max\{N_n(g(y)), f(x)\}.$$

The approach (ii) is clearly a generalization of the use of material implication when $h(x) = 1 - x$. For other works on fuzzy implication operators, see Yager (1983), Sembi and Mamdani (1979), Mattila (1986), and Smets (1990).

Goodman and Stein (1989) attempted a definition for fuzzy conditioning, based upon the fuzzy set analogue of the basic characterization of conditional events as the solution set of a Boolean linear equation, that is, $\{x : x \in R, xb = ab\}$. Specifically, if S is a generalization of Zadeh's classical $(min,\ max,\ 1 - (\cdot))$ system over the set of all membership functions of fuzzy subsets of Ω (called there a semi-Boolean algebra, being a complete, bounded distributive DeMorgan lattice) with conjunction $*$ and partial order relation $\leq$ then, for $f, g \in S$, the conditional form $(f|g)$ is given by

$$(f|g) = \{h : h \in S,\ h*g = f*g\}.$$

This led to, for $x \in \Omega$,

$$(f|g)(x) = \begin{cases} f(x) & if\ f(x) < g(x)\,, \\ [g(x),\ 1] & if\ f(x) \geq g(x)\,. \end{cases}$$

Unfortunately, unlike the Boolean counterpart, closure of functionally extended operations did not hold.

In this section, we propose a new approach to fuzzy conditioning using random set representations of fuzzy set membership functions. Let X be a set, and for simplicity, let the Boolean ring R be $\mathcal{P}(X)$. For $a, b \in R$, the syntax representation of the conditional "a given b" is the coset $(a|b) = a + Rb'$, while its semantic representation (DeFinetti, 1964; Schay, 1968) is its "generalized" indicator function

$$\varphi(a|b) : X \to \{0,\ 1,\ u\}$$

defined by

$$\varphi(a|b)(x) = \begin{cases} 1 \text{ if } x \in ab \\ 0 \text{ if } x \in a'b \\ u \text{ if } x \in b' \end{cases},$$

where u stands for "undefined."

It is time to say a little more about the symbol u. In view of Lukasiewicz's three-valued logic, the logical operations on the truth space $\{0, 1, u\}$ are defined by

$$0' = 1, \quad 1' = 0, \quad u' = u;$$

$$0 \wedge 1 = 0 \wedge 0 = 0 \wedge u = 0, \quad 1 \wedge 1 = 1, \quad u \wedge 1 = u \wedge u = u;$$

$$0 \vee 0 = 0, \quad 0 \vee 1 = 1 \vee 1 = u \vee 1 = 1, \quad 0 \vee u = u \vee u = u.$$

For concreteness, u can be taken to be a number in $(0, 1)$, say $u = 1/2$, so that for $x, y \in \{0, 1, 1/2\}$,

$$x' = 1 - x, \quad x \vee y = max(x, y), \quad x \wedge y = min(x, y) .$$

Consider now the case of fuzzy sets. Our approach to defining fuzzy conditionals is based upon a relationship between membership functions of fuzzy sets and canonical random sets which are induced by uniformly distributed random variables. See, for example, Goodman and Nguyen (1985). Specifically, let $f : X \to [0, 1]$, let $(\Omega, \mathcal{A}, P)$ be a probability space, and U be a random variable defined on it and uniformly distributed over the unit interval $[0, 1]$. The random variable U is thought as a device for randomizing the α-level sets of f. Thus for $x \in X$,

$$f(x) = P(\omega : U(\omega) \leq f(x))$$

$$= P(U \leq f(x)) = P(U^{-1}[0, f(x)]).$$

In this way, f is the *one-point coverage function* of the canonical random set S_U defined by

$$S_U(\omega) = \{x : x \in X, \ U(\omega) \leq f(x)\} = f^{-1}[U(\omega), 1].$$

That is, $f(x) = P(\omega : x \in S(\omega)), x \in X$.

The logical operations among membership functions can be defined using this relationship. First, the set complement of $U^{-1}[0, f(x)]$ is $U^{-1}[0, f(x)]^c$, and

$$P(U^{-1}[0, f(x)]^c) = P(x \notin f^{-1}[U, 1]) = 1 - P(U \leq f(x)) = 1 - f(x),$$

since U is uniformly distributed. Thus, the negation of f is $1 - f$. Next, let $f, g : X \to [0, 1]$, and U, V be their corresponding uniformly distributed random variables, both defined on $(\Omega, \mathscr{A}, P)$. The joint distribution function of U, V is a *copula* F (or more precisely a 2-copula, see Schweizer and Sklar, 1983, p. 82-83).

To define the conjunction of f and g, we look at the set intersection of their random set representations, namely

$$U^{-1}[0, f(x)] \cap V^{-1}[0, g(x)].$$

We have

$$P(U^{-1}[0, f(x)] \cap V^{-1}[0, g(x)]) = P(x \in f^{-1}[U, 1] \cap g^{-1}[V, 1])$$

$$= P(U \leq f(x),\ V \leq g(x)) = F(f(x), g(x))$$

Thus the conjunction $f \wedge g$ is defined for $x \in X$ by $(f \wedge g)(x) = F(f(x), g(x))$.

For disjunction $\vee$ among membership functions, we look at the set union of their random set representations, namely

$$U^{-1}[0, f(x)] \cup V^{-1}[0, g(x)].$$

We have

$$P(U^{-1}[0, f(x)] \cup V^{-1}[0, g(x)])$$

$$= P(U^{-1}[0, f(x)]) + P(V^{-1}[0, g(x)])$$

$$= P(U^{-1}[0, f(x)] \cap V^{-1}[0, g(x)])$$

$$= P(x \in f^{-1}[U, 1]) + P(x \in g^{-1}[V, 1])$$

$$= P(x \in f^{-1}[U, 1] \cap g^{-1}[V, 1])$$

$$= f(x) + g(x) - F(f(x), g(x))$$

$$= F^{*}(f(x), g(x)),$$

where F^{*} is the *dual copula* of F. Of course, the logical system above of membership functions depends upon the copula F.

The procedure above is carried over to the conditional case as follows. The conditional counterpart of "f given g," denoted as $(f|g)$, is the conditional event $(U^{-1}[0, f(x)] \,|\, V^{-1}[0, g(x)])$ in the conditional space $\mathscr{A} \,|\, \mathscr{A}$. Thus, it is natural to define

$$(f|g)(x) = P(U^{-1}[0, f(x)] \,|\, V^{-1}[0, g(x)])$$

$$= P(U \leq f(x) \,|\, V \leq g(x))$$

$$= \frac{F(f(x), g(x))}{g(x)}$$

when $g(x) \neq 0$.

For $(f|g)(\cdot)$ to reduce to $\varphi(a|b)(\cdot)$ when $f = 1_a$, $g = 1_b$, with a, b being elements of a field of subsets of X, a third value u (undefined) has to be assigned to $(f|g)(x)$ when $g(x) = 0$.

We consider first a special case in which the copula F is taken to be *min*, that is $F(x, y) = min\{x, y\}$. We use also the symbol $\wedge$ for minimum. Also, in the sequence $u = [0, 1]$.

Definition. *Let $f, g : X \to [0, 1]$. The semantic part of the fuzzy conditional*

$$(f|g) : X \to [0, 1] \cup \{u\},$$

is defined by

$$(f|g)(x) = \begin{cases} \frac{f(x) \wedge g(x)}{g(x)} & when\ \ g(x) \neq 0 \\ u & when\ \ g(x) = 0\,. \end{cases}$$

As in the case of ordinary conditionals, the abstract symbol u has to be clarified. This is because the range of $(f|g)(\cdot)$ involves u, implying that the nature of fuzziness of $(f|g)$ depends on u. Of course, one can simply imagine that u is an abstract symbol, and define the logical operations $'$, $\wedge$, $\vee$ on $[0, 1] \cup \{0, 1\}$. A concrete candidate for u is the whole unit interval $[0, 1]$. This choice turns out to be convenient and also consistent with interval analysis. Taking u as $[0, 1]$, fuzzy conditionals are interval-valued fuzzy sets.

Before proceeding further, let us specify the (Lukasiewicz) logical operations on the space $[0, 1] \cup \{[0, 1]\}$, where real numbers x in $[0, 1]$ are considered as intervals $[x, x]$. Two intervals $[a_1, a_2]$, $[b_1, b_2]$ in $[0, 1]$ are equal if and only if $a_1 = b_1$ and $a_2 = b_2$. The logical operations on $[0, 1]$ are

$$x' = 1 - x,$$

$$x \wedge y = min(x, y),$$

$$x \vee y = max(x, y).$$

As in interval analysis (for example, Moore, 1966, 1979; Alefeld and Herzberger, 1983), logical operations on the set $I([0, 1])$ of intervals of $[0, 1]$ are set-extension operations, so that, using the same notation,

$$[a, b]' = \{x' : a \leq x \leq b\} = [1 - b, 1 - a] = 1 - [a, b] .$$

Note that $[a, b]'' = [a, b]$. In particular

$$u' = [0, 1]' = [0, 1] = u,$$

$$[a, b] \wedge [c, d] = \{x \wedge y : a \leq x \leq b, c \leq y \leq d\},$$

$$= [a \wedge c, b \wedge d],$$

$$[a, b] \vee [c, d] = \{x \vee y : a \leq x \leq b, c \leq y \leq d\}$$

$$= [a \vee c, b \vee d].$$

Note that $'$ is not a true complement (so that the law of excluded middle does not hold) since, in general,

$$[a, b]' \wedge [a, b] \neq 0,$$

$$[a, b]' \vee [a, b] \neq 1.$$

However, it is easy to check that DeMorgan's laws do hold, that is,

$$([a, b] \wedge [c, d])' = [a, b]' \vee [c, d]',$$

$$([a, b] \vee [c, d])' = [a, b]' \vee [c, d]'.$$

Moreover, both $\wedge$ and $\vee$ are commutative and associative. Also, the following distributive laws hold:

$$[a, b] \wedge ([c, d] \vee [e, f]) = ([a, b] \wedge [c, d]) \vee ([a, b] \wedge [e, f]) ,$$

$$[a, b] \vee ([c, d] \vee [e, f]) = ([a ,b] \vee [c, d]) \wedge ([a, b] \vee [e, f]).$$

This last fact follows by the distributivity of $\wedge$ over $\vee$ on real numbers. Finally, the order relation on $I([0, 1])$ is defined by setting $[a, b] \leq [c, d]$ if and only if $[a, b] = [a, b] \wedge [c, d]$, which is the same as $a \leq c$ and $b \leq d$. The smallest and greatest elements of $I([0, 1])$ are $[0, 0] = 0$, $[1, 1] = 1$, respectively.

The logical operations on $[0, 1] \cup \{[0, 1]\}$ are the restrictions of the above operations on $I([0, 1])$ to its subset $[0, 1] \cup \{[0, 1]\}$. Thus, for example, $x \in [0, 1]$, we have

$$x \wedge [0, 1] = [x, x] \wedge [0, 1] = [0, x],$$

$$1 \vee [0, 1] = 1, \; [0, 1] \wedge [0, 1] = [0, 1],$$

$$x \vee (y \wedge [0, 1]) = [x, x \vee y], \ldots .$$

In particular, these restrictions to $\{0, 1, u\}$, with $u = [0, 1]$, form a Lukasiewicz three-valued logic.

In the sequel, $u = [0, 1]$. For $f, g : X \to [0, 1]$, define

$$\left(\frac{f\wedge g}{g}\right)_0(x) = \begin{cases} \frac{f\wedge g}{g}(x) & \text{when } g(x) \neq 0 \\ 0 & \text{when } g(x) = 0 \end{cases},$$

and let $1_{(g\neq 0)}$ denote the indicator function of the set $\{x : g(x) \neq 0\}$. We can write

$$(f|g)(x) = \left(\frac{f\wedge g}{g}\right)_0(x)1_{(g\neq 0)}(x) \vee (1_{(g=0)}(x) \wedge u) .$$

Note that $\left(\frac{f\wedge g}{g}\right)_0$ can be replaced by $(f\wedge g)/(g\vee 1_{(g=0)})$ in the above equality. Also, multiplication $\cdot$ on $I([0, 1])$ is the set-extension operation of $\cdot$ on real numbers, so that, for $x \in [0, 1]$,

$$x \cdot u = x \cdot [0, 1] = \{xy : y \in [0, 1]\} = [0, x] = x \wedge [0, 1] = x \wedge u .$$

Thus, it is convenient to use the form

$$(f|g) = F \vee Gu = [F, F \vee G]$$

for membership functions of fuzzy conditionals, where

$$F = \left(\frac{f\wedge g}{g}\right)_0 1_{(g\neq 0)}, \quad G = 1_{(g=0)}.$$

Note that G takes only values in $\{0, 1\}$, and if $G(x) = 1$, then $F(x) = 0$.

Theorem 1. *Let* $f_1, g_1, f_2, g_2 \in \mathcal{F}(X)$. *Then* $(f_1|g_1) = (f_2|g_2)$ *if and only if there is a positive function* K *on* X *such that* $g_1 = Kg_2$ *and* $f_1 \wedge g_1 = K(f_2 \wedge g_2)$.

Proof. For sufficiency, let K be a positive function on X such that $g_1 = Kg_2$ and $f_1 \wedge g_1 = K(f_2 \wedge g_2)$. From $g_1 = Kg_2$ and $K > 0$, we see that $g_1 = 0$ if and only if $g_2 = 0$, so that

$$(f_1|g_1) = (f_2|g_2) \text{ on } (g_1 = 0) = (g_2 = 0).$$

Next, on $(g_1 \neq 0) = (g_2 \neq 0)$,

$$\begin{aligned}(f_1|g_1)(x) &= \frac{f_1 \wedge g_1}{g_1}(x) \\ &= \frac{K(x)(f_2(x) \wedge g_2(x))}{g_1(x)} \\ &= \frac{f_2 \wedge g_2}{g_2}(x) \\ &= (f_2|g_2)(x) .\end{aligned}$$

For necessity, suppose that $(f_1|g_1) = (f_2|g_2)$. Define

$$K(x) = \begin{cases} \dfrac{g_1(x)}{g_2(x)} & on\ (g_2 \neq 0) = (g_1 \neq 0) . \\ c > 0 & on\ (g_2 = 0) = (g_1 = 0) . \end{cases}$$

We then have $g_1 = Kg_2$ on X, and

$$\frac{f_1 \wedge g_1}{g_1} = \frac{f_2 \wedge g_2}{g_2} \quad \text{on } (g_1 \neq 0) = (g_2 \neq 0)$$

implies that

$$f_1 \wedge g_1 = \frac{g_1}{g_2}(f_2 \wedge g_2) = K(f_2 \wedge g_2) .$$

On $(g_1 = 0) = (g_2 = 0)$, we always have

$$g_1 \wedge g_1 = K(f_2 \wedge g_2) .$$

□

As in the case of fuzzy sets, let us specify the syntax representation of fuzzy sets, let us specify the syntax representation of fuzzy conditionals. First, let's look again at conditional events. For $a, b \in \mathcal{P}(X)$, we have seen in Chapter 3 that $(a|b)$ is equal to the interval $[ab, b \rightarrow a]$ in $\mathcal{P}(X)$, where $b \rightarrow a = b' \vee a$ (material implication).

Thus, $(a|b)$ is equivalent to $\{ab, b \rightarrow a\}$ or to $\{ab, b \rightarrow a, X\}$. $A = \{ab, b \rightarrow a, X\}$ can be viewed as a *finite* flou set with characteristic function of the form

$$\psi_{A^{(t)}}(x) = \begin{cases} 1 & \text{if } x \in ab \\ 0 & \text{if } x \in a'b \\ t & \text{if } x \in b' \end{cases}$$

for some $t \in [0, 1]$, where in flou set form,

$$A^{(t)} = \{A_\alpha, \alpha \in [0, 1]\}$$

with $A_0 = X$, $A_\alpha = a \vee b'$ for $0 < \alpha \leq t$, and $A_\alpha = ab$ for $t < \alpha \leq 1$.

For each $t \in [0, 1]$, define $\varphi_t(a|b) : X \rightarrow \{0, 1, t\}$ by

$$\varphi_t(a|b)(x) = \begin{cases} 1 & \text{if } x \in ab \\ 0 & \text{if } x \in a'b \\ t & \text{if } x \in b' \end{cases} .$$

Then, since $u = [0, 1]$, we have, for $x \in X$,

$$\varphi(a|b)(x) = \{\varphi_t(a|b)(x) : t \in [0, 1]\}.$$

That is, the generalized indicator function $\varphi(a|b)$ is precisely the collection of real-valued functions $\varphi_t(a, b)$, $t \in [0, 1]$.

In the general case, let $f, g \in \mathcal{F}(X)$. Define $(f|g)_t : X \rightarrow [0, 1]$ for each $t \in [0, 1]$ by

$$(f|g)_t(x) = \begin{cases} \frac{f \wedge g}{g}(x) & \text{when } g(x) \neq 0 \\ \\ t & \text{when } g(x) = 0 . \end{cases}$$

Then $(f|g) = \{(f|g)_t$ and $t \in [0, 1]\}$. Let $A^{(t)}$ be the flou set associated with $(f|g)_t$. Then the syntax representation of $(f|g)$ is the family of flou sets $\{A^{(t)} : t \in [0, 1]\}$.

We turn now to logical operators among fuzzy conditionals. Since fuzzy conditionals are interval-valued fuzzy sets, operations among them are defined pointwise, that is, by logical operations on $I([0, 1])$. First,

$$(f|g)'(x) = 1 - (f|g)(x) = \begin{cases} 1 - \frac{f \wedge g}{g}(x) & \text{on } (g \neq 0) \\ \\ 1 - [0, 1] = [0, 1] & \text{on } (g = 0) . \end{cases}$$

Now,

$$1 - \frac{f(x) \wedge g(x)}{g(x)} = \frac{g(x) - f(x) \wedge g(x)}{g(x)} = \frac{((g(x) - f(x)) \vee 0) \wedge g(x)}{g(x)} .$$

Thus $(f|g)' = ((g - f) \vee 0|g)$. The situation for $\wedge$ and $\vee$ is not that simple, in the sense that *compound fuzzy conditionals* are arbitrary interval-valued fuzzy sets. Using interval representations,

$$\begin{aligned}(f_1|g_1) \wedge (f_2|g_2) &= [F_1, F_1 \vee G_1] \wedge [F_2, F_2 \vee G_2] \\ &= [F_1 \wedge F_2, (F_1 \vee G_1) \wedge (F_2 \vee G_2)] \\ &= (F_1 \wedge F_2) \vee (((F_1 \vee G_1) \wedge (F_2 \vee G_2))u),\end{aligned}$$

and

$$\begin{aligned}(f_1|g_1) \vee (f_2|g_2) &= [F_1 \vee F_2, (F_1 \vee G_1) \vee (F_2 \vee G_2)] \\ &= (F_1 \vee F_2) \vee (((F_1 \vee G_1) \vee (F_2 \vee G_2))u).\end{aligned}$$

Thus, compound fuzzy conditionals are of the form $f \vee gu = [f, f \vee g]$ with $f, g : X \to [0, 1]$. Simple fuzzy conditionals are of very special form, namely g takes only values in $\{0, 1\}$, and when $g = 1$, we have $f = 0$. However,

Theorem 2. *If $f, g : X \to [0, 1]$, then $f \vee gu = \alpha \cdot (e|h) \vee \beta$ where $\alpha, \beta, e, h : X \to [0, 1]$.*

Proof. Let

$$\alpha(x) = g(x) \vee 1_{(g=0)}(x).$$

Then

$$f \vee gu = \alpha(\frac{f}{\alpha} \vee \frac{g}{\alpha} u)$$

$$= \alpha(\frac{f}{\alpha}(1_{(g=0)} \vee 1_{(g \neq 0)}) \vee 1_{(g \neq 0)}u)$$

$$= \alpha(\frac{f}{\alpha} 1_{(g=0)} \vee 1_{(g \neq 0)}u) \vee f \cdot 1_{(g \neq 0)}$$

$$= \alpha(\frac{f}{\alpha}|1_{(g=0)}) \vee f \cdot 1_{(g \neq 0)}.$$

□

Remark.

An alternative approach to defining logical operations among fuzzy conditionals is this. Instead of using arithmetic of intervals, we will explore the connection between fuzzy sets and random sets. Let $f_i, g_i : X \to [0, 1]$, $i = 1, 2$ with corresponding uniform random variables U_i and V_i, respectively, all defined on a probability space $(\Omega, \mathcal{A}, P)$. Let F be the joint distribution function of (U_1, V_1, U_2, V_2), that is F is a 4-copula. Let $*$ be a binary operator on $\mathcal{A}\,|\,\mathcal{A}$, for example, conjunction or disjunction. The corresponding operator among fuzzy conditionals is determined by

$$((f_1|g_1)*(f_2|g_2))(x) = P((a|b)*(c|d))$$

where

$$a = U_1^{-1}[0, f_1(x)], \quad b = V_1^{-1}[0, g_1(x)],$$

$$c = U_2^{-1}[0, f_2(x)], \quad d = V_2^{-1}[0, g_2(x)].$$

Now $(a|b)*(c|d) = (\alpha|\beta)$, say, so that

$$((f_1|g_1)*(f_2|g_2))(x) = P(\alpha|\beta) = \frac{P(\alpha\beta)}{P(\beta)},$$

in which $P(\alpha\beta)$ and $P(\beta)$ can be computed in terms of F, the f_i's, g_i's, and x. If we let $P(\alpha\beta) = h(x)$, $P(\beta) = \ell(x)$, then for $x \in X$,

$$((f_1|g_1)*(f_2|g_2))(x) = (h|g)(x).$$

To illustrate this approach, consider negation and conjunction in the case where F is *min*. The situation for negation is simple, involving only a unary operator. Let $f, g : X \to [0, 1]$ with corresponding U, V. Let $a = U^{-1}[0, f(x)]$, $b = V^{-1}[0, g(x)]$ for an arbitrary $x \in X$. Then $(a|b)' = (a'|b)$ and $P((a|b)') = P(a'|b) = 1 - P(a|b)$, so that $(f|g)' = 1 - (f|g)$.

For conjunction with $F = min$, using the same notation in the procedure described above, we have

$$(a|b)(c|d) = (ab|a'b \vee c'd \vee bd),$$

and

$$P(abcd) = min\{f_1(x), g_1(x), f_2(x), g_2(x)\},$$

$$P(a'b \vee c'd \vee bd) = P(a'b \vee c'd \vee abcd)$$

$$= P(a'b \vee c'd) + P(abcd),$$

$$P(a'b \vee c'd) = P(a'b) + P(c'd) - P(a'bc'd),$$

$$P(a'b) = P(b) - P(ab)$$

$$= g_1(x) - min\{f_1(x), g_1(x)\},$$

$$P(c'd) = g_2(x) - min\{f_2(x), g_2(x)\},$$

$$P(a'bc'd) = P(bd(a \vee c)')$$

$$= P(bd) - P(abd \vee cbd)$$

$$= P(bd) - (abd) - P(cbd) + P(abcd)$$

$$= min\{g_1(x), g_2(x)\} - min\{f_1(x), g_1(x), g_2(x)\} - min\{g_1(x), f_2(x), g_2(x)\} + P(abcd).$$

Thus,

$$P(a'b \vee c'd \vee bd) = g_1(x) - min\{f_1(x), g_1(x)\} + g_2(x) - min\{f_2(x), g_2(x)\}$$

$$- min\{g_1(x), g_2(x)\} + min\{f_1(x), g_1(x), g_2(x) + min\{g_1(x), f_2(x), g_2(x)\}.$$

Therefore,

$$((f_1 | g_1) \wedge (f_2 | g_2))(x) = (h | \ell)(x),$$

where

$$h(x) = min\{f_1(x), g_1(x), f_2(x), g_2(x)\},$$

$$\ell(x) = g_1(x) - min\{f_1(x), g_1(x)\} + g_2(x) - min\{f_2(x), g_2(x)\}$$

$$- min\{g_1(x), g_2(x)\} + min\{f_1(x), g_1(x), g_2(x)\} + min\{g_1(x), f_2(x), g_2(x)\}.$$

7.5 Probability qualification

If we view fuzzy conditionals as uncertain rules in expert systems, then, according to fuzzy logic (Zadeh, 1988), there are three possible modes of qualification of these rules, truth-qualification, probability-qualification, and possibility-qualification. In this section, we address only the numerical aspect of probability qualification of fuzzy conditionals; we lay down the mathematical framework for semantic evaluations of fuzzy conditionals in Probability Logic. Other modes of qualification as well as fuzzy probabilities are not treated here.

Let (X, R) be a measurable space. At the semantic level, following Zadeh (1968), a fuzzy event is defined to be a measurable map from X to $[0, 1]$ (where $[0, 1]$ is regarded as a measurable space with its induced Borel σ-field). A probability measure P on (X, R) is viewed as a model, and $\|\cdot\|_P$ denotes the semantic evaluation map in the model P. Thus, if f is a fuzzy event, then , as proposed by Zadeh (1968), $\|f\|_P$ is defined as follows. Let ξ be a random variable with values in X, having P as its probability law.

$$\|f\|_P = E_P f(\xi) = \int_X f(x)dP(x).$$

Next, we look at the case of ordinary conditional events. For $a, b \in R$, the syntax part $(a|b)$ was derived (Chapter 2) in a compatible manner with conditional probability. That is, if P is a probability on R, then $P((a|b)) = P(ab)/P(b)$, when $P(b) > 0$, is well-defined. Thus, the probability evaluation of its "generalized" indicator function (or its semantic part) $\varphi(a|b)$ is taken to be $P(a|b)$, that is,

$$\|\varphi(a|b)\|_P = P(a|b).$$

This evaluation of $\varphi(a|b)$ with respect to a model P is sometimes referred to as a third value for $\varphi(a|b)$. See Chapter 5, also Coletti et al., 1990. Now, with the notation of Section 7.4,

$$\varphi(a|b) = \{\varphi_t(a|b) : t \in [0, 1]\},$$

and $E_P\varphi_t(a|b)(\xi) = P(ab) + tP(b')$. It is easy to check that $P(a|b)$ is the *fixed point* of the map

$$t \in [0, 1] \to E_P\varphi_t(a|b)(\xi).$$

This observation suggests an extension of Zadeh's concept of probabilities of fuzzy events (Zadeh, 1968) to the case of probabilities of fuzzy conditional events. Specifically, let f, g be two fuzzy events. From Section 7.4, we have

$$(f|g) = \{(f|g)_t : t \in [0, 1]\}.$$

Define $\|(f|g)\|_P$ to be the fixed point of the map $t \to E_P(f|g)_t(\xi)$. Then

$$\begin{aligned} E_P(f|g)_t(\xi) &= E_P((f|g)_t(\xi)\,|\,g(\xi) > 0)P(g(\xi) > 0) \\ &\quad + E_P((f|g)_t(\xi)\,|\,g(\xi) = 0)P(g(\xi) = 0) \\ &= E_P(\tfrac{f\wedge g}{g}(\xi)\,|\,g(\xi) > 0)P(g(\xi) > 0) + tP(g(\xi) = 0). \end{aligned}$$

Thus the fixed point is

$$E_P(\tfrac{f\wedge g}{g}(\xi)\,|\,g(\xi) > 0)\,,$$

and

$$\|(f|g)\|_P = E_P(\tfrac{f\wedge g}{g}(\xi)\,|\,g(\xi) > 0)\,.$$

Obviously, this evaluation generalizes those in the two previous special cases. Since

$$(f|g) = (\tfrac{f\wedge g}{g})_0 1_{(g>0)} \vee 1_{(g=0)}u,$$

$(f|g)$ takes values in $[0, 1]$ on $(g > 0)$ (on $(g = 0)$, $(f|g) = u$), namely $\frac{f\wedge g}{g}$. This observation is used to define evaluation of compound fuzzy conditionals as follows.

A compound fuzzy conditional is of the form $f \vee gu$ where $f, g : X \to [0, 1]$. Since $f \vee gu = [f, f \vee g]$, we see that $f \vee gu$ takes values in $[0, 1]$ only on $(g \le f)$, that is, $x \in (g \le f)$ if and only if $(f \vee gu)(x) = f(x) \in [0, 1]$. Thus, by analogy with the simple fuzzy conditionals case, we define

$$\|f \vee gu\|_P = E_P(f|g \le f).$$

This evaluation is well-defined, since if $f \vee gu = h \vee ku$ then $[f, f \vee g] = [h, h \vee k]$. Thus $f = h, f \vee g = h \vee k$, and $(g \le f) = (k \le h)$. Hence $E_P(f|g \le f) = E_P(h|k \le h)$.

7.6 Iterated fuzzy conditionals

The topic of iterated conditioning will be treated in Section 8.1 of Chapter 8, from a syntactic viewpoint. Here, to be complete, we discuss this concept in the setting of fuzzy sets, but from a semantic viewpoint, that is, using generalized indicator functions of conditional objects rather than the objects themselves. Let R be a field of subsets of a set X. For $a, b \in R$, the generalized indicator function of $(a|b)$ is defined as

$$\varphi(a|b) : X \to I([0, 1]),$$

where $I([0, 1])$ denotes the set of all closed sub-intervals of $[0, 1]$, equipped with the arithmetic of intervals, and

$$\varphi(a|b)(x) = \begin{cases} 1 & \text{for } x \in ab \\ 0 & \text{for } x \in a'b \\ u = [0, 1], & \text{for } x \in b'. \end{cases}$$

$\varphi(a|b)$ is a special fuzzy conditional, since $\varphi(a|b) = (1_a|1_b) = 1_a \wedge 1_b = 1_{ab}$ on b and is u on b'. Also, if

$$\varphi_t(a|b) = \begin{cases} 1 \text{ for } x \in ab \\ 0 \text{ for } x \in a'b \\ t \text{ for } x \in b' \end{cases}$$

then

$$\varphi(a|b) = \{\varphi_t(a|b) : t \in [0, 1]\}.$$

Each $\varphi_t(a|b)$ can be viewed as an element in $\varphi(a|b)$. Observe that $\varphi_t(a|b)\cdot 1_b = 1_a\cdot 1_b$, for $t \in [0, 1]$. Thus, as a natural approximation, we can view $\varphi(a|b)$ as

$$\{f : X \to [0, 1] : f\cdot 1_b = 1_a\cdot 1_b\}.$$

In a similar way, for $f, g : X \to [0, 1]$ one can approximate a fuzzy conditional $(f|g)$ as

$$\{h : X \to [0, 1] : h\cdot g = f\cdot g\}.$$

The heuristic considerations above lead to an approximate form of iterated fuzzy conditionals. For $f_i, g_i : X \to [0, 1], i = 1, 2,$ define

$$\nabla((f_1|g_1)|(f_2|g_2)) = \bigcup_{f,g} \{(f|g) : (f|g)(f_2|g_2) = (f_1|g_1)(f_2|g_2)\},$$

where operations among fuzzy conditionals are those in $I([0, 1])$. Note that by a union of the form $\bigcup_{f,g} \{(f|g)\}$, we mean the union of the set $(f|g)(x)$ whose members are either $\{t\}$, for some $t \in [0, 1]$, or $[0, 1]$, for each $x \in X$. In other words, ∇ is a map from X to $\mathcal{P}[0, 1]$. The main result of this section is the proof of the fact that ∇ is an operator on the space of fuzzy conditionals.

For this purpose, we proceed as follows. Consider the equation

$$(f|g)(f_2|g_2) = (f_1|g_1)\cdot(f_2|g_2)\,. \tag{1}$$

Let $h = \frac{f\wedge g}{g}$ on $(g > 0)$. We write

$$(f|g) = h1_{(g>0)} \vee u1_{(g=0)}.$$

Similarly, let $h_i = \frac{f_i\wedge g_i}{g_i}$ on $(g_i > 0), i = 1, 2.$ The equation (1) is rewritten as

$$(h1_{(g>0)} \vee u1_{(g=0)})(h_2 1_{(g_2>0)} \vee u1_{(g_2=0)}) =$$

$$(h_1 1_{(g_1>0)} \vee u1_{(g_1=0)})(h_2 1_{(g_2>0)} \vee u1_{(g_2=0)}). \tag{2}$$

After multiplying out terms, we get

$$\alpha \vee \beta u = \gamma \vee \xi u, \tag{3}$$

where

$$\alpha = h1_{(g>0)}h_2 1_{(g_2>0)},$$

$$\beta = h1_{(g>0)}1_{(g_2=0)} \vee h_2 1_{(g_2>0)}1_{(g=0)} \vee 1_{(g=0)}1_{(g_2=0)},$$

$$\gamma = h_1 1_{(g_1>0)}h_2 1_{(g_2>0)},$$

$$\xi = h_1 1_{(g_1>0)}1_{(g_2=0)} \vee h_2 1_{(g_2>0)}1_{(g_1=0)} \vee 1_{(g_1=0)}1_{(g_2=0)}.$$

Since (3) is precisely

$$[\alpha, \alpha \vee \beta] = [\gamma, \gamma \vee \xi],$$

we have

$$\alpha = \gamma \text{ and } \alpha \vee \beta = \gamma \vee \xi. \tag{4}$$

To solve (4), we consider the partition of X consisting of $(g_2 = 0)$, $(g_2 > 0, h_2 > 0)$ and $(g_2 > 0, h_2 = 0)$.

On $(g_2 = 0)$, (4) becomes

$$h1_{(g>0)} \vee 1_{(g=0)} = h_1 1_{(g_1>0)} \vee 1_{(g_1=0)}. \tag{5}$$

Thus,

$$(f|g) = h1_{(g>0)} \vee u1_{(g=0)}$$

$$= (h_1 1_{(g_1>0)} \vee 1_{(g_1=0)})1_{((1-h_1)g_1>0)} \vee u1_{((1-h_1)g_1)=0)} \tag{6}$$

since on $(g > 0)$, (5) yields

$$h = h_1 1_{(g_1>0)} \vee 1_{(g_1=0)},$$

and on $(g = 0)$,

$$1 = h_1 1_{(g_1>0)} \vee 1_{(g_1=0)}.$$

This is equivalent to $h_1 = 1$ or $g_1 = 0$, that is, to

$$(1 - h_1)(g_1) = 0.$$

On $(g_2 > 0, h_2 > 0)$, we have from (4) that

$$h1_{(g>0)} = h_1 1_{(g_1>0)},$$

and

$$h_1 1_{(g_1>0)} \vee 1_{(g=0)} = h_1 1_{(g_1>0)} \vee 1_{(g_1=0)}. \tag{7}$$

From (7) we see that $(g = 0) = (g_1 = 0)$. Indeed, if $g_1(x) = 0$, then $g(x) = 0$. Conversely, if $g(x) = 0$, then either $g_1(x) = 0$ or $h_1(x) = 0$. But the case where $g_1(x) > 0$ and $h_1(x) = 0$ is impossible in view of (7).

Next, on $(g > 0)$, we have $h = h_1$. Thus

$$(f|g) = h1_{(g>0)} \vee u1_{(g=0)} = h_1 1_{(g_1>0)} \vee u1_{(g_1=0)} = (f_1|g_1).$$

On $(g_2 > 0) \cap (h_2 = 0)$, (4) supplies no constraint on f and g, so that $(f|g)$ is a solution. But for $x \in X$,

$$\bigcup_{f,g} \{(f|g)(x)\} = [0, 1] = u.$$

Thus, we have

Theorem 1. *For* $f_i, g_i : X \to [0, 1], i = 1, 2,$

$$\nabla((f_1|g_1)|(f_2|g_2)) = ((\frac{f_1 \wedge g_1}{g_1})_o 1_{(C\neq 0)} | 1_{(D\neq 0)},$$

where

$$C = g_1 h_2 g_2 \vee g_1(1 - h_1)1(g_2 = 0),$$

$$D = g_1 h_2 g_2 \vee g_1(1 - h_1)(h_2 \vee 1_{(g_2=0)}),$$

and

$$h_i = \frac{f_i \wedge g_i}{g_i}, i = 1, 2.$$ □

As an example, consider $f_1 = 1_a, \ f_2 = 1_c, \ g_1 = 1_b, \ g_2 = 1_d$. We have

$$(\frac{f_1 \wedge g_1}{g_1})_o = 1_{ab},$$

$$(C \neq 0) = bcd \vee b(ab)'d',$$

$$(D \neq 0) = bcd \vee b(ab)'(cd \vee d') = b(a'd' \vee cd),$$

so that

$$\nabla(\varphi(a|b)|\varphi(c|d)) = (1_{abcd}|1_{b(a'd' \vee cd)}) = \varphi(abcd|b(a'd' \vee cd))$$
$$= \varphi(ab|b(a'd' \vee cd)).$$

This should be compared with Theorem 3 of Section 8.1 of Chapter 8.

CHAPTER 8

ITERATED CONDITIONING AND MISCELLANEOUS ISSUES

This last chapter is concerned with some topics related to measure-free conditioning. In Section 8.1, an investigation of iterated conditioning is carried out. In Section 8.2, some aspects of non-monotonic logic on conditionals are discussed. In Section 8.3, we generalize some of the results concerning operations on cosets of Booleans rings to commutative von Neumann regular rings. Finally, in Section 8.4, we close by suggesting open problems for future research.

8.1 Iterated conditioning

In Section 7.6, we have touched upon the concept of conditionals of conditionals in the fuzzy case. In this section, we return to the Boolean case and formulate the basic concepts of higher-order conditioning. This investigation of iterated conditioning is a first attempt. We hope that this will trigger further work in this area.

By Lewis' Triviality Result (Chapter 1), there is no binary operation $\Diamond$ on a Boolean ring that is compatible with conditional probabilities. That is, there is no binary operation $\Diamond$ on R such that for all $a, b \in R$, and all probability measures P on R such that $P(b) \neq 0$, the equation

$$P(a \Diamond b) = P(a|b) = P(ab)/P(b).$$

holds. Thus, to define conditional events compatible with probabilites, one is forced to go outside R, and we enlarged R to $R|R$ for that purpose. Now, having the conditional space $R|R$, we wish to consider conditionals on it. But, again, $R|R$ will not accomodate conditionals between its elements that are compatible with probability. More precisely, the situation is this.

Theorem 1. *(The Triviality Result for* R|R*) Let R be a Boolean algebra with at least sixteen elements. Then there does not exist a binary operation $\Diamond$ on $R|R$ such that*

$$P((a|b) \Diamond (c|d)) = \frac{P((a|b)(c|d))}{P(c|d)}$$

for all $a, b, c, d \in R$ and for all probability measures P on R such that

$P(b) \neq 0 \neq P(c|d)$.

Proof. If R is not atomic, then there exist four mutually disjoint non-zero elements of R. Just take $a \in R$ where a has no atom, and let $a > b > c > d$, all non-zero. Then d, cd', bc', and ab' are four such elements of R. If R is atomic, let a be an atom of R, b be an atom of a', c be an atom of $(a \vee b)'$, and d be an atom of $(a \vee b \vee c)'$. This is possible else R has fewer than sixteen elements. Thus, in any case, R has four mutually disjoint non-zero elements r, s, t, and u. Now, by the Stone Representation Theorem, R is a subalgebra of $\mathcal{P}(\Omega)$ for some set Ω, and so viewing R, let v, w, x, and y be elements of r, s, t, and u respectively. Define a probability measure P on $\mathcal{P}(\Omega)$ via

$$P(v) = 0.1, \ \ P(w) = P(x) = P(y) = 0.3.$$

Then, P is a probability measure on R. Let

$$\begin{aligned} a &= r \vee s, \\ b &= r \vee s \vee t, \\ c &= r \vee t, \\ d &= r \vee s \vee t \vee u. \end{aligned}$$

A solution

$$(x|y) = (a|b) \lozenge (c|d)$$

so that

$$P(x|y) = \frac{P((a|b))(c|d))}{P(c|d)}$$

yields

$$\begin{aligned} P(x|y) &= \frac{P(ac|a'b \vee c'd \vee bd)}{P(c|d)} \\ &= \frac{P(ac)P(d)}{P(a'b \vee c'd \vee bd)P(c)} \\ &= \frac{(0.1)(1)}{P(t \vee (r \vee t) \vee (r \vee s \vee t))(0.6)} \\ &= \frac{0.1}{(0.7)(0.6)} = \frac{P(x)}{P(y)}, \end{aligned}$$

taking $x \leq y$. But there is not such a pair $x, y \in R$. □

There are some special cases for which solutions exist.

(i) For $b = d = 1$, we have $(x|y) = (a|c) \in R|R$.

(ii) More generally, for $b = d$, we have

$$P[(a|b)(c|b)]/P(c|b) = P(ac|b)/P(c|b) = P(abc)/P(bc) = P(a|bc),$$

so that a solution $(x|y)$ is $(a|bc)$.

(iii) Generalizing in a different direction, letting only $d = 1$, we have

$$P[(a|b)c]/P(c) = P(abc|b \vee c')/P(c) = P(ac)/P(c)$$

if $c \leq b$, so that when $c \leq b$, $((a|b)|c) = (a|c)$. In particular, $((a|b)|b) = (a|b)$.

The interpretation of all the above is plausible from a rule deduction viewpoint. (See Dubois and Prade, 1990, and also Calabrese, 1987). For iterated conditionals of conditionals with the same antecedent (that is, $b = d$), see also Pfanzagl (1971, p. 200). In this case, for fixed $b \in R$, iterated conditionals of the form $((a|b)|(c|b))$ are nothing more than conditionals on the (quotient) Boolean ring $R|Rb'$. The operations on the ring R/Rb' are

$$(a|b) + (c|b) = (a + c|b),$$

$$(a|b)\cdot(c|b) = (ac|b),$$

$$(a|b)' = (a'|b).$$

Thus

$$((a|b)|(c|b)) = (a|b) + (R/Rb')(c|b)' \in ((R/Rb')/(R/Rb')(c'|b)).$$

We are going to show that $((a|b)|(c|b))$ can be identified with $(a|bc) \in R|R(bc)'$. For this purpose, consider the map

$$\lambda : R|Rb' \to R|R(bc)'$$

defined by

$$\lambda(x + Rb') = x + R(bc)'.$$

First, this map is well-defined. Indeed, changing x to $x + rb'$, the image under λ is $x + rb' + R(bc)'$. But $rb' \leq b' \vee c'$ so that $rb' \in R(bc)'$, that is, $rb' + R(bc)' = R(bc)'$. It is obvious that λ is a ring homomorphism and is onto. It remains to verify that the kernel of λ is precisely the principal ideal $(R|Rb')(c'b)$ of $R|Rb'$. We have

$$\lambda((x + Rb')(c' + Rb')) = \lambda(xc' + Rb') = xc' + R(bc)' = R(bc)'$$

(since $xc' \leq (bc)'$), which is the zero in $R/R(bc)'$. Thus, $((R/Rb')/(R/Rb')(c'|b))$ is isomorphic to $R/R(bc)'$. □

However, the identification of $((a|b)|(c|b))$ with $(a|bc)$ does nothing toward getting a general definition of conditionals for conditionals. Of course, one can argue from some logical viewpoint, and then *define*, in an ad-hoc or plausible manner, an iterated conditional $(a|b)|(c|d)$ in such a way that the above intuitive (and compatible) special cases hold.

Our approach here is this. We cannot proceed in exactly the same manner as we did to get $R|R$ from R. The space $R|R$ consists of all cosets of all principal ideals of R. The space $R|R$ is not even a ring, and so we cannot make a totally analogous construction. However, in R, a coset $a + Rb' = \{a + rb' : r \in R\}$ is the set of all solutions x to the equation $xb = ab$. In $R|R$, we can carry out the construction analogous to that. So we are led to the following definition.

Definition 1. *For* $(a|b)$, $(c|d) \in R|R$, *the iterated conditional* $((a|b)|(c|d))$ *is the set*

$$\{(x|y) : (x|y)(c|d) = (a|b)(c|d)\}.$$

The collection of these sets is denoted $(R|R)|(R|R)$ *and is called the space of iterated conditionals.*

Now $((a|b)|(c|d))$ is not empty since it contains $(a|b)$ as well as $((a|b)(c|d))$. In the case of ordinary events, the set $\{x : xb = ab\}$ is the interval $[ab, a \vee b']$. That is, solutions x to the equation $xb = ab$ are exactly those x such that $ab \leq x \leq a \vee b'$. So a conditional event is also an interval in R. This was discussed in Chapter 2. One might expect that $((a|b)|(c|d))$ is an interval in $R|R$ under the partial order we defined by $(a|b) \leq (c|d)$ if $(a|b) = (a|b)(c|d)$. In fact, $R|R$ is a pseudo-complemented lattice with respect to this order, as expounded upon in Chapter 4. Now $((a|b)|(c|d))$ does have a smallest element, namely $(a|b)(c|d)$. Furthermore, this is the counterpart to the smallest element ab in the interval $[ab, b' \vee a]$. However, various counterparts to $b' \vee a = b \rightarrow a$ (material implication), such as $(a|b) \vee (c|d)'$ and Lukasiewicz's implication are not solutions to $(x|y)(c|d) = (a|b)(c|d)$, that is, are not in $((a|b)|(c|d))$. However, $R|R$ has a property that we have not yet exploited. It is *relatively pseudo-complemented.* It turns out that $b \rightarrow a$ is a relative pseudo-complement in R of b with respect to a since $x \leq b' \vee a$ if and only if $xb \leq a$. (See the definition below.) So there is another counterpart in $R|R$ to $b' \vee a$, and it is that element that is a maximum solution to $(x|y)(c|d) = (a|b)(c|d)$ and guarantees that $((a|b)|(c|d))$ is indeed an interval in $R|R$.

Definition 2. *A lattice* L *is relatively pseudo-complemented if for every* $a, b \in L$, *there is an element* $a*b \in L$ *with the property that* $x \le a*b$ *if and only if* $a \wedge x \le b$.

Clearly there is only one such $a*b$, and it is called the *pseudo-complement of* a *relative to* b. The element $a*b$ satisfies $a \wedge a*b \le b$, and is the supremum of the set of all such elements. The relative pseudo-complement of a with respect to 0 is called the *pseudo-complement* of a, and that notion played an important role in Chapter 4.

The relevance of relative pseudo-complements to our problem is this. Suppose that $R|R$ is relatively pseudo-complemented. Then applying that property to the pair of elements $(c|d)$ and $(a|b)(c|d)$, $R|R$ has an element $e|f = (c|d)*((a|b)(c|d))$ such that $(e|f)(c|d) \le (a|b)(c|d)$ and such that $(x|y) \le (e|f)$ if and only if $(x|y)(c|d) \le (a|b)(c|d)$. But there are *solutions* to $(x|y)(c|d) = (a|b)(c|d)$. Hence the pseudo-complement $(e|f)$ of $(c|d)$ relative to $(a|b)(c|d)$ satisfies $(e|f)(c|d) = (a|b)(c|d)$. Thus if

$$(x|y)(c|d) = (a|b)(c|d)$$

then

$$(a|b)(c|d) \le (x|y) \le (e|f).$$

Conversely, if

$$(a|b)(c|d) \le (x|y) \le (e|f),$$

then

$$(a|b)(c|d)(c|d) = (a|b)(c|d) \le (x|y)(c|d) \le (e|f)(c|d) = (a|b)(c|d),$$

and so

$$(x|y)(c|d) = (a|b)(c|d).$$

Therefore

$$(a|b)|(c|d) = [(a|b)(c|d), (c|d)^*((a|b)(c|d))].$$

Thus we need two things. We need that $R|R$ is relatively pseudo-complemented, and we need a formula for the relative pseudo-complement $(c|d)^*((a|b)(c|d))$.

Theorem 2. *$R|R$ is relatively pseudo-complemented, and the pseudo-complement of $(a|b)$ relative to $(c|d)$ is*

$$(a|b)*(c|d) = (cd \vee a'b \vee b'd' \,|\, d \vee a'b \vee b'd')$$

Proof. Let $e = cd \vee a'b \vee b'd'$, and $f = d \vee a'b \vee b'd'$. We need that $(a|b)(x|y) \leq (c|d)$ if and only if $(x|y) \leq (e|f)$. Now $(a|b)(x|y) \leq (c|d)$ if and only if

$$(ax|a'b \vee x'y \vee by) \leq (c|d)$$

if and only if

$$ax(a'b \vee x'y \vee by) \leq cd$$

and

$$c'd \leq (ax)'(a'b \vee x'y \vee by),$$

if and only if

$$abxy \leq cd$$

and

$$c'd \leq (a' \vee x')(a'b \vee x'y \vee by) = a'b \vee x'y.$$

So we have that $(a|b)(x|y) \leq (c|d)$ if and only if

$$abxy \leq cd \text{ and } c'd \leq a'b \vee x'y.$$

Conversely, $(x|y) \leq (e|f)$ if and only if

$$\begin{aligned} xy \leq ef &= (cd \vee a'b \vee b'd')(d \vee a'b \vee b'd') \\ &= cd \vee a'b \vee b'd', \end{aligned}$$

and

$$\begin{aligned} x'y \geq e'f &= (cd \vee a'b \vee b'd')'(d \vee a'b \vee b'd) \\ &= (c' \vee d')(a \vee b')(b \vee d)(d \vee a'b \vee b'd) \\ &= (c' \vee d')(a \vee b')(d \vee a'b) \\ &= (c' \vee d')(ad \vee b'd) \\ &= (ac'd \vee b'c'd) \\ &= c'd(a \vee b') \end{aligned}$$

Thus we have $(x|y) \leq (e|f)$ if and only if

$$xy \le cd \vee a'b \vee b'd' \text{ and } x'y \ge c'd(a \vee b').$$

We need

$$xy \le cd \vee a'b \vee b'd' \text{ and } x'y \ge c'd(a \vee b')$$

if and only

$$abxy \le cd \text{ and } c'd \le a'b \vee x'y.$$

But $x'y \ge c'd(a \vee b')$ implies that

$$a'b \vee x'y \ge a'b \vee c'd(a \vee b') \ge c'd,$$

and $c'd \le a'b \vee x'y$ implies that

$$c'd(a \vee b') \le (a'b \vee x'y)(a \vee b') = x'y(a \vee b') \le x'y.$$

From $xy \le cd \vee a'b \vee b'd'$ we get

$$abxy \le ab(cd \vee a'b \vee b'd') = abcd.$$

Finally, $abxy \le cd$ and $c'd \le a'b \vee x'y$ imply

$$xy \le (ab)'cd$$

and

$$c'd(a \vee b') \le (a'b \vee x'y)(a \vee b')$$

$$= x'y(a \vee b') \le x'y \le x' \vee y',$$

from which we get

$$x' \vee y' \ge ab(c' \vee d')$$

and

$$x' \vee y' \ge c'd(a \vee b').$$

Thus

$$x' \vee y' \ge ab(c' \vee d') \vee c'd(a \vee b').$$

But $xy \le cd \vee a'b \vee b'd'$ is equivalent to

$$x' \vee y' \ge (c' \vee d')(a \vee b')(b \vee d)$$

$$= ab(c' \vee d') \vee c'd(a \vee b'). \qquad \square$$

The relative pseudo-complement $(a|b)*(c|d)$ can be written in an apparently simpler form, namely

$$(a|b)*(c|d) = (cd \vee a'b \vee b'd' | a' \vee b' \vee d).$$

One should note the special case $(c|d) = (0|1)$. We have

$$(a|b)*(0|1) = (a|b)^* = (a'b|1),$$

the pseudo-complement in $R|R$ of $(a|b)$.

The relative pseudo-complement of $(a|b)$ with respect to $(c|d)$ is a form of an "implication operator"

$$(a|b) \Rightarrow (c|d) = (a|b)*(c|d) = (cd \vee a'b \vee b'd' | a' \vee b' \vee d)$$

on $R|R$ extending material implication on R. It can be viewed as the counterpart of material implication in $R|R$. The truth table of $(a|b) \Rightarrow (c|d)$ follows. Let $x = (a|b)$ and $y = (c|d)$.

$x \Rightarrow y$

$x \backslash y$	$0|1$	$1|1$	$0|0$
$0|1$	$0|1$	$0|1$	$0|1$
$1|1$	$1|1$	$1|1$	$1|1$
$0|0$	$1|1$	$0|1$	$1|1$

Corollary 1. $((a|b)|(c|d)) = [(a|b)(c|d), (c|d)*((a|b)(c|d))]$

$$= [(a|b)|(c|d), (abcd \vee c'd \vee ad' \vee b'd' | b \vee c'd \vee ad' \vee b'd')].$$

Proof. We need only to show that

$$(c|d)^*((a|b)(c|d)) = (abcd \vee c'd \vee ad' \vee b'd' | b \vee c'd \vee ad' \vee b'd').$$

Let $e = a'b \vee c'd \vee bd$. By the formula in Theorem 2,

$$\begin{aligned}(c|d)*((a|b)(c|d)) &= (c|d)*(ac|a'b \vee c'd \vee bd)\\ &= (ace \vee c'd \vee d'e' | e \vee c'd \vee d'e')\\ &= (abcd \vee c'd \vee ad' \vee b'd' | a'b \vee c'd \vee bd \vee ad' \vee b'd')\\ &= (abcd \vee c'd \vee ad' \vee b'd' | b \vee c'd \vee ad' \vee b'd').\end{aligned}$$

□

The right hand endpoint

$$(c|d)*((a|b)(c|d)) = (abcd \vee c'd \vee ad' \vee b'd' | b \vee c'd \vee ad' \vee b'd')$$

can also be written in the somewhat simpler form

$$(ab \vee c'd \vee ad' \vee b'd' | b \vee c' \vee d').$$

Corollary 1 gives a way to identify $R|R$ with a subset of $(R|R)|(R|R)$. An easy calculation shows that $((a|b)|(1|1)) = [(a|b), (a|b)]$. Thus the map $(a|b) \rightarrow ((a|b)|(1|1))$ is one-to-one.

Corollary 2. $((a|b)|(c|d)) = (R|R)((c|d)*((a|b)(c|d))) \vee (a|b)(c|d)$.

Proof. An element $(x|y)$ in the interval $[(a|b)|(c|d), (c|d)*((a|b)(c|d))]$ is

$$(x|y)((c|d)*((a|b)(c|d))) \vee (a|b)(c|d),$$

which is in

$$(R|R)((c|d)*((a|b)(c|d))) \vee (a|b)(c|d).$$

The converse is equally clear. □

Now $((a|b)|(c|d))$ is an interval in $R|R$, and it would be nice to have simple criteria for the equality $((a|b)|(c|d)) = ((e|f)|(g|h))$. Two conditional events $(a|b)$ and $(c|d))$ are equal if and only if $ab = cd$ and $c = d$. The analogous condition here is that $(a|b)(c|d) = (e|f)(g|h)$ and $(c|d) = (g|h)$. This does not seem to be the case, however, and the best we can do at the moment is to say that $(a|b)(c|d) = (e|f)(g|h)$, and $(c|d)*((a|b)(c|d)) = (g|h)*((e|f)(g|h))$, that is, that the end points be the same. For example, there does not seem to be a way to recover $(c|d)$ from $(a|b)(c|d)$ and $(c|d)*((a|b)(c|d))$. This precludes making the definition

$$P((a|b)|(c|d)) = P((a|b)(c|d))/P(c|d)$$

since $(c|d)$ is not available. However, in the conditional case,

$$P(a|b) = P(ab)/P(b) = P(ab)/(1 + P(ab) - P(a \vee b')).$$

This last expression affords a way to define P on $(R|R)|(R|R)$, namely by the equation

$$P((a|b)|(c|d)) = P((a|b)(c|d))/(1 - P(a|b)(c|d) + P((c|d)*((a|b)(c|d)))).$$

Furthermore,

$$P((a|b)|(1|1)) = P((a|b)(1|1))/(1 - P((a|b)(1|1)) + P((1|1)*((a|b)(1|1))))$$

$$= P(a|b)/(1 - P(a|b) + P(a|b)) = P(a|b),$$

and this definition extends the definition of P on $R|R$, viewing $R|R$ as embedded in $(R|R)|(R|R)$ by $(a|b) \to ((a|b)|(1|1))$.

An element $((a|b)|(c|d))$ of $(R|R)|(R|R)$ contains some special elements besides its endpoints $(a|b)(c|d)$ and $(c|d)*((a|b)(c|d))$. It is a subset of $R|R$, and so consists of a set of subsets of R. As the latter, its *point set union* can be taken, yielding a subset of R. It is rather remarkable that doing so yields a coset, that is, an element of $R|R$, and moreover that coset is in $(a|b)|(c|d)$. We proceed now to verify all this.

Let $(c|d)*((a|b)(c|d)) = (\alpha|\beta)$. Since $(a|b)(c|d) = (ac|a'b \vee c'd \vee bd)$, we have

$$\alpha = abcd \vee \gamma$$

$$\beta = (a'b \vee c'd \vee bd) \vee \gamma$$

where

$$\gamma = (a'b \vee c'd \vee bd)'d' \vee c'd$$

$$= (ab \vee b')d' \vee c'd.$$

We also have

$$((a|b)|(c|d)) = (R|R)\gamma \vee (a|b)(c|d).$$

Indeed,

$$(\alpha|\beta) = (a|b)\cdot(c|d) \vee \gamma.$$

Thus,

$$((a|b)|(c|d)) = (R|R)((a|b)(c|d) \vee \gamma) \vee (a|b)(c|d)$$

$$= (R|R)(a|b)(c|d) \vee (R|R)\gamma \vee (a|b)(c|d)$$

$$= (R|R)\gamma \vee (a|b)(c|d).$$

The point of the equality $((a|b)|(c|d)) = (R|R)\gamma \vee (a|b)(c|d)$ is that there is a special element $\gamma \in R$ such that

$$(R|R)\gamma \vee (a|b)(c|d) = (R|R)(c|d)*((a|b)(c|d)) \vee (a|b)(c|d).$$

Of course we are identifying γ with $(\gamma|1)$. Now, for any set of subsets S of R, let

$\cup(S)$ denote the union of all the sets in S. Noting that

$$\begin{aligned} \cup((R|R)(x|1)) &= \cup\{(a|b)(x|1) : (a|b) \in R|R\} \\ &= \cup\{(a + Rb')(x + R0) : (a|b) \in R|R\} \\ &= \cup\{(a + Rb')x : (a|b) \in R|R\} \\ &= Rx, \end{aligned}$$

and that for two sets S and T, $\cup(S \vee T) = \cup(S) \vee \cup(T)$, makes the proof of the following theorem transparent.

Theorem 3. For $a, b, c, d \in R$,

$$\cup[(a|b)|(c|d)] = (ab|b(a'd' \vee cd)) \in ((a|b)|(c|d)).$$

Proof. We have

$$\begin{aligned} \cup[(a|b)|(c|d)] &= \cup[(R|R)\gamma \vee (a|b)(c|d)] \\ &= \cup(R|R)\gamma \vee (a|b)(c|d) \\ &= R\gamma \vee (a|b)(c|d) \\ &= (0|\gamma') \vee (a|b)(c|d) \\ &= (abcd|abcd \vee \gamma'(a'b \vee c'd \vee bd)) \\ &= (abcd|b(a'd' \vee cd) \\ &= (ab|b(a'd' \vee cd)). \end{aligned}$$

To see that $(ab|b(a'd' \vee cd)) \in ((a|b)|(c|d))$, simply verify that

$$(ab|b(a'd' \vee cd))(c|d) = (a|b)(c|d). \qquad \square$$

One may view $\cup$ as a binary operation on $R|R$, with

$$\cup((a|b), (c|d)) = (ab|b(a'd' \vee cd)).$$

Now Calabrese (1987) has defined a binary "conditioning" operation on $R|R$ which is his candidate for iterated conditioning. His operation is given by

$$(a|b)\circ(c|d) = (ab|b(d' \vee c)) = (ab|b(d \to c)).$$

As a simple check shows, it is not true that

$$((a|b)(c|d))\circ(c|d) = (a|b)\circ(c|d),$$

while it is the case that

$$((a|b)(c|d)|(c|d)) = ((a|b)|(c|d)),$$

and hence that

$$\cup((a|b)(c|d)|(c|d)) = \cup((a|b)|(c|d)).$$

However, when $(a|b) \leq (c|d)$,

$$\cup[(a|b)|(c|d)] = (ab|b(a'd' \vee cd))$$

$$= (ab|b(d' \vee c)) = (ab|b(d \rightarrow c)).$$

Therefore, the binary operations $\cup$ and Calabrese's "conditioning" operator on $R|R$ agree on pairs $((a|b),(c|d))$ with $(a|b) \leq (c|d)$.

If $b = d = 1$, then $\cup(a|c) = (a|c)$, so $\cup$ is onto. If $b = d$, then

$$\cup[(a|b)|(c|b)] = (ab|bc) = (a|bc)\ .$$

If $d = 1$ and $b = c$, then

$$\cup[(a|b)|b] = (ab|b) = (a|b).$$

Thus $\cup$ produces "compatible" solutions, at least in the special cases considered at the beginning of this section. It is obvious that $\cup$ preserves logical operations. Moreover, the restriction of $\cup : (R|R)|(R|R) \rightarrow R|R$ to $\{(a|b)|(c|d) : a, b \in R\}$ is an isomorphism for each pair $c, d \in R$. Also the restriction to $\{(a|b)|(c|b) : a, c \in R\}$ is an isomorphism for each $b \in R$. To prove these facts, only injectivity needs to be verified. For the first, suppose

$$\cup[(a_1|b_1)|(c|d)] = \cup[(a_2|b_2)|(c|d)].$$

By Theorem 3, we then have:

$$(a_1b_1|b_1(a_1'd' \vee cd)) = (a_2b_2|b_2(a_2'd' \vee cd)),$$

that is,

$$(1) \qquad \begin{cases} a_1b_1cd = a_2b_2cd \\ b_1(a_1'd' \vee cd) = b_2(a_1'd' \vee cd)\ . \end{cases}$$

Since $b_icd = a_ib_icd \vee a_i'b_icd$, (1) is equivalent to

(2)
$$\begin{cases} a_1b_1cd = a_2b_2cd \\ a_1'b_1(d \to c) = a_2'b_2(d \to c) . \end{cases}$$

Now

$$\begin{aligned}(a_i|b_i)(c|d) &= (a_ib_icd|a_i'b_i \vee c'd \vee b_id) \\ &= (a_ib_icd|a_i'b_i(d \to c) \vee c'd \vee b_id) ,\end{aligned}$$

since

$$\begin{aligned}a_i'b_i \vee c'd \vee b_id &= (a_i'b_i)(c'd)' \vee c'd \vee b_id \\ &= a_i'b_i(d' \vee c) \vee c'd \vee b_id \\ &= (a_i'b_i)(d \to c) \vee c'd \vee b_id.\end{aligned}$$

Also, observe that

$$\begin{aligned}&[(a_i'b_i)(d \to c) \vee c'd] \vee b_id \\ &= [(a_i'b_i)(d \to c) \vee c'd] \vee (b_id)[(a_i'b_i)(d \to c) \vee c'd]' \\ &= [a_i'b_i(d \to c) \vee c'd] \vee a_ib_icd,\end{aligned}$$

with the last union being a disjoint one.

Thus, (2) implies

$$(a_1|b_1)(c|d) = (a_2|b_2)(c|d),$$

and hence

$$((a_1|b_1)|(c|d)) = ((a_2|b_2)|(c|d)) .$$

To prove the second fact, suppose that

$$\cup[(a_1|b)|(c_1|b)] = \cup[(a_2|b)|(c_2|b)],$$

that is,

$$a_1bc_1|bc_1) = (a_2bc_2|bc_2),$$

or

(3)
$$\begin{cases} a_1bc_1 = a_2bc_2 \\ bc_1 = bc_2 \end{cases} .$$

Now, $(a_i|b)(c_i|b) = (a_ibc_i|b)$. Thus (3) implies that

$$(a_1|b)(c_1|b) = (a_2|b)(c_2|b).$$

Also, (3) implies that

$$(c_1 | b) = (c_2 | b),$$

and hence

$$((a_1 | b) | (c_1 | b)) = ((a_2 | b) | (c_2 | b)). \qquad \square$$

8.2 Non-monotonic logics on conditionals

This section discusses non-monotonic entailment relations in conditional logic (CL). In Chapter 6, the building block for a conditional probability logic (CPL) is the base space $R|R$ together with Lukasiewicz's three-valued logic. Conditional probabilities were introduced into the analysis mainly for purpose of reasoning under uncertainty. Of course, other uncertainty measures could be used instead of probability (see for example, Goodman, Nguyen and Rogers, 1990). This is basically a *numerical approach* to reasoning with uncertainty in the sense that the uncertainty involved is taken into account in a quantitative way. However, a *qualitative approach* to reasoning can be carried out at the level of CL. In view of the structure of $R|R$, qualitative notions will be compatible with quantitative ones. The need to manipulate conditionals qualitatively is apparent in problems such as combination of rules in expert systems. Our concern here is to extract some non-monotonic aspects of CL as well as to discuss the possibility for building non-monotonic entailment relations on $R|R$.

In the case of classical two-valued logic (C_2), truth is the only primitive notion. As stated in Section 6.3, the logical entailment relation $\models$ in C_2 is defined in terms of models (homomorphisms Ω from R to $\{0, 1\}$, or equivalently, maximal filters of R). In turn, $\models$ is expressed in terms of the order relation $\leq$ on R by $b \models a$ if and only if $b \leq a$. Now, since for $c \in R$, $bc \leq b$, we see that if $b \models a$ then for $c \in R$, $bc \models a$. This property of $\models$ is referred to as "monotonicity," that is, roughly speaking, additional evidence will not affect the validity of previous logical conclusions. In this sense, C_2 is called a monotonic deduction system, or the logic C_2 is monotonic. In this case, the monotonicity of $\models$ is due to the transitivity property of $\leq$. From an axiomatic approach to entailment relations (for example, Gabbay, 1985), the monotonic $\models$ satisfies

(i) reflexivity: for $a, b \in R$, $ab \models a$,
(ii) monotonicity: if $b \models a$, then for $c \in R$, $bc \models a$, and
(iii) transitivity (or cut): if $ab \models c$ $b \models a$, then $b \models c$.

PL is also monotonic since probability is compatible with the order relation $\leq$ on R. To capture common sense reasoning, some form of "non-monotonic" deduction is desirable. Roughly speaking, an entailment relation $\models$, in some logic, is non-monotonic if in light of new evidence, previous logical conclusions may fail. Specifically, $\models$ is non-monotonic if the monotonicity property (ii) above does not hold. The non-monotonicity of a logical system refers precisely to an entailment relation in it. Thus, a logic can have both a monotonic entailment and a non-monotonic one.

Examine again the $\models$ in C_2. In applications, given a set of data $\{b_1, ..., b_n\}$, the relation $\models$ is used to express the fact that some a follows logically from the data, written

$$\{b_1, ..., b_n\} \models a.$$

There are two procedures in this deduction process. First, combination of evidence is taken as "conjunction" which is the ring multiplication. Second, $\models$ is defined as $\leq$. This (partial) order relation on R is defined precisely in terms of $\wedge$, via $a, b \in R$, $b < a$ if, by definition, $a \wedge b = b$. Thus, in order to break the monotonicity of a system, one can either consider combination of evidence differently or define $\models$ independently of $\leq$. We will return to this issue shortly.

We proceed now to clarify the statement that "probabilistic reasoning captures a form of non-monotonic reasoning." We know that PL is monotonic. What makes "probabilistic reasoning" non-monotonic depends on the framework of inference. Suppose we consider the (partial, quantitative) entailment of an event a from a collection of events $\{b_1, ..., b_n\}$ as a conditional probability $P(a|b_1 \wedge ... \wedge b_n)$, denoted $\{b_1, ..., b_n\} \models a$ with degree $P(a|b_1 \wedge ... \wedge b_n)$. In other words, this partial entailment relation is non-monotonic. Note that the two primitive notions involved here are truth and probability.

It is possible to express the above aspect of non-monotonicity in a qualitative fashion. Indeed, in the CL (Chapter 6), we have $(a|b) \leq (c|d)$ if and only if $ab \leq cd$ and $c'd \leq a'b$, and

$$(a|b) \overset{CL}{\models} (c|d)$$

is defined as

$$(a|b) \leq (c|d).$$

Now, $(a|b)$ and $(a|bc)$ are not comparable in general, since we always have $abc \leq ab$, but not $a'b \leq a'bc$, in general. On the other hand, the structure of $R|R$ is such that probabilities are compatible with operations on $R|R$, in particular P preserves the (partial) order relation $\leq$ on $R|R$. Note also that, for the purpose of automation, syntactic representation of $\models$ is desirable.

Now from R (base space of C_2), we go to $R|R$ (base space of CL). The truth space of $R|R$ is $\{0, u, 1\}$. Note that, in the analysis of reasoning processes in AI, three-valued logics often surface, for example, in Computation Theory (McCarthy, 1967), in the semantics of non-monotonic entailment (Sandewall, 1989), and in modeling of default rules (Dubois and Prade, 1990).

Since the truth-space of $R|R$ is a three-element set, one can consider various logics on $R|R$. In other words, the class of all possible logics for $R|R$ is that of all three-valued logics. Theorem 2 of Section 3.4 established their correspondences with logical operators and relations on $R|R$, that is, at the syntax level. Depending upon interpretations of conditional objects and intuitive logical aspects of problems at hand, different choices of three-valued logics can be made. For example, for reasoning in the theory of computable partial recursive functions, *non-commutative* three-valued logics might be appropriate (for example, Guzman and Squier, 1990; see also Section 3.4). In a direction related to quantum logic, *non-distributive* systems can be looked for (for example, Schay, 1968).

As far as *commutative* three-valued logics are concerned, the standard literature is summarized in Rescher (1969). As stated earlier, different choices of connectives for three-valued logics (that is, truth tables) lead to different logical operators on $R|R$. Thus, for example, Lukasiewicz, Sobocinski and Bochvar's logics correspond respectively to our operators in Chapter 3, Adams-Calabrese-Schay's operators, and an alternative system of Schay. (See Section 3.5; also Dubois and Prade, 1989, 1990).

Consider first the case of Lukasiewicz' logic on $R|R$, corresponding to operators $'$, $\wedge$, $\vee$ of Chapter 3. Suppose data consist of conditional information, or conditionals are viewed as production rules in expert systems. A simple way to express the fact that the conditional information (or rule) $(e|f)$ follows logically from the data $\{a|b, (c|d)\}$ is to define $\models$ as $\{(a|b), (c|d)\} \models (e|f)$ if and only if

$$(a|b) \wedge (c|d) \leq (e|f) .$$

This deduction process is exactly the same as in the case of C_2, and hence is monotonic. As suggested by Dubois and Prade (1989), one way to destroy the monotonicity of $\models$ is to modify it at the combination of evidence level. Instead of using Lukasiewicz' conjunction $\wedge$, one might replace it by another one, for example, Sobocinski's, (See Chapter 3.) The reason is this. Since $\leq$ on $R|R$ is defined as

$$(a|b) \leq (c|d) \text{ if and only if } (a|b) \wedge (c|d) = (a|b),$$

as on R, the transitivity of $\leq$, coupled with this definition, is responsible for the monotonicity of $\models$. If $\wedge$ is replaced by Adams-Calabrese-Schay's conjunction $\wedge_o$ then $\models_o$ is non-monotonic, where

$$\{(a|b), (c|d)\} \vdash_o (e|f)$$

if, by definition,

$$(a|b) \wedge_o (c|d) \leq (e|f);$$

and where

$$a|b) \wedge_o (c|d) = ((b' \vee a)(d' \vee c)|b \vee d).$$

Indeed, suppose $(a|b) \leq (e|f)$. By inspection, we see that

$$(a|b) \wedge_o (c|d) \leq (a|b)$$

does not hold, so that, in general, $(e|f)$ might not follow from $\{(a|b), (c|d)\}$.

Note that, in view of Theorem 1, Section 3.3, the order relation $\leq$ on $R|R$ can be defined by $(a|b) \leq (c|d)$ if $ab \leq cd$ and $c'd \leq a'b$, that is, by using only the ring structure of R, without calling upon $\wedge$. For other order relations on $R|R$, see the recent work of Calabrese (1990).

Another way to modify $\models$ to obtain non-monotonicity is suggested by Sandewall (1989). First, to define "partial interpretations," Sandewall considered Kleene three-valued base logic. By base logic, we mean truth tables of the three basic connectives "not," "and" and "or". This is the same as Lukasiewicz's three-valued base logic (Rescher, 1969, p. 34). The main difference between the two logics lies in the concept of implication. Thus, in our setting, $R|R$ is equipped with operators $'$, $\wedge$, and $\vee$ of Chapter 3. The logical entailment relation is next defined by introducing a *preference order* on the set of models (partial interpretations). For details, see Sandewall (1989). This is in line with the general methodology advocated by Hawthorne (1988) for building non-monotonic logics. To achieve non-monotonicity, one should generalize the classical concept of models by taking more primitive notions than just "truth." In Hawthorne's words "there is more to the meaning of a sentence than the determination of truth-values at possible worlds." One should also take "entailment" as a primitive notion. That means an entailment relation should be autonomous with respect to truth-values semantics. Then, as in the case of "truth" as a primitive notion, once an entailment concept has been taken, one will specify its "semantic rules" (in the same way that truth tables of logical connectives specify how truth values of compound formulae are assigned) governing deduction processes. For an axiomatic approach to non-monotonic entailment relations, see Gabbay (1985). Recent relevant papers on non-monotonic logics include Grosof (1988), Bibel (1986), McLeish (1988).

8.3 Operations on cosets of regular rings.

The algebraic sturctures more general than Boolean rings that are pertinent for our considerations of conditional events, iterated conditionals, and so on, seem to be lattices of some sort rather than more general rings. For example, in Chapter 4, we extended R, which is both a Boolean ring and equivalently a Boolean lattice, to the space $R|R$ of conditional events. This space is a Stone algebra, which is a lattice more general than a Boolean lattice. It is not a ring. That is, $R|R$ generalizes R as a lattice, not as a ring. There is the possibility, however, of generalizing this process of going from R to $R|R$ by starting with a ring more general than a Boolean one. Now $R|R$ is the set of all cosets of principal ideals of R, and the operations between its elements were defined to be those induced by the operations on R. That is, if A and B are subsets of R, and $*$ is any binary operation on R, then, by definition, $A*B = \{ab : a \in A, b \in B\}$. In the Boolean ring case, addition and multiplication between cosets yielded cosets. In fact, for $a, b \in R$, and ideals I and J of R,

$$(a + I) + (b + J) = (a + b) + (I + J),$$

and

$$(a + I)\cdot(b + J) = ab + Ib + aJ + IJ.$$

These facts were thoroughly discussed in Chapter 3. These operations on $R|R$ were the basis of its development. While the set addition of cosets is a coset holds in any ring and is easily verified, the fact that the set product of cosets is a coset is unexpected and non-trivial. The question naturally arises as to the generality of this phenomenon. In particular, for what rings does it hold? In this section, we will show that it holds for commutative von Neumann regular rings. In Boolean rings, every element is an idempotent, and these regular rings are good candidates for such an extension because of the abundance of idempotents in them. Our principal result is Theorem 4, the extension of Theorem 1 of Section 3.2 to these more general rings.

Definition 1. *A commutative ring R is (von Neumann) regular if it has an identity, and if for each $x \in R$, there is a $y \in R$ such that $xyx = x$.*

We will call these commutative von Neumann regular rings simply regular rings. Here are some examples of regular rings:

(*1*) Any Boolean ring is a regular ring.

(2) Any field is a regular ring.

(3) The Cartesian product of any family of regular rings is a regular ring. The ring operations in such a product are, of course, componentwise.

(4) Quotients of regular rings are regular rings. That is, if R is a regular ring and I is an ideal of R, then R/I is a regular ring.

(5) p-rings are regular rings. These are rings such that for some prime p and every element x, $px = 0$ and $x^p = x$. Boolean rings are those for which $p = 2$.

For an element x in a regular ring, the element y such that $xyx = x^2y = x$ is not unique since, for example, in a Boolean ring one may take y to be x or 1. The element xy is unique. We denote it x°.

Lemma 1. *Let R be a regular ring. Then for all $x \in R$,*

(*i*) *x° is unique,*

(*ii*) *x° is an idempotent, and*

(*iii*) *$Rx = Rx^\circ$.*

Proof. For (i), if $(xy)x = (xz)x = x$, then $xy = xzxy = xz$. For (ii), $(xy)(xy) = (xyx)x = xy$. Finally, for (iii), clearly $R(xy) \subseteq Rx$. If $a = rx \in Rx$, then $a = (rx)(xy) \in R(xy)$, whence $Rx^\circ = Rx$. □

Theorem 1. *Let R be a regular ring. The following hold.*

(*i*) *For any principal ideal Ra of R, $Ra = Re$ for a unique idempotent e.*

(*ii*) *$I^2 = I$ for any ideal I of R.*

(*iii*) *$Ra^2 = Ra$ for any $a \in R$.*

(*iv*) *Finitely generated ideals of R are principal.*

(*v*) *For ideals I and J of R, we have $IJ = \{ij : i \in I, j \in J\}$ is an ideal.*

Proof. To prove (i), $Ra = Ra^\circ$ with a° idempotent by Lemma 1. If $Re = Rf$, with e, f idempotents, then $e = rf$ and $f = se$ for suitable elements r and s of r, and

$$e = rf = rse = rsef = f.$$

For (ii), clearly $I^2 \subseteq I$. If $i \in I$, then $i = i^\circ i \in I^2$. Now (iii) follows since $Ra^2 = RaRa = Ra$ by (ii). To get (iv), we need that $Ra_1 + Ra_2 + \ldots + Ra_n = Ra$ for some $a \in R$. We may assume that each a_i is idempotent. Now,

$$Ra_1 + Ra_2 = R(a_1 + a_2 - a_1a_2)$$

since $a_i(a_1 + a_2 - a_1a_2) = a_i$ for $i = 1, 2$, whence $Ra_1 + Ra_2 \subseteq R(a_1 + a_2 - a_1a_2)$. The other inclusion is easy.

Finally, to prove (v), IJ is closed under multiplication by any element a of R since $a(ij) = (ai)j$ with $ai \in I$, $j \in J$. We need $i_1j_1 + ... + i_nj_n \in IJ$ for $i_k \in I$, $j_k \in J$, $k = 1, ..., n$. From (iv), let $Rj_1 + Rj_2 + ... + Rj_n = Rj$. Then

$$i_1j_1 + ... + i_nj_n = (i_1r_1)j + ... + (i_nr_n)j$$

for suitable r_k, $k = 1, ..., n$. □

The following is a characterization of regular rings in terms of products of cosets of the same ideal.

Theorem 2. *Let R be a commutative ring with identity. Then R is regular if and only if the set product of any two cosets of an ideal I is the product of those two cosets as elements of the quotient ring R/I. That is, R is regular if and only if*

$$(a + I)(b + I) = ab + I$$

for each ideal I of R, and a, b $\in R$.

Proof. If the equality above holds, then taking $a = b = 0$ yields $I^2 = I$ for all ideals I. Taking $I = Rx$ gets $RxRx = Rx = Rx^2$, so that $x = yx^2$ for some y in R. Thus R is regular. Now assume that R is regular. We need

$$(a + I)(b + I) = ab + I,$$

or that

$$\{(a + i)(b + j) : i, j \in I\} = \{ab + ib + aj + ij : i, j \in I\}$$
$$= \{ab + k : k \in I\}.$$

Clearly, $(a + I)(b + I) \subseteq ab + I$. We need to write $ab + k$ in the form $ab + ib + aj + ij$. Letting $i = k^\circ(1 - a)$ and $j = k^\circ a(k - b + ab)$ accomplishes that.

□

Note that Theorem 2 yields an ideal theoretic characterization of regular rings, namely that a ring is regular if and only if $I^2 = I$ for all ideals I.

We now turn to the problem of showing that the set product of two cosets of ideals of a regular ring is again a coset. Specifically, we will show that

$$(a + I)(b + J) = ab + aJ + bI + IJ.$$

To do this, we investigate the quantity

$$K(a, b, I, J) = \{aj + bi + ij : i \in I, j \in J\}.$$

We have

$$(a + I)(b + J) = ab + aJ + bI + IJ = ab + K(a, b, I, J).$$

In general, $K(a, b, I, J)$ is not an ideal of R. However, if we let

$$K_o(a, b, I, J) = \sigma(IJ) + aJ + bI$$

where $\sigma(IJ)$ denotes the ideal generated by IJ, then $K_o(a, b, I, J)$ is always an ideal, and, moreover, we have:

Lemma 2. *Let R be a commutative ring with unit 1. Then for $a, b \in R$, and I, J ideals of R,*

$$aJ \cup bI \subseteq K(a, b, I, J) \subseteq K_o(a, b, I, J) = \sigma(IJ) + K(a, b, I, J).$$

Proof. Since 0 is in any ideal, it follows that

$$aJ \cup bI \subseteq K(a, b, I, J).$$

Next, if $i \in I$ and $j \in J$, then $ij \in \sigma(IJ)$, hence

$$K(a, b, I, J) \subseteq K_o(a, b, I, J).$$

Clearly

$$\sigma(IJ) + K(a, b, I, J) \subseteq K_o(a, b, I, J).$$

Conversely, let $k \in \sigma(IJ)$. We have

$$aj + bi + k = (k - ij) + (aj + bi + ij) \in \sigma(IJ) + K(a, b, I, J).$$

□

Lemma 3. *Let R be a commutative ring with identity. The following are equivalent.*

(i) *For $a, b \in R$, and I, J ideals of R, $K(a, b, I, J)$ is an ideal.*

(ii) *For $a, b \in R$, and I, J ideals of R, $K(a, b, I, J) = K_o(a, b, I, J)$.*

(iii) *$\sigma(IJ) \subseteq K(a, b, I, J)$, for $a, b \in R$, and I, J ideals of R.*

Proof. That (ii) implies (i) is obvious. Assume (i). Then for $i \in I, j \in J$, we have

$$ij = (ij + ja + bi) - (ja + bi) \in K(a, b, I, J),$$

since

$$aJ \subseteq K(a, b, I, J), \ bI \subseteq K(a, b, I, J).$$

Thus (i) implies (iii). Assume (iii). In view of Lemma 2, it suffices to show that,

$$K_o(a, b, I, J) \subseteq K(a, b, I, J).$$

For this purpose, let $aj + bi + k \in K_o(a, b, I, J)$. Then

$$aj + bi + k = (aj + bi + ij) + (k - ij)$$

with $k - ij \in \sigma(IJ)$. Now, by hypothesis, (iii) holds for any a, b in R. Thus taking $c = a + i, d = b + j$, we have $\sigma(IJ) \subseteq K(c, d, I, J)$. That is, $k - ij$ is of the form

$$(a + i)j_1 + (b + j)i_1 + i_1j_1$$

for some $i_1 \in I, j_1 \in J$. Hence

$$aj + bi + k = aj + bi + ij + (a + i)j_1 + (b + j)i_1 + i_1j_1$$

$$= a(j + j_1) + b(i + i_1) + ij + ij_1 + ji_1 + i_1j_1$$

$$= a(j + j_1) + b(i + i_1) + (i + i_1)(j + j_1) \in K(a, b, I, J).$$ □

Theorem 3. *Let R be a regular ring. Then for $a, b \in R$, and ideals I, J of R,*

$$(a + I)(b + J) = ab + aJ + bI + IJ \in \mathcal{I}(R).$$

Proof. By Theorem 1, $\sigma(IJ) = IJ$. But $IJ = I \cap J$. Clearly, $IJ \subseteq I \cap J$. Conversely, if $a \in I \cap J$, then since R is regular,

$$a = (aa^\circ)a \in IJ.$$

Thus if $r \in I \cap J$,

$$i = (rr^\circ)(1 - a) \in I,$$

$$j = (rr^\circ)(r - b + ab) \in J,$$

and

$$r = aj + bi + ij \in K(a, b, I, J),$$

so that

$$\sigma(IJ) \subseteq K(a, b, I, J).$$

In view of Lemma 3, we then have

$$K(a, b, I, J) = K_o(a, b, I, J) = aJ + bI + IJ. \qquad \square$$

We have just seen that if R is a regular ring, then set-extenstion operations of addition and multiplication are operators on the space $\mathcal{I}(R)$ of all cosets of R, extending coset operations on each fixed quotient ring. Of course, by Theorem 2, this property is unique to regular rings. However, it is not known which rings have the property that products of cosets are cosets, or indeed if having this property is unique to commutative regular rings.

To extend Theorem 1 of Section 3.2 to regular rings, we define analogs of $'$ and $\vee$ for regular rings. For $a, b \in R$, let

$$a' = 1 - a,$$

and

$$a \vee b = a + b - ab.$$

These operations are extended to subsets of R as usual. For $A, B \subseteq R$,

$$A' = \{1 - a : a \subset R\},$$

$$A \vee B = \{a + b - ab : a \in A, b \in B\}.$$

One should note that $A \vee b$ is not

$$A + B - AB = \{a + b - cd : a, c \in A, b, d \in B\}.$$

However, DeMorgan laws do hold.

$$(AB)' = A' \vee B',$$

$$(A \vee B)' = A'B'.$$

The following theorem is a generalization of Theorem 1 in Section 3.2.

Theorem 4. *Let R be a regular ring. Then for $a, b \in R$, and ideals I, J of R,*

(*i*) $(a + I) + (b + J) = (a + b) + (I + J)$,

(*ii*) $(a + I)\cdot(b + J) = ab + aJ + bI + IJ$,

(*iii*) $(a + I)' = a' + I$,

(*iv*) $(a + I) \vee (b + J) = a \vee b + (a'J + b'I + IJ)$.

Proof. (i) and (ii) have been proved previously. (iii) is easy. For (iv), we use one of the DeMorgan laws above and (ii). We have

$$\begin{aligned}(a + I) \vee (b + J) &= ((a' + I)(b' + J))' \\ &= 1 - (a'b' + a'J + b'I + IJ) \\ &= a \vee b + a'J + b'I + IJ.\end{aligned}$$

□

The difficult part of Theorem 4 is (ii). It was proved by inspecting the quantity $K(a, b, I, J)$. There is a more direct proof, which goes as follows. First, assume that I and J are principal ideals. Let $I = Re$, $J = Rf$ with e, f idempotents. It suffices to solve the equation

$$ij + ib + ja = i_1j_1 + i_2b + j_2a$$

for $i \in I, j \in J$, where $i_1, i_2 \in I; j_1, j_2 \in J$. Letting

$$i = (x - a)ef + i_2(1 - f)e,$$

$$j = (1 - b)ef + j_2(1 - e)f$$

where $x = i_1j_1 + i_2b + j_2a + ab$, yields a solution.

For the general case, by Theorem 1, the ideal $Ri + Ri_1 + Ri_2$ is a principal ideal Re, and $R_j + Rj_1 + Rj_2$ is Rf with e, f idempotents. Thus, the principal ideal case finishes the proof. □

There are analogs for $'$ and $\vee$ on a regular ring other than the ones we defined above. An alternative is this. In analogy with the Boolean case, define, for $a, b \in R$,

$$(a|b) = \{x \in R : xb = ab\}.$$

Then, assuming throughout that R is regular,

$$(a|b) = a + R(1 - b^{\circ}).$$

Indeed, first observe that

$$a + R(1 - b^\circ) = ab^\circ + R(1 - b^\circ).$$

This can be seen as follows. If $x \in a + R(1 - b^\circ)$, then

$$\begin{aligned} x &= a + r(1 - b^\circ) \\ &= a(1 - b^\circ + b^\circ) + r(1 - b^\circ) \\ &= ab^\circ + (a + r)(1 - b^\circ) \end{aligned}$$

which is in $ab^\circ + R(1 - b^\circ)$. Conversely, for $x = ab^\circ + s(1 - b^\circ)$,

$$\begin{aligned} x &= a(1 - 1 + b^\circ) + s(1 - b^\circ) \\ &= a + (s - a)(1 - b^\circ) \end{aligned}$$

which is in $a + R(1 - b^\circ)$.

Now let $x \in (a|b)$, that is, $xb = ab$. Multiplying through by b° yields $xb^\circ = ab^\circ$. Thus

$$\begin{aligned} x &= x(1 - b^\circ + b^\circ) \\ &= x(1 - b^\circ) + xb^\circ \\ &= x(1 - b^\circ) + ab^\circ, \end{aligned}$$

which is in

$$ab^\circ + R(1 - b^\circ) = a + R(1 - b^\circ).$$

Conversely, if $x = ab^\circ + r(1 - b^\circ)$ for some $r \in R$, then

$$xb = ab^\circ b + r(1 - b^\circ)b = ab. \qquad \square$$

The fact that $\{x \in R : xb = ab\} = a + R(1 - b^\circ)$ rather than $a + R(1 - b)$ suggests that one might want to define $'$ on regular rings by $a' = 1 - a^\circ$ rather than $1 - a$. In that case, in order for DeMorgan's laws to hold, and in analogy with the Boolean case, one should define $\vee$ by

$$\begin{aligned} a \vee b &= (a' \wedge b')' = (a'b')' \\ &= ((1 - a^\circ)1 - b^\circ))' \\ &= 1 - (1 - a^\circ - b^\circ + a^\circ b^\circ)^\circ \\ &= a^\circ + b^\circ - a^\circ b^\circ. \end{aligned}$$

With respect to these operations, a regular ring R satisfies the following properties.

(1) $a(ab \vee ac) = ab \vee ac$ *and* $ab \vee c = (a \vee c)(b \vee c)$.

(2) $(a \vee b)' = a'b'$ *and* $(ab)' = (a' \vee b')$.

(3) $a(a \vee b) = a$ *and* $a \vee ab = a^{\circ}$.

The verification of these properties is completely routine. The upshot of property *(1)* is that $\vee$ distributes over products, but not the other way around. Property (2) asserts that DeMorgan's laws hold. Property (3) is one absorption law, and the failure of the other.

There does not seem to be a way to define a partial order $\leq$ on R in terms of these operations so that R is a lattice. In fact, defining $\leq$ by $a \leq b$ if $a = ab$, or if $a = ab^{\circ}$ does not yield a partial order. Anti-symmetry is not achieved. For example, for the case $a \leq b$ if and only if $a = ab^{\circ}$, if $a \leq b$, and $b \leq a$, then $a^{\circ} = b^{\circ}$, but $a \neq b$ unless a and b are idempotents. Thus, this alternate definition of $'$, and consequently of $\vee$, on R, utilizing more heavily the idempotent part of R, does not result in a particularly tractable algebraic system on which to base a logic.

It is instructive to see what Theorem 4 becomes with these alternate definitions of $'$ and $\vee$. Of course parts (i) and (ii) do not change. Some properties of these new operations when extended to cosets follow. Properties *(5)* and *(6)* are the analogs of parts (iii) and (iv) of Theorem 4 are these.

(4) $(a + Rb)^{\circ} = R^{\circ}b^{\circ} + a^{\circ}b'$.

(5) $(a + Rb)' = R^{\circ}b^{\circ} + a'b'$.

(6) $(a + Rb) \vee (c + Rd) = R^{\circ}(bd \vee a'd \vee bc') + (ab' \vee cd')$.

To give a better appreciation of the analogs, we present a proof of *(5)*. If x is an element of a regular ring R, then there is an element y such that $xyx = x$. Denote such an element by $x^{\dagger}$. Thus $x^{\dagger}x = x^{\circ}$. Now note the following equalities.

$$a + b = a + bb^{\circ} = ab^{\circ\prime} + (a + b)b^{\circ},$$

$$(a^{\dagger}b^{\circ\prime} + (a + b)^{\dagger}b^{\circ})(ab^{\circ\prime} + (a + b)b^{\circ}) = a^{\circ}b^{\circ\prime} + (a + b)^{\circ}b^{\circ},$$

and

$$(a^{\circ}b^{\circ\prime} + (a + b)^{\circ}b^{\circ})(ab^{\circ\prime} + (a + b)b^{\circ}) = a + b.$$

Thus

$$(a + b)^{\circ} = a^{\circ}b^{\circ\prime} + (a + b)^{\circ}b^{\circ}.$$

Now to the equality

$$(a + Rb)' = R^\circ b^\circ + a'b'.$$

We have

$$\begin{aligned}(a + rb)' &= 1 - (a + rb)^\circ \\ &= 1 - a^\circ(r^\circ b^\circ)' - (a + rb)^\circ r^\circ b^\circ \\ &= 1 - a^\circ + a^\circ r^\circ b^\circ - (a + rb)^\circ r^\circ b^\circ \\ &= 1 - a^\circ - b^\circ + a^\circ b^\circ + b^\circ - a^\circ b^\circ + a^\circ r^\circ b^\circ - (a + rb)^\circ r^\circ b^\circ \\ &= (1 - a^\circ)(1 - b^\circ) + (1 - a^\circ - a^\circ r^\circ - (a + rb)^\circ r^\circ)b^\circ.\end{aligned}$$

It is readily checked that the quantity

$$1 - a^\circ - a^\circ r^\circ - (a + rb)^\circ r^\circ)$$

is idempotent, so we have the inclusion

$$(a + Rb)' \subseteq R^\circ b^\circ + a'b'.$$

Now let $eb^\circ + a'b' \in R^\circ b^\circ + a'b'$, with e idempotent of course. It suffices to solve the equation

$$eb^\circ + a'b' = 1 - (a + sb)^\circ,$$

or the equation

$$(a + sb)^\circ = a^\circ(1 - b^\circ) + (1 - e)b^\circ$$

for s. Setting

$$x = a^\circ(1 - b^\circ) + (1 - e)b^\circ,$$

and noting that x is idempotent, means that we need s such that

$$x(a + sb) = (a + sb),$$

and

$$x = y(a + sb)$$

for some y. Letting

$$s = - ab^\dagger + (1 - e)b^\dagger$$

and

$$y = a^{\dagger}(1 - b^{\circ}) + b^{\circ}(1 - e)$$

does the trick.

For further work on developing algebraic properties for conditionals on regular rings, see Goodman and Nguyen (1990).

8.4 Miscellaneous issues and open problems

It is time to summarize our work and to discuss open problems.

The topic of conditioning is perhaps very old since it is central to empirical sciences. However, the concept of "measure-free" conditioning has not been studied seriously due to a lack of motivation. It is the fundamental aspect of probabilistic inference in expert systems that motivated us to look again at this topic and to formulate a rigorous theory of conditioning.

The subject of probabilistic inference in expert systems has attracted considerable attention among researchers in artificial intelligence and has caused much discussion. Several fundamental ideas and methodologies relating to conditioning have been proposed, most of which were highly appealing on common sense grounds. However, serious foundational problems have been encountered, as has been the case in many other areas of science. Accordingly, clarification of conditioning at the basic level is necessary. The purpose of this monograph is to introduce a rigorous theory of measure-free conditioning which can be utilized in inference procedures in intelligent machines. The theory developed here concerns mainly basic mathematical objects such as ordinary sets and probability measures. It can be regarded as a first step that will lead to extensions in various directions of interest.

Basically, this work is an effort to provide a better understanding of the logics of conditionals. It is an attempt to bring conditional logic closer to the level of understanding as that of classical logic. Such an understanding is needed since more and more AI techniques rely on formal methods in logic to guide programming in intelligent machines. Logics can be viewed as knowledge representation languages in which facts, rules, and inference can be stated and manipulated.

Uncertainty modeling is a tricky business in AI. Unlike the term "conditionals" used in classical two-valued logic, where "conditional" is referred to as material implication, conditionals or implicative statements used in this text need to be modeled properly in the context of reasoning with uncertainty. A "measure-free" approach seems to be the most objective way to lay the first brick. However, we are all biased by the popular approach to uncertainty modeling, namely, probability theory, in which there is a fundamental concept of probabilistic conditioning. We have tried to revise the work of others concerning conditioning concepts to be compatible with probability theory. We provided an axiomatic approach to conditionals, and built a conditional logic. This should stimulate further work to improve it and to extend it towards applications. In a time of fast advances in AI technologies, we hope that it is useful to have a monograph on the subject, even at a tentative level. Many issues remain to be re-examined and much further work is needed. We now discuss some of these issues and some open problems.

A. Conditionals on more general algebraic systems

The axiomatic approach in Chapter 2 led to the coset form for conditionals on Boolean rings. This mathematical representation of implicative statements is satisfactory in the sense that it reflects earlier thoughts on the concept of conditioning in logic, and coincides with that derived from other work on the subject. There are a number of elegant characterizations of conditional probability without any reference to conditional events, such as Aczel's generalization of Renyi axioms (Aczel, 1966) or Cox's approach (Cox, 1961). However, DeFinetti (1974) and, more generally, Lindley (1982), characterized conditional probability via the "Dutchbook," or equivalently, uncertainty decision game. This does use (tacitly) DeFinetti's conditional event indicator function (see also Goodman et al (1990) for a modification of certain of Lindley's conclusions concerning the inadmissibility of uncertainty measures). In connection with these results, it is of some interest to attempt to relate all of these characterizations with the standard probability evaluation we use, namely $P((a|b)) = P(a|b)$.

The next problem has been: once the concrete conditional space $R|R$ is obtained, what are the logical operators on it? From a "syntax" viewpoint, this is an extension problem. The operations on the Boolean ring R need to be extended to operators on $R|R$ which capture, in some reasonable sense, aspects of combination of evidence in ruled-based systems. In Chapter 3, the approach is algebraic. It is motivated by an interesting problem in ring theory, namely, how to extend appropriately coset operations on each quotient ring of R to $R|R$? It turns out that set-extension operations provide a natural solution to this extension problem. In this way, $R|R$ becomes a Stone algebra (Chapter 4).

All that was done for Boolean rings, for mathematical interest as well as for applications. Conditionals on more general algebraic structures now need to be investigated. In Chapters 7 and 8, we have touched upon two generalizations: fuzzy sets and regular rings.

In carrying out the construction of $R|R$ from R for R a commutative regular ring rather than a Boolean ring, problems arise. We can certainly let $R|R$ be the set of all cosets of principal ideals of R. But just how they should manipulated in order to provide suitable generalizations of the Boolean case is not settled. It is not totally clear what a conditional event should be in this context. Should $(a|b)$ be the coset $a + R(1 - b)$ or the coset $a + R(1 - b^{\circ})$? In either case, since products and sums of cosets are cosets, they can be added and multiplied, but there are choices to be made for $'$ and $\vee$. As mentioned earlier, it seems not to be known which rings have the property that products of cosets are cosets, and of course we are only considering the commutative case.

B. Three-valued logics of conditionals

Various open issues have been suggested by Schay (1968, p. 343-344) as far as $R|R$, viewed as the space of generalized (three-valued) indicator functions, is concerned. Viewing the conditional space $R|R$ as some specific algebraic structure, for example, as a Stone algebra, "probability-like" measures on it should be formulated in a more thorough measure theoretical basis. This is somewhat similar to the situation in quantum probability (see, for exmple, Gudder, 1988) in which the domain of a generalized measure is an algebraic structure slightly more general than the usual concept of σ-algebra, namely a σ-additive class.

On the other hand, one might ask what would $R|R$ be, as an algebraic structure, if instead of using Lukasiewicz's three-valued logic (corresponding to logical operations on $R|R$ as developed in Chapter 3), one started with either Schay's first or second system, or with Sobocinski's or Bochvar's three-valued logic?

In Chapter 3 we established the connection between logical operations on the conditional space $R|R$ and truth tables in three-valued logic. It might be interesting to explore the situation in n-valued logics (n > 3). Logical operators on $R|R$, as developed in Chapter 3, lead to a well-known three-valued logic, namely that of Lukasiewicz. The algebraic structure of $R|R$ so obtained is a modification of Koopman's non-totally comparable conditional qualitative probability structure, (Koopman (1940, 1964)). Referring to the excellent analysis and summary by Fine (1973, p. 183-196), the order relation on $R|R$, as defined in Chapter 3, can be seen to satisfy essentially all but two of Koopman's axioms. Additional work should be carried out for this aspect of conditional event algebra, and should focus on the basic equivalence (not just implication) between the partial order on $R|R$ and the numerical partial order on corresponding conditional probabilities. By proving that there is a bijection between the class of all three-valued logics and logical operators on $R|R$, the search for operators on $R|R$ might begin by examining the class of all possible three-valued logics. For example, it turns out that

Schay's system of logical operators on $R|R$ corresponds to Bochvar's three-valued logic, while Schay's other system (as well as Adams' and Calabrese's) corresponds to Sobocinski's. This should be viewed as a healthy situation in reasoning with uncertain conditional data, rather than a divergence of opinion. This is similar to the debate about choices of uncertainty measures in AI, and to the choice of logical systems in fuzzy logic. For the latter, each logical system in fuzzy logic is modeled by a triple (N, T, S), where N is a negation operator, T is a t-norm and S its dual t-conorm. The art of modeling is delicate. For example, based on three-valued logic, a form of fuzzy conditionals was adopted in Chapter 7. The choice of the copula *min* was suggested since it is the simplest one. Other choices can be motivated on an empirical basis. For example, if "conjunction" is to be modeled mathematically in a given problem, and if there is some randomness involved in the gathered data, one can pick a copula for a t-norm, and view the modeling problem as a non-parametric statistical estimation problem, and estimate the joint distribution function from data on marginals.

This seems appropriate in problems such as modeling activation functions in neural networks. Indeed, basically the architecture of an artificial neural network can be placed within the theory of approximations of functions of several variables. (See, for example, Lorentz, 1966.) More specifically, it is related to Kolmogorov's theorem on representation of functions of several variables by superpositions of functions of fewer variables (Lorentz, 1966, chapter 11, or Vitushkin, 1978). As such, statistical estimation procedures for semi-parametric models can be used as learning rules. It is interesting to note that the popular back-propagation training algorithm in neural networks bears some close relationship with backfitting procedures in projection pursuit regression (see Huber, 1985). It seems that a fundamental question in the field of neural networks is this. Given a class of functions, not necessarily completely specified, how to design an efficient artificial neural network to "process" any member of this class?

In a recent personal communication, Hestir (1990) showed that extreme points of the space of copulas (identified as probability measures on the unit square with uniform marginals) can be characterized, so that the above estimation problem might be feasible. The space of copulas is a compact, convex space with the topology of weak convergence of measures.

C. Non-monotonic entailments on conditionals

When probability is used as a quantification of uncertainty, an extension of Probability Logic is needed for $R|R$. The resulting logic is called a Conditional Probability Logic (Chapter 6). Conditional Probability Logic should be extended from the sentential level to first order predicate calculus.

At a pragmatic level, non-monotonic entailment relations need to be specified as far as common sense reasoning is concerned. Some aspects of this problem were discussed in this Chapter. This difficult and important issue in mathematical logic should be further investigated. See also the recent work of the group Lea Sombe (1990).

D. Higher order conditioning

Once conditionals on Boolean rings are defined, it is natural, at least from a mathematical standpoint, to consider conditionals of conditionals. See, however, Pfanzagl (1971). In Chapters 7 and 8, we have touched upon this problem, both from the syntactic and semantic viewpoints.

The material in Section 8.1 is incomplete. Conditionals are defined on the space $R|R$, yielding the space of iterated conditionals $(R|R)|(R|R)$. The basic result is Corollary 1 in that section, asserting that they consist of intervals in the Stone Algebra $R|R$. This relied heavily on the fact that $R|R$ is relatively pseudo-complemented. This relative pseudo-complementation played the role of material implication. There, we also touched on a way to assign "probabilities" to these iterated conditionals. An algebra of these iterated conditionals has not been developed. No binary or unary operations on $(R|R)|(R|R)$ were defined and investigated. Much work remains to be done to clarify the issue and to obtain a more satisfying theory of higher-order conditioning. Doing so could be rewarding, and result in a tractable and important algebraic system, not only for its modeling of higher order conditioning, but for its possible connections with higher order logics.

E. Fuzzy conditionals and probability qualification

In view of the success of fuzzy logic in AI, we have devoted the entire Chapter 7 to the extension of ordinary conditionals to the fuzzy case. Our semantic approach to fuzzy conditionals is novel. It is motivated by a connection between membership functions and random sets, namely randomization of level-sets associated with membership functions of fuzzy sets. The simplest copula min was chosen to define membership functions of fuzzy conditionals, which turn out to be interval-valued fuzzy sets. As in fuzzy logic, other choices of copulas are possible. It might be of interest to compare fuzzy conditionals as perceived here with various concepts of conditional possibility distributions in the literature. Also, inference with fuzzy conditionals, for example fuzzy implication operators, should be investigated further for applications. See, for example, Smets, 1990, and Goodman, 1990.

F. Modal logic

Since conditioning was shown not to be just a primitive concept, best described by axioms, but rather can be analyzed further through the power class structure of the space of conditionals, one can inquire whether other model forms (see, for example, Rescher, 1968) can be analogously investigated, including deontic, alethic and voluntative modes, among others. This could use reduction of such forms to conditionals, using a synthetic approach as in the analysis of the latter, not a top-down external approach as taken by Palmer (1986) nor the formal logical stand of Searle and Vanderveken (1985). See also Ruspini (1989).

G. Non-additive uncertainty measures

As mentioned several times in this monograph, especially in Chapter 5, conditionals, as cosets of Boolean rings, were derived under the condition of compatibility with conditional probability. Here, Lewis' Triviality Result plays an important role. It is clear that this result depends heavily on the additivity property of probability measures. If probability measures, viewed as set functions, are generalized to, say, Dempster-Shafer's belief functions, which are non-additive set functions, then Lewis' Triviality Result does not hold. Indeed, as pointed out in Chapter 5, material implication on Boolean rings is compatible with conditional belief assignments. Belief functions are not the only non-additive set functions considered in the literature of artificial intelligence. Fuzzy measures (for example, Sugeno, 1974), or decomposable measures with respect to t-conorms (for example, Weber, 1984) are non-additive set functions. Although, in many cases, non additive measures can be transformed into additive ones, in the spirit of Lindley's admissibility (Lindley, 1982; Goodman, Nguyen, and Rogers, 1990), an analysis of conditional events compatible with a given class of uncertainty measures might be of interest, as a way to specify, at the syntex level, the "non-standard" logics underlying the semantic aspects when reasoning with various types of uncertainty.

REFERENCES AND BIBLIOGRAPHY

A

Aczel, J. (1966). *Lectures on Functional Equations and their Applications.* Academic Press, N.Y.

Adams, E. (1975). *The Logic of Conditionals.* D. Reidel, Dordrecht, Netherlands.

Aigner, M. (1979). *Combinatorial Theory.* Springer-Verlag, N.Y.

Alefeld, G. and Herzberger, J. (1983). *Introduction to Interval Computations.* Academic Press, N.Y.

Appiah, A. (1985). *Assertion and Conditionals.* Cambridge Univ. Press, London.

Arnold, B.C. and Press, S.J. (1989). Compatible conditional distributions. *J. Amer. Statist. Assoc.* 84 (405), 152-156.

B

Balbes, R. (1970). On implicative boolean algebras. *Math. Zeitschr.* (114), 170-172.

Barigelli, B. and Scozzafava, R. (1984). Remarks on the role of conditional probability in data exploration. *Statist. and Prob. Letters* (2), 15-18.

Bellman, R. (1978). *Artificial Intelligence.* Boyd & Fraser, Boston.

Belluce, L.P. (1986). Semisimple algebras of infinite valued logic and bold fuzzy set theory. *Can. J. Math.* 38(6), 1356-1379.

Berge, C. (1956). *Espaces Topologiques.* Dunod, Paris.

Berge, C. (1973). *Graphs and Hypergraphs.* North-Holland, Amsterdam.

Berge, C., Christensen, J. and Ressel, P. (1984). *Harmonic Analysis on Semi-Groups.* Springer, New York.

Bibel, W. (1986). Methods of automated reasoning. In *Fundamentals of AI*, Lecture Notes in C.S. #232, (G. Goos, J. Hartmanis, Eds.) Springer-Verlag, 171-217.

Birkhoff, G. (1948). *Lattice Theory.* Amer. Math. Soc., N.Y.

Bonissone, P. and Decker, K.S. (1985). Selecting uncertainty calculi and granularity: an experiment in trading-off precision and complexity. In *UCLA Workshop on Uncertainty and Probability in Artificial Intelligence*, 57-66.

Boole, G. (1854). *The Laws of Thought.* Dover, 1958 (originally published by McMillan in 1854).

Bouchon, B. (1987). Fuzzy inferences and conditional possibility distributions. *Fuzzy Sets and Systems* (23), 23-41.

Bruno, G. and Gilio, A. (1985). Comparison of conditional events of zero probability in Bayesian statistical inference (Italian). *Rivisita di Matematica per le Scienze economiche e sociali* (Milan) 8(2), 141-152.

Burton, D.M. (1967). *Introduction to Modern Abstract Algebra.* Addison-Wesley, Reading, Mass.

C

Calabrese, P. (1975). The probability that p implies q. Abstract in *Amer. Math. Soc. Notices*, 22(3), A 430-431.

Calabrese, P. (1987). An algebraic synthesis of the foundations of logic and probability. *Information Science*, (42), 187-237.

Calabrese, P. (1990). Deduction and Inference using conditional logic and probability. To appear in *Conditional Logic in Expert System* (I.R. Goodman, M.M. Gupta, H.T. Nguyen and G.S. Rogers, Eds.), North-Holland, Amsterdam.

Chang, C.C. (1958). Algebraic analysis of many-valued logics. *Trans. Amer. Math. Soc.* (88), 467-490.

Chang, C.C. (1959). A new proof of the completeness of the Lukasiewicz axioms. *Trans. Amer. Math. Soc.* (93), 74-80.

Choquet, G. (1953). Theory of capacities. *Ann. Inst. Fourier* (5), 131-292.

Coletti, G., Gilio, A. and Scozzafava, R. (1990). Conditional events with vague information in expert systems. *Proceedings Third IPMU*, Paris, 37-39.

Copeland, A.H. (1941). Postulates for the theory of probability. *Amer. J. Math.* (63), 741-762.

Copeland, A.H. (1945). A postulational characterization of statistics. *Proc. First Berk Symp. Math. Stat. and Prob.*, (J. Neyman, Ed.), Univ. Calif. Press, L.A. 51-61.

Copeland, A.H. (1950). Implicative boolean algebras. *Math. Zeitschr.* 53(3), 285-290.

Copeland, A.H. and Harary, F. (1953a). A characterization of implicative boolean rings. *Cand. J. Math.* (5), 465-469.

Copeland, A.H. and Harary, F. (1953b). The extension of an arbitrary boolean algebra to an implicative boolean algebra. *Proc. Amer. Math. Soc.* (4), 751-758.

Copeland, A.H. (1954), Probabilities, observations and predictions. *Proc. Third Berkeley Symp. Math. Stat. and Prob.*, 41-47.

Copeland, A.H. (1967). Formal logic as a tool for mathematics. In *Lectures on Calculus* (K.O. May, Ed.) Holden-Day, San Francisco, 3-12.

Cox, R.T. (1961). *The Algebra of Probable Inference.* J. Hopkins Press, Baltimore.

D

Dale, A.I. (1980). Probability, vague statements and fuzzy sets. *Philos. Sci.*, (47), 38-55.

DeFinetti, B. (1964). Forsight: its logical laws, its subjective sources. *Studies in Subjective Probability.* (H.E. Kyburg, H.E. Smokler, Eds.) J. Wiley, N.Y., 93-158.

DeFinetti, B. (1972). *Probability, Induction and Statistics.* J. Wiley, N.Y.

DeFinetti, B. (1974). *Theory of Probability.* Vol. 1 & 2, J. Wiley, N.Y.

Delgrande, J.P. (1987). A first-order conditional logic for prototypical properties. *Artificial Intelligence* (33), 105-130.

Dempster, A.P. (1967). Upper and lower probabilities induced by a multi-valued mapping. *Ann. Math. Statist.* (38), 325-339.

Domotor, Z. (1969). Probabilistic relational structures and their applications. Tech. Report #144, Psych. Series, Stanford Univ.

Dubois, D. and Prade, H. (1980). *Fuzzy Sets and Systems: Theory and Applications.* Academic Press, N.Y.

Dubois, D. and Prade, H. (1987). *Théorie des Possibilités* (2nd Edition). Masson, Paris.

Dubois, D. and Prade, H. (1990). The logical view of conditioning and its application to possibility and evidence theories. *Int. J. Approximate Reasoning* 4(1), 23-46.

Dubois, D. (1986). Generalized probabilistic independence and its implications for utility. *Operations Research Letters* 5(5), 255-260.

Dubois, D. and Prade, H. (1991). Conditioning, non-monotonic logic and non-standard uncertainty models. In *Conditional Logic in Expert Systems* (Goodman, I.R., et al. Eds.) To appear.

F

Fine, T.L. (1973). *Theories of Probability.* Academic Press, N.Y.

Fishburn, P.c. (1983). Ellsberg revisited: a new look at comparative probability. *Ann. Statist.* 11(4), 1047-1059.

Foulis, D.J. and Randall, C.H. (1971). Conditioning maps on orthomodular lattices. *Glasgow Math.* (12), 35-42.

Foulis, D.J. and Randall, C.H. (1974). The stability of pure weights under conditioning. *Glasgow Math.* (15), 5-12.

French, S. (1990). Comment on Lindley's paper "The present position in Bayesian statistics," *Statistical Science* 5(1), 78-80.

G

Gabbay, D.M. (1985). Theoretical foundation for non-monotonic reasoning in expert systems. In *Logics and Models of Concurrent Systems* (K.R. Apt., Ed.) Springer-Verlag, 439-457.

Genest, C. and MacKay, J. (1986a). The Joy of Copulas: bivariate distributions with uniform marginals. *J. Amer. Statist. Assoc.* 40(4), 280-283.

Genest, C. and MacKay, J. (1986b). Copules archimédiennes et familles de lois bidimensionnelles dont les marges sont donneés. *Can. J. Statist.* 14(2), 145-159.

Gentilhomme, Y. (1968). Les ensembles flous en linguistique. *Cahier Linguist. Theor. Appl.* (Bucarest), (5), 47-63.

Goodearl, K.R. (1979) *Von Neumann Regular Rings.* Pitman.

Goodman, I.R. and Stein. D.W. (1989). Extension of the measure-free approach to conditioning of fuzzy sets and other logics. *Proc. 3rd Intern. Fuzzy Syst. Assoc.*, Univ. of Wash., Seattle, 361-364.

Goodman, I.R. and Nguyen, H.T. (1985). *Uncertainty Models for Knowledge Based Systems.* North-Holland, Amsterdam.

Goodman, I.R. and Nguyen, H.T. (1988). Conditional objects and the modeling of uncertainties.In *Fuzzy Computing, Theory, Hardware and Applications* (M.M. Gupta, T. Yamakawa, Eds.), North-Holland, Amsterdam, 119-138.

Goodman, I.R. (1987). A measure-free approach to conditioning. *Proc. 3rd AAAI Workshop on Uncertainty in AI*, Seattle, July 1987, 270-277.

Goodman, I.R., Nguyen, H.T. and Rogers, G.S. (1990). On scoring approach to the admissibility of uncertainty measures in expert systems. To appear *J. Math. Anal. and Appl.*

Goodman, I.R. (1990). Algebraic and probabilistic bases for fuzzy sets and development of fuzzy conditioning. To appear in *Conditional Logic in Expert Systems* (I.R. Goodman, M.M. Gupta, H.T Nguyen, and G.S. Rogers, Eds.), North-Holland, Amsterdam.

Goodman, I.R. and Nguyen, H.T. (1990). Foundations for an algebraic theory of conditioning. *J. Fuzzy Sets and Systems.* To appear.

Grätzer, G. (1968). *Universal Algebra.* Van Nostrand, N.Y.

Grätzer, G. (1978). *General Lattice Theory.* Birkhauser, Basel.

Grosof, B. (1988). Non-monotonicity in probability reasoning. In *Uncertainty in AI 2* (J.F. Lemmer, L.N. Kanal, Eds.), North-Holland, 237-249.

Grosof, B.N. (1988). Non-monotonicity in probabilistic reasoning. *Uncertainty in Artificial Intelligence 2*, (J.F. Lemmer, L.N. Kanal, Eds.), North-Holland, 237-249.

Gudder, S.P. (1988). *Quantum Probability.* Academic Press, San Diego, CA.

Guzman, F. and Squier, C.C. (1990). The algebra of conditional logic. *Algebra Universalis* (27) 88-110.

H

Hailperin, T. (1965). Best possible inequalities for the probability of a logical function of events. *Amer. Math. Monthly* (72), 343-359.

Hailperin, T. (1976). *Boole's Logic and Probability.* North-Holland, Amsterdam.

Hailperin, T. (1984). Probability logic. *Notre Dame J. Formal Logic* 25(3), 198-212.

Halmos, P.R. (1961). *Measure Theory.* Van Nostrand, N.Y.

Halmos, P.R. (1962). *Algebraic Logic.* Chelsea, N.Y.

Halmos, P.R. (1963). *Lectures in Boolean Algebras.* Springer-Verlag, N.Y.

Hamilton, A.G. (1978). *Logic for Mathematicians.* Cambridge Univ. Press, Cambridge.

Hawthorne, J. (1988). A semantic approach to non-monotonic entailment. In *Uncertainty in AI2* (J.F. Lemmar, L.N. Kanal, Eds.), North-Holland, 251-262.

Henkind, S.J. and Harrison, M.C. (1988). An analysis of four uncertainty calculi. *IEEE Trans. Syst., Man and Cybern.* 18(5), 700-714.

Hestir, K., Nguyen, H.T. and Rogers, G.S. (1990). A random set formalism for evidential reasoning. To appear in *Conditional Logic in Expert Systems* (I.R. Goodman et al, Eds.), North-Holland, Amsterdam.

Hestir, K. (1990). Extreme points of the set of copulas. (Personal communication.)

Hisdall, E. (1978). Conditional possibilities, independence and non-interaction. *J. Fuzzy Sets and Systems,* (1), 283-297.

Huber, P. J. (1985). Projection persuit. *Annals of Statistics* 13(2), 435-475.

Hungerford, T.W. (1974). *Algebra.* Holt, Rinehart, and Winston, N.Y.

J

Jevons, W.S. (1879). *The Principles of Science, A Treatise on Logic and Scientific Method* (3rd Ed.), McMillan, London (Dover, reprint, 1958).

Jónsson, B. (1954). Review of the "extension of an arbitrary boolean algebra to an implicative boolean algebra." *Math. Reviews* 15(1), 193-194.

K

Klein, E. and Thompson, A.C. (1984). *Theory of Correspondences*, Wiley, New York.

Klir, G.J. and Folger, T.A. (1988). *Fuzzy Sets, Uncertainty and Information.* Prentice Hall, New Jersey.

Kolmogorov, A.N. (1950). *Foundations of the Theory of Probability.* Chelsea, N.Y.

Koopman, B.O. (1940). The bases of probability. *Amer. Math. Soc. Bull.* 2(46), 763-774.

Koopman, B.O. (1940). The axioms and algebra of intuitive probability. *Ann. of Math.* 41(2), 269-292.

Koopman, B.O. (1941). Intuitive probabilities and sequences. *Ann. of Math.* 42(1), 169-187.

Kosko, B. (1986). Fuzzy entropy and conditioning. *Information Science* (40), 165-174.

L

Lambek, J. (1966). *Lectures on Rings and Modules.* Blaisdell, Watham, Mass.

Lauritzen, S.L. and Spiegelhalter, D.J. (1988). Local computations of probabilities in graphical structures and their application to expert systems. *J. Roy. Statist. Soc.* B 50(2), 157-224.

Levi, I. (1973). *Gambling with Truth.* MIT Press.

Lewis, C.I. (1960). *A Survey of Symbolic Logic.* Dover, N.Y.

Lewis, C.I. and Langford, C.H. (1956). History of Symbolic Logic, in *The World of Mathematics* (J.R. Newman, Ed.). Simon and Schuster, N.Y., 1859-1877.

Lewis, D. (1976). Probabilities of conditionals and conditional probabilities. *Phil. Rev.* 85, 297-315.

Lindley, D.V. (1982). Scoring rules and the inevitability of probability. *Intern. Statist. Rev.* (50), 1-26.

Ling, C.H. (1965). Representation of associative functions. *Publ. Math. Debrecen* (12), 189-212.

Lorentz, G. G. (1966). *Approximation of Functions.* Holt, Rinehart, and Winston, N. Y.

M

MacFarlane, A. (1879). *Principles of the Algebra of Logic.* Douglas, Edinburgh.

Maclennan, B. (1988). Logic for the new AI. In *Aspects of Artificial Intelligence* (J.H. Fetzer, Ed.), Kluwer Academic, 163-192.

Marshall, A.W. and Olkin, I. (1988). Families of multivariate distributions. *J. Amer. Statist. Assoc.* 83(403), 834-841.

Matheron, G. (1975). *Random Sets and Integral Geometry.* J. Wiley, N.Y.

Mattila, J.K. (1986). On some logical points of fuzzy conditional decision making. *Fuzzy Sets and Systems* (20), 137-145.

Mazurkiewicz, S. (1956). *Podstawy Rachunku Prawdopodobienstwa* (Foundations of the calculus of probability), Paustwowe Wydawnictwo Naukawe, Polska, Akademia Nauk, Warsaw, (32), (prepared by J. Las, from posthumous notes of Mazurkiewicz, in Polish with Math. Review, Vol. 18, 1957).

McCarthy, J. (1967). A basis for a mathematical theory of computation. In *Computer Programming and Formal Systems* (P. Braffort and D. Hirschberg, Eds.), North-Holland, Amsterdam, 33-70.

McCarthy, J. (1980). Circumscription: a form of non-monotonic reasoning. *Art. Intell.* (13), 27-39.

McCoy, N.H. (1962). *Rings and Ideals.* Carus Math. Monograph #8, Math. Assoc. of Amer.

McDermott, D.V. and Doyle, J. (1980). Non-monotonic logic 1. *Art. Ingell.* (13), 41-72.

McLeish, M. (1988). Probabilistic logic: some comments and possible use for non-monotonic reasoning. In *Uncertainty in AI2*. (J.F. Lemmer, L.N. Kanal, Eds.), North-Holland, 55-62.

Mendelson, E. (1970). *Boolean Algebra and Switching Circuits.* Schaum's Outline Series, McGraw Hill, N.Y.

Moore, R.E. (1966). *Interval Analysis.* Prentice-Hall, Englewood Cliffs, N.J.

Moore, R.E. (1979). *Methods and Applications of Interval Analysis.* SIAM Studies in Applied Mathematics, SIAM, Philadelphia, Pennsylvania.

Mosteller, F. and Youtz, C. (1990). Quantifying probabilistic expressions. *Statistical Science* 5(1), 2-12 (discussions: 12-34).

N

Neapolitan, R.E. (1990). *Probability Reasoning in Expert Systems. Theory and Algorithms.* J. Wiley, N.Y.

Negoita, C.V. and Ralescu, D.A. (1975). Representation theorems for fuzzy concepts. *Kybernetes* (4), 169-174.

Nguyen, H.T. (1978). On conditional possibility distributions. *J. Fuzzy Sets and Systems,* (1), 299-309.

Nguyen, H.T. (1978). On random sets and belief functions. *J. Math. Anal. and Appl.* (65), 531-542.

Nilsson, N. (1986). Probabilistic logic. *Art. Intell.*, (28), 71-87.

Nothcott, D. G. (1968). *Lessons on Rings, Modules, and Multiplicities.* Cambridge Univ. Press, Cambridge, G. B.

Nute, D. (1980). *Topics in Conditional Logic.* D. Reidel, Dordrech, Holland.

P

Palmer, F.R. (1986). *Mood and Modality.* Cambridge Univ. Press, Cambridge, G.B.

Pearl, J. (1988). *Probabilistic Reasoning in Intelligent Systems. Networks of Plausible Inference.* Morgan Kaufmann, San Mateo, California.

Pearl, J. (1990). Reasoning with belief functions: an analysis of compatibility. *International Journal of Approximate Reasoning* (4), 363-389.

Peirce, C.S. (1880). On the algebra of logic. *Amer. J. Math.* (3), 15-57.

Peirce, C.S. (1967). On an improvement in Boole's calculus of logic. *Proc. Amer. Acad. Arts and Sci.* (7), 250-261.

Pfanzagl, J. (1967). Subjective probability derived from the Morgenstern-Von Neumann utility concept. In *Essays in Mathematical Economics* (M. Shubik, Ed.). Princeton Univ. Press, 237-251.

Pfanzagl, J. (1971). *Theory of Measurement.* Physica-Verlag, Würzburg-Wien.

Popper, K.R. (1961). *The Logic of Scientific Discovery.* Science Editions, N.Y.

Popper, K.R. (1963). *Conjectures and Refutations.* Routledge and Kegan Paul, London.

Popper, K.R. and Miller, D.W. (1987). Why probabilistic support is not inductive? *Philo. Trans. Royal Soc.* Series A (321), 569-591.

R

Radecki, T. (1977). Level fuzzy sets. *J.* (7), 189-198.

Ramer, A. (1989). Conditional possibility measures. *Cybern. and Systems* (20), 233-147.

Ramsay, A. (1988). *Formal Methods in Artificial Intelligence.* Cambridge Univ. Press.

Randall, C.H. and Foulis, D.J. (1970). An approach to empirical logic. *Amer. Math. Monthly* (77), 363-374.

Randall, C.H. (1967). Empirical logic and orthomodular lattices. (1967). Unpublished Notes, Univ. Mass., Amherst.

Rasiowa, H. (1974). *An Algebraic Approach to Non-Classical Logics.* North-Holland, Amsterdam.

Redder, W. (1982). Conditions for probabilities of conditionals to be conditional probabilities. *Synthese* (53), 439-443.

Reichenbach, H. (1948). *Philosophical Foundations of Quantum Mechanics.* Univ. Calif. Press, Berkeley.

Reichenbach, H. (1940). *The Theory of Probability* (2nd Ed.). Univ. of Calif. Press, Berkeley.

Reiter, R. (1980). A logic for default reasoning. *Art. Intell.* (13), 81-132.

Rényi, A. (1970). *Foundations of Probability Theory.* Holden-Day, San Francisco.

Rescher, N. (1968). *Topics in Philosophical Logic.* D. Reidel, Dordrecht, Holland.

Rescher, N. (1969). *Many-valued Logics.* McGraw-Hill, N.Y.

Rose, G.F. (1953). Review of "axiomatization of a partial system of three-valued calculus of propositions" by B. Sobocinski. *J. Symbolic Logic* (18), 283.

Rota, G.C. (1964). On the foundation of combinatorial theory: I, Theory of Möbius functions. *Z. Works* (2), 340-368.

Rudeanu, S. (1974). *Boolean Functions and Equations.* North-Holland, Amsterdam.

Rumelhart, D.E. and McClelland, J.L. (1986). *Parallel Distributed Processing,* Vol. 1, MIT Press.

Ruspini, E.H. (1989). On the semantics of fuzzy logic. Technical report No. 475, SRI International, Menlo Park, CA.

S

Sandewall, E. (1989). The semantics of non-monotonic entailment defined using partial interpretations. In *Non-Monotonic Reasoning* (M. Reinfrank, J. de Kleer, M.L. Ginsberg, E. Sandewall, Eds.), Springer-Verlag, Lecture Notes in AI #346, 27-41.

Savage, L.J. (1972). *The Foundations of Statistics.* Dover, N.Y.

Schay, G. (1968). An algebra of conditional events. *J. Math. Anal. and Appl.* (24), 334-344.

Scholz, H. (1959). Review of "axiomatization of a partial system of three-valued calculus of propositions" by B. Sobocinski. *Zentral. für Math. und Ihre Grenz* (49), 242.

Schroder, E. (1890, 1891, 1895) *Vorlesungen uber die Algebra der Logik* (3 volumes). Teubner, Leipzig.

Schweizer, B. and Sklar, A. (1983). *Probabilistic Metric Spaces.* North-Holland, Amsterdam.

Scott, D. and Krauss, P. (1966). Assigning probabilities to logical formulae. In *Aspects of Inductive Logic.* (Hintikka, J. and Suppes, P. Eds.) North Holland, 219-264.

Scozzafava, R. (1984). A survey of some common misunderstanding concerning the role and meaning of finitely additive probabilities in statistical inference. *Statistica* (Bologna, Italy) XLIV(1), 21-45.

Searle, J.R. and Vanderveken, D. (1985. *Foundations of Illocutionary Logic.* Cambridge Univ. Press, Cambridge, G.B.

Semi, B.S. and Mamdani, E.H. (1979). On the nature of implication in fuzzy logic. *Proc. 9th Intern. Symp. Multi. Logic.*, Bath, G.B., 143-149.

Shafer, G. (1976). *A Mathematical Theory of Evidence.* Princeton Univ. Press, Princeton, N.J.

Shafer, G., Shenoy, P.P. and Mellouli, K. (1987). Propagation of belief functions in qualitative Markov trees. *Intern. J. Approx. Reasoning* 1(4), 349-400.

Shafer, G. (1990). Perspectives on the theory and practice of belief functions. *International Journal of Approximate Reasoning*(4), 323-362.

Sklar, A. (1959). Fonctions de répartition à n variables et leurs marges. *Publ. Inst. Statist. Univ. Paris*, (8), 229-231.

Sklar, A. (1973). Random variables, joint distribution functions and copulas. *Kybernetika* (9), 449-460.

Smets, P. (1990). Implication and modus ponens in fuzzy logic. To appear in *Conditional Logic in Expert Systems*, (Goodman, I.R., et al., Eds.) North Holland, Amsterdam.

Sobocinski, B. (1952). Axiomatization of a partial system of three-valued calculus of propositions. *J. Computing Systems* 1(1), 23-55.

Sombé, L. (1990). *Reasoning Under Incomplete Information in Artificial Intelligence.* J. Wiley, N.Y.

Spies, M. (1990). Application aspects of qualitative conditional independence. *Proceedings Third IPMU*, Paris, 233-235.

Stalnaker, R.C. (1970). Probability and conditionals. *Philos. Sci.* (37), 64-80.

Sugeno, M. (1974). Theory of fuzzy integral and its applications. Ph.D. Thesis, Tokyo Inst. Techn., Tokyo.

Suppes, P. (1973). New foundations of objective probability: axioms for propensities. *Logic, Methodology of Science IV* (P. Suppes, L. Henkin, G. Moisil, A. Joja, Eds.), North-Holland, Amsterdam, 515-529.

T

Tarski, A. (1956). *Logic, Semantics, Metamathematics.* Clarendon Press.

Thoma, H.M. (1989). Factorization of belief functions. Ph.D. Thesis, Harvard Univ., Dept. of Statistics.

Thoma, H.M. (1991). Belief function computations. In *Conditional Logic in Expert Systems* (I.R. Goodman, et. al., Eds.) North-Holland. To appear.

Traugott, E.C., Meulen, A., Reilly, J. and Ferguson, C. (1986), Eds. *On Conditionals.* Cambridge Univ. Press, N.Y.

Turner, R. (1984). *Logics for Artificial Intelligence.* Ellis Harwood, England.

U

Uzkov, A.I. (1949). On rings of quotients of commutative rings. *Amer. Math. Soc. Transl.* (3), 1-5 (Amer. Math. Soc., N.Y.).

V

Van Fraasen, B.C. (1976). Probabilities of conditionals. *Foundations of Probability Theory, Statistical Inference and Statistical Theories of Science* I, (W.L. Harper, C.A. Hooker, Eds.), D. Reidel, Dordrecht, Holland, 261-308.

Villegas, C. (1967). On qualitative probability. *Amer. Math. Monthly* 74(2), 661-668.

Vitushkin, A. G. (1978). On representation of functions by means os superpositions and related topics. *Enseignement Mathematique.* Monogr. No. 25.

Von Neumann, J. and Morgenstern, O. (1947). *Theory of Games and Economic Behavior.* Princeton Univ. Press, Princeton, N.J.

W

Walker, E.A. (1987). *Introduction to Abstract Algebra.* Random House, N.Y.

Weber, S. (1984). Decomposable measures and integrals for archimedean t-conorms. *J. Math. Anal. and Appl.* (101), 114-138.

White, H. (1989). Some asymptotic results for learning in single hidden-layer feedforward network models. *J. Amer. Stat. Assoc.* 84(408). 1003-1013.

Whitt, W. (1976). Bivariate distributions with given marginals. *Ann. Statist.* 4(6), 1280-1289.

Williams, R.J. (1986). The logic of activation functions. In *Parallel Distributed Processing*, Vol. 1 (D.E. Rumelhart and J.L. McClelland), MIT Press, 423-443.

Y

Yager, R.R., Ovchinnikov, S., Tong, R.M. and Nguyen, H.T. (Eds.) (1987). *Fuzzy Sets and Applications: Selected papers by L.A. Zadeh.* J. Wiley, N.Y.

Yager, R.R. (1983). On the implicative operator in fuzzy logic. *Information Science* (31), 141-164.

Z

Zadeh, L.A. (1985). A computational theory of dispositions. Berkeley Cognitive Science Report No. 30, Univ. of Calif., Berkeley.

Zadeh, L.A. (1968). Probability measures of fuzzy events. *J. Math. Anal. Appl.* (23), 421-427.

Zadeh, L.A. (1978). Fuzzy sets as a basis for a theory of possibility. *Fuzzy Sets and Systems* (1), 3-28.

Zadeh, L.A. (1983). The role of fuzzy logic in the management of uncertainty in expert systems. *Fuzzy Sets and Systems* (11), 199-227.

Zadeh, L.A. (1988). Fuzzy logic. *Computer*, 21(4), 83-92.

SUBJECT INDEX

A

B

C